ARCTIC AIR POLLUTION

Studies in Polar Research

This series of publications reflects the growth of research activity in and about the polar regions, and provides a means of disseminating the results. Coverage is international and interdisciplinary: the books will be relatively short (about 200 pages), but fully illustrated. Most will be surveys of the present state of knowledge in a given subject rather than research reports, conference proceedings or collected papers. The scope of the series is wide and will include studies in all the biological, physical and social sciences.

Editorial Board

R. J. Adie, British Antarctic Survey, Cambridge
T. E. Armstrong, Scott Polar Research Institute, Cambridge
D. J. Drewry, Scott Polar Research Institute, Cambridge
S. W. Greene, Department of Botany, University of Reading
B. Stonehouse, Scott Polar Research Institute, Cambridge
P. Wadhams, Scott Polar Research Institute, Cambridge
D. W. Walton, British Antarctic Survey, Cambridge
I. Whitaker, Department of Anthropology, Simon Fraser University, British Columbia

Other titles in this series:

The Antarctic Circumpolar Ocean
Sir George Deacon

The Living Tundra
Yu. I. Chernov, transl. D. Löve

Transit Management in the Northwest Passage
edited by C. Lamson and D. Vanderzwaag

The Antarctic Treaty Regime
edited by Gillian D. Triggs

ARCTIC AIR POLLUTION

EDITED BY B. STONEHOUSE

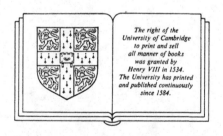

The right of the
University of Cambridge
to print and sell
all manner of books
was granted by
Henry VIII in 1534.
The University has printed
and published continuously
since 1584.

CAMBRIDGE UNIVERSITY PRESS

Cambridge

London   New York   New Rochelle

Melbourne   Sydney

CAMBRIDGE UNIVERSITY PRESS
Cambridge, New York, Melbourne, Madrid, Cape Town, Singapore, São Paulo, Delhi

Cambridge University Press
The Edinburgh Building, Cambridge CB2 8RU, UK

Published in the United States of America by Cambridge University Press, New York

www.cambridge.org
Information on this title: www.cambridge.org/9780521330008

© Cambridge University Press 1986

First published 1986
This digitally printed version 2008

A catalogue record for this publication is available from the British Library

ISBN 978-0-521-33000-8 hardback
ISBN 978-0-521-09339-2 paperback

CONTENTS

CONTRIBUTORS TO THIS VOLUME

ACKERMAN, Dr T. P. Space Science Division, NASA Ames Research Center, Moffett Field, CA 94035 USA

BARRIE, Dr L. A. Atmospheric Environment Service, 4905 Dufferin Street, Downsview Ont., Canada M3H 5T4

BEAN, M. Wildlife Liason, Association of Village Council Presidents, Bethel, AK 99559 USA

BENSON, Prof C. S. Geophysical Institute, University of Alaska at Fairbanks, Fairbanks, AL 99701 USA

BLANCHET, J-P. Canadian Climate Centre, 4095 Dufferin Street, Downsview, Ontario, M3H 5T4 Canada

BODHAINE, Dr B. A. Office of Geophysical Monitoring for Climatic Change, Air Resources Laboratory, NOAA-US Dept of Commerce, 325 Broadway, Boulder, CO 80303 USA

COVERT, D. S. Department of Environmental Health, University of Washington, Seattle, WA 98195 USA

DELMAS, Dr R. J. Laboratoire de Glaciologie du CNRS, 2 rue Tres-Cloitres, 38031 Grenoble-Cedex, France

FRIDAY, D. Alaska Department of Fish and Game, Division of Boards, Bethel, AK 99559 USA

HANSEN, A. D. A., Lawrence Berkeley Laboratory, University of California, Berkeley, CA 94720 USA

HANSEN, Prof J. C. Hygiejnisk Institut, Aarhus Universitet, 800 Aarhus, Denmark

HANSON, Dr W. C. Hanson Environmental Research Services, 1902 Yew St Road, Bellingham, WA 98226 USA

HANSSON, J-C. Department of Meteorology, University of Stockholm, 10691 Stockholm, Sweden

HASHIMOTO. Y. Faculty of Science and Technology, Keio University, 3–14–1 Hioshi, Kohokuku, Yokohama 223 Japan

HEIDAM, Prof N. Z. National Agency of Enviromental Protection, Air Pollution Laboratory, RisP National Laboratory, DK-4000 Roskilde, Denmark

HEINTZENBERG, Prof J. Department of Meteorology, University of Stockholm, 10691 Stockholm, Sweden

HENDERSON, Dr M. M. Department of Epidemiology, SC-36 University of Washington, Seattle, WA 98195 USA

HOV, Dr Ø. Norwegian Institute for Air Research, PO Box 130, Elvegaten 52, N-2001 Lillestrøm, Norway

LOWENTHAL, D. H. Center for Atmospheric Chemistry Studies, Graduate School of Oceanography, University of Rhode Island, Narragansett, RI 02882–1197 USA

MING-XING WANG, Institute of Atmospheric Physics, Chinese Academy of Sciences, Beijing, People's Republic of China

NEVÉ, R. A. Office of the Governor of Alaska, Juneau, AK USA.

OGREN, J. A. Department of Meteorology, University of Stockholm, 10691 Stockholm, Sweden

O'ROURKE, Dr P. J. Office of the Chancellor, University of Alaska at Fairbanks, Fairbanks, AK 99701, USA

OTTAR, Prof B. Norwegian Institute for Air Research, PO Box 130, N-2001, Lillestrøm, Norway

PACYNA, J. M. Norwegian Institute for Air Research, PO Box 130, N-2001, Lillestrøm, Norway

RAHN, Prof K. A. Center for Atmospheric Chemistry Studies, Graduate School of Oceanography, University of Rhode Island, Narragansett, RI 02882–1197 USA

REY, Prof L. Comité Arctique Internationalc, La Jaquiere, CH-1066 Epalinges, Switzerland

ROEDERER, Dr J. G. Geophysical Institute, University of Alaska at Fairbanks, Fairbanks, AK 99701 USA

ROSEN, Dr H. IBM Research Laboratories, San José, CA USA

SCHNELL, Dr R. C. CIRES, University of Colorado, ERL-ARL, 325 Broadway, Boulder, CO 80309 USA

SHAW, Prof G. E. Geophysical Institute, University of Alaska at Fairbanks, Fairbanks, AK 99701 USA

SPARCK, H. Nunam Kitlutsisti, PO Box 206, Bethel, AK 99559 USA

SPENGLER, Prof J. D. Dept of Environmental Science & Physiology, Harvard University School of Public Health, Building 1, Room 1305, 665 Huntington Avenue, Boston, MA 02115 USA

STENBACK, J. M. Space Science Division, NASA Ames Research Center, Moffett Field, CA 94035 USA

STONEHOUSE, Dr B. Scott Polar Research Institute, Lensfield Road, Cambridge CB2 1ER UK

STUTZMAN, Dr C. D. Radiation Oncology, School of Medicine, Vanderbilt University, Nashville, TE 37232 USA

TURNER, W. A. Harvard University School of Public Health, Boston MA 02115 USA

UPTON, Dr A. C. Institute of Environmental Medicine, New York University Medical Centre, 550 First Avenue, MSB 213, New York, NY 10016 USA

VALERO, Dr F. P. J. Astrophysical Experiment Branch, NASA Ames Research Center, Moffett Field, CA 94035 USA

WALLÉN, C. C. World Meteorological Organization, Secretariat, CH- 1211 Geneva 20, Switzerland

WINCHESTER, Prof J. W. Department of Oceanography, Florida State University, Tallahassee, FL 32306–3048 USA

Front Row: left to right: G. Shaw (Convener); L. Gorman (Secretary); J. G. Roederer; L. Rey; A. Kohler; P. J. O'Rourke; D. J. Drewry; C. C. Wallén; L. Barrie; B. Ottar; K. Rahn

Second Row: D. Szepesi; N. Heidam; W. C. Hanson; J. Middaugh; H. Sparck; T. P. Ackerman; R. Delmas; W. Turner; A. C. Upton; B. Bodhaine

Third Row: S. Pilkington; D. Gray; J. Heintzenberg; R. A. Nevé; J. C. Hansen; R. Schnell; J. Harte; H-C. Hansson; H. Rosen.

Back Row: M. Bean; J. W. Winchester; B. Stonehouse (Editor); C. S. Benson; Ø. Hov; E. Wolff; M. Scott.

FOREWORD

PATRICK J. O'ROURKE

Chancellor, University of Alaska-Fairbanks

In May 1984 Senator Frank Ferguson, a senior member of the Alaska State Legislature, asked the University of Alaska-Fairbanks to sponsor two international symposia on the phenomenon known as Arctic haze. An appropriation was provided, and Prof Glenn Shaw began immediately to implement this request, convening a session in October 1984 in cooperation with the northwest meeting of the American Association for the Advancement of Science. Dr David Drewry attended this meeting and graciously offered the services and facilities of the Scott Polar Research Institute of the University of Cambridge as the site for the 1985 international symposium on Arctic Air Pollution.

The Arctic is one of those regions of the world that has long fascinated both men and women. It has invited the explorer, it has stimulated the soul of the poet, it has intrigued the scientist, it has challenged the industrialist, and it has engendered awe in those populations unfamiliar with its character and essence. At the same time, for centuries the Arctic has served as home to a hardy population of indigenous people who have learned through the years that it offers them a very special quality of life. These people, although separated artificially by our geopolitical boundaries, have evolved a culture and value system that operates in harmony with the land they inhabit. They have come to understand the fragility of its nature and that, because of its intense extremes, small infusions of non-native substances may have larger-than-normal impacts of wide reach. To some extent the Cambridge symposium alleviated the acute anxiety that we are, in the short-term, facing major arctic destruction and health risks. However, as an observer at the conference, it seemed to me clear that, as much as we know, we are still missing many pieces of the puzzle that will allow us to better assess long-term, chronic impact.

From a public policy perspective, the interest of Senator Ferguson and his colleagues in the Alaska State Senate is quite refreshing. Too often issues of scientific concern and impact do not receive the focus of public policy attention until chronic conditions become acute. Too often state policy makers have not taken the global perspective necessary to understand potential impacts in their home regions. Not so in this case. Thus, it seems a most appropriate time to continue to expand investigations so that the base of knowledge required to make sound international policies is well grounded in scientific thought and understanding. Our scientific understanding of the elements in which we live evolves and moves in slow iterations. The 1985 Cambridge symposium brought together the collective knowledge of a group of talented scientists who, as in almost all scientific endeavours, answered many questions, but stimulated far more.

Any reader of these proceedings will see that good baseline data, against which change in levels of arctic air pollution may be measured, are missing. The effects of carbon deposition on the albedo of the polar cap are very preliminary, and it is apparent that these effects must be looked at additively and cumulatively with algae and other heat absorbing materials. One will observe the need for the expansion and broadening of ecological monitoring and analysis efforts. The information contained in this volume brings to focus a problem that has received intermittent attention for 40 years. It summarizes quite concisely the existing knowledge, indentifies potentially fruitful areas of further investigation and research, and raises some interesting questions for international problem resolution.

INTRODUCTION: INTERNATIONAL SYMPOSIUM ON ARCTIC AIR POLLUTION

GLENN E. SHAW

An International Symposium on Arctic Air Pollution was held at the Scott Polar Research Institute, Cambridge, England, 2–5 September 1985. The conference, sponsored by the State of Alaska through the University of Alaska at Fairbanks, involved participants from the circumpolar region in discussions about the location of pollution sources and transport pathways, climatic influences of Arctic air pollution, and possible effects on human health. On the final day participants considered issues of international scientific cooperation and state responsibility; the latter topic included the difficult subject of international liabilities regarding the transfer of pollutants across international boundaries. In view of its multinational, even multi-continental character, the problem of Arctic air pollution is perhaps the quintessential example of such transferral and its attendant problems. This symposium volume contains edited versions of the papers presented at the four-day meeting in Cambridge, and the conclusions and recommendations adopted by consensus among the participants.

This was not the first symposium devoted to the phenomenon of Arctic air pollution. Indeed, the first conference during which the subject of Arctic haze was discussed was held in Norway in 1977. At that conference it was agreed to try and set up a loosely coordinated, informal Arctic Chemical Network to share data on atmospheric chemistry. In retrospect this informal agreement represented a superb example of unselfish international cooperation. The data obtained were of sufficient quality that, three years later, it was possible to convene a meeting of meteorologists and atmospheric chemists at the University of Rhode Island, to obtain a consensus on what had become a paradigm regarding the Arctic as a 'chemical sink'. The validity of this concept became even more evident at the Arctic Air Chemistry Conference held in Toronto in 1984.

These earlier meetings focused narrowly on atmospheric chemistry and meteorological processes relating to Arctic air pollution. The Cambridge symposium built on the bases established at those meetings, and expanded discussion into frontier areas of public health, ecological impact and possible climatic influences. Thus the present volume examines the problem of Arctic air pollution in an integral, comprehensive and multi-disciplinary fashion, an approach indeed warranted by the considerable attention that the media have given recently to the problem of Arctic haze.

Meditating on Arctic air pollution, one cannot help but be struck with a certain sense of irony. We are after all only a few generations from the 'romantic era' of polar exploration that so strongly gripped the world in late Victorian times. In those days the Arctic was literally and figuratively at the end of the world. Such remote places were supposed to be wild and pristine! Sipping tea during the breaks between conference sessions in the museum of the Scott Polar Research Institute, surrounded by relics of

xvi

the expeditions of Scott, Amundson and Franklin, all the delegates were powerfully reminded of the brief span of time elapsed since those triumphant days.

Before being over-pessimistic about man-made pollution reaching remote parts of the globe, we should keep in mind that the spate of current research into Arctic air pollution does not necessarily mean that the Arctic has just suddenly become polluted. Recent papers in many instances reflect recent technological advances in analytical chemistry. In fact we know very little about the secular behavior of Arctic air pollution. It is even conceivable that pollution levels could already be declining, due to pollution control legislation adopted by several countries in the past decade: I must emphasize that long term trends are at present unknown.

It became clear during the Cambridge discussions that, despite spectacular news stories, Arctic air pollution at its present level appears not to have any significant effects on human health. However, the presence of a sooty haze over the reflecting surface of Arctic ice cannot be ignored, for it may play a significant role in the machinery of global climate. Germane to discussing the climatic influences is learning more about secular trends in air pollutants, and improving numerical climatic modelling in the polar zones.

One thing is cause for celebration: Arctic haze has been caught *impedimentia in actu*—caught in the act *before* many industrial concerns have have migrated poleward. This gives us the chance to understand an important environmental problem before it becomes serious, and in good time for industry, universities and government to take intelligent planning action, rather than expensive after-the-fact remedial action.

I would like to express appreciation to Frank Ferguson, Senator in the Alaskan Legislature, for making it possible to hold this conference and bring these proceedings to light. I thank also the Director and staff of the Scott Polar Research Institute for their several roles in helping to make this conference successful. The conclusions and recommendations agreed in the meetings were sculptured into a self-contained whole by C. C. Wallén, to whom we all owe a debt of gratitude.

PART 1.

COMPOSITION, SOURCE AREAS AND TRANSPORT PATHWAYS

Back: B. Ottar; J. Heintzenberg; K. Rahn; C. Benson
Front: L. Barrie; G. Shaw (Convener); N. Heidam

PART 1. INTRODUCTION

The symposium began with addresses of welcome by Dr David Drewry (Director, Scott Polar Research Institute) and Dr Patrick O'Rourke (Chancellor, University of Alaska-Fairbanks), and opening statements by representatives of the sponsoring bodies. The first day's work was then devoted to six papers that between them outlined the size and scope of the Arctic air pollution problem.

Dr Leonard Barrie, of the Canadian Atmospheric Environment Service, provided the initial overview. The arctic atmosphere is no longer a pristine environment untouched by man. It is continuously polluted by industrial by-products that drift in patchily with air masses from lower latitudes. The pollution becomes most apparent in winter, when stable conditions encourage layering and accumulation, and cleansing precipitation is minimal. Pollutants include soot and other particulate matter, and such gases as sulphur dioxide, perfluorocarbons, pesticides and oxides of nitrogen. Interactions with other atmospheric components give rise to arctic haze, a phenomenon first documented 30 years ago and since studied in detail by chemists and climatologists. The history of atmospheric pollution can be followed from glaciological records: studies of packed snow and ice indicate that arctic atmospheric pollution has increased during the present century, most markedly with the surge of industrial activity of the last few decades.

The following three papers reported Scandinavian approaches to arctic aerosol research. Heintzenberg's Swedish group began with a useful summary of what needs to be known for a full understanding of arctic air pollution. What are the local and long-distance sources, and what quantities of pollution are they contributing? Can we distinguish primary and secondary sources, and their influence on the size, shape and composition of pollutant particles? How are pollutants transported to the Arctic, and what physical and chemical processes affect them in transit? What scavenging processes operate during transport, and in the Arctic itself? How long do pollutants remain in suspension; where and in what form are they deposited? Can we identify sources from the chemical composition ('signature') of a pollution mass? Finally, and perhaps most important of all, what effects have pollutants on the arctic radiation budget? The group's paper summarized the results of six years' Swedish research into some of these problems, reporting a consistent picture of the composition and radiative effects of the arctic aerosol, good understanding of seasonal variations and ageing processes, and progress with

3

modelling. Unresolved issues include variations in concentration between years, chemical processes in the aerosol during both dark and light phases of the arctic year, and identification of regional and remote sources.

Heidam reported on Danish research into the elemental composition of particulate matter in the arctic aerosol, based at four ground stations in Greenland. Chemical analyses of particulates have identified three to five independent components from distant sources, both natural and anthropogenic; fossil fuel combustion in mid-latitudes is a major contributor to particles of anthropogenic origin, and metallic composition indicates that some at least of the polluted Greenland aerosol originates in the industrial Urals area of the USSR. Norwegian investigations, outlined by Ottar and Pacyna of the Norwegian Institute for Air Research, used aircraft to explore the atmosphere over Svalbard. Polluted layers of air were found at altitudes above 2,000 m in summer and 2,500 in winter; for these and lower layers, particle size distributions, chemical composition and wind patterns indicated that industry in the northern USSR contributes substantially to winter pollution over the Norwegian Arctic; the much lower levels of summer pollution include contributions also from Eurasia and North America.

Carl Benson of the University of Alaska-Fairbanks drew attention to underlying reasons for the accumulation and persistence of airborne pollution in the Arctic. Strong, stable air temperature inversions at ground level characterize the whole region, especially in winter. Extending upward for one or two km, they encourage incoming air masses to form thin layers with negligible dispersion above or below. On a regional scale the result is persistent Arctic haze. On a city scale, for example over Fairbanks, domestic and industrial pollutants (especially from gasoline and other fuels) and their secondary products accumulate visibly in the atmosphere, undisturbed by convection and unleached by precipitation. In winter they build up to pollution levels that match those of much larger industrial communities in lower latitudes. Rahn and Lowenthal, of the Center for Atmospheric Chemistry Studies, University of Rhode Island, returned to the question of origins, reporting on a new regional element tracer system that, in combination with meteorological data, enables them to pinpoint by chemical tracers the origins of aerosols. Their contribution includes an analysis of a pollution episode of March 1983, based on early samples from ground level in Svalbard and later samples from the atmosphere above Barrow, Alaska.

ARCTIC AIR CHEMISTRY: AN OVERVIEW

L. A. BARRIE

ABSTRACT. From December to April the arctic air mass is polluted by man-made mid-latitudinal emissions from fossil fuel combustion, smelting and industrial processes. For the rest of the year pollution levels are much lower. This is due to less efficient processes of pollutant removal and better south-to-north transport during winter. In winter, the arctic air mass covers much of Eurasia and North America. Meteorological flow fields, and the distribution of anthropogenic SO_2 emissions in the northern hemisphere favour northern Eurasia as the main source of the visibility-reducing haze. Observations of sulphate concentrations in the atmosphere throughout the Arctic yield a mean for January-April of 1.5–3.9 μgm^{-3} in the Norwegian Arctic to 1.2–2.2 μgm^{-3} in the North American Arctic. An estimate of the mean vertical profile of fine particle aerosol mass shows that, on average, pollution is concentrated in the lower 5 km of the atmosphere. Not only are anthropogenic particles present in the arctic atmosphere, but also gases such as SO_2, perfluorocarbons and pesticides. The acidic nature and seasonal variation of arctic pollution is reflected in precipitation, the snowpack and glacier snow in the Arctic. A pH of 4.9–5.2 in winter and around 5.6 in summer is expected where calcareous windblown soil is absent. Glacial records indicate that arctic air pollution has undergone a marked increase since the mid 1950s, paralleling a marked increase in SO_2 and NO_x emissions in Europe. Real and potential effects of arctic pollution include reduction in visibility, perturbation of the solar radiation budget in April-June, and the acidification and toxification of sensitive ecosystems.

Contents

Introduction

The first documented report of arctic air pollution (coining the term 'arctic haze') was published nearly 30 years ago by Mitchell (1956). Based on the qualitative observations during routine 'Ptarmigan' weather reconnaissance flights in the Alaskan Arctic, he made several observations: (1) haze is comprised of particles no larger than 2 μm diameter that are not ice crystals; (2) it occurs in patches some 800–1,300 km across; (3) it has been observed at all levels in the atmosphere below 9 km; and (4) curiously, it is seldom found over Greenland.

For the next 20 years these observations lay dormant, until unexpectedly high values of total atmospheric turbidity, measured in the Alaskan Arctic during spring of 1971 and 1972 (Shaw and Wendler 1972), prompted a renewed interest in the nature and origin of arctic haze. A subsequent airborne study (Holmgren and others 1974) found layers of anomalously high turbidity at altitudes of a few kilometers. About 40% of the total turbidity was above 4 km. Next came an aircraft study of the chemical composition of haze particles comprising 15 flights during April and May 1976 (Rahn and others 1977). Although particulate matter in air between distinct haze layers was due to pollution (as indicated by enrichment in oil-derived vanadium) the haze layers themselves were of crustal composition, originating presumably from Asian deserts between 40 and 50° N. Many thus believed that arctic haze consisted mainly of windblown dust, and not until the first routine ground-level observations of the chemical composition of suspended particulate matter in the winter of 1976–77 (Rahn and McCaffrey 1979a, 1980) did another picture emerge. Arctic haze particles in the lower troposphere of the Alaskan Arctic were found to vary seasonally: they were 20–40 times more abundant in winter than in summer. Furthermore, during winter they were mostly of man-made origin, while in summer what little there was consisted of wind-blown dust and sea salt.

The investigations of Rahn and McCaffrey (1979a, 1980) and the first symposium on arctic aerosols held in April 1977 in Norway initiated a surge of research into arctic air pollution. In this overview, the current state of knowledge of arctic air pollution is summarized. Previous reviews of arctic haze have been published (Rahn and McCaffery 1979a, Rahn 1982a, 1982b) and three important compendia of arctic research results are found in Arctic Symposium (1981, 1985) and in AGASP (1984).

Arctic pollution meteorology

The arctic air mass is uniquely characterized by: (1) sub-zero temperatures much of the year; (2) little precipitation (most of it falling in the warmer half of the year); (3) stable stratification preventing strong vertical mixing; and (4) low levels of solar radiation, especially in winter. During winter it extends from the polar region over much of snow-covered North America and Eurasia (Figure 1). Three-dimensionally the winter air mass consists of a dome 7–8 km deep over the pole with shallow tongues of air 0–5 km deep spilling southward over the land masses. In summer it is confined to the northern polar region. There is more cloud in summer than in winter, mainly stratus clouds in the lower troposphere. This is well illustrated by the annual variation in occurrence of lower (0–2 km), middle (2–4 km) and upper (4–7 km) level clouds, as well as clear skies in the western Eurasian Arctic (Figure 2). While low cloud varies seasonally from about 75% between May and September to about 35% between December and March, middle and high cloud frequencies vary less. Clear skies are very rare in summer (5–8%) and more frequent in winter (30–40%). These results are typical of the arctic region. Since low summer stratus is often formed by warm air advection over cold ice pack, it is frequently associated with drizzle, which is important in the removal of air pollution. Indeed, it is observed that the annual variation of low level cloud cover is in anti-phase to the annual variation of anthropogenic SO_4^- concentration in the lower arctic troposphere (Barrie and others 1981). An investigation of arctic cloudiness since 1920 (Raatz 1981) has revealed no significant trends that might be ascribed to increasing levels of air pollution.

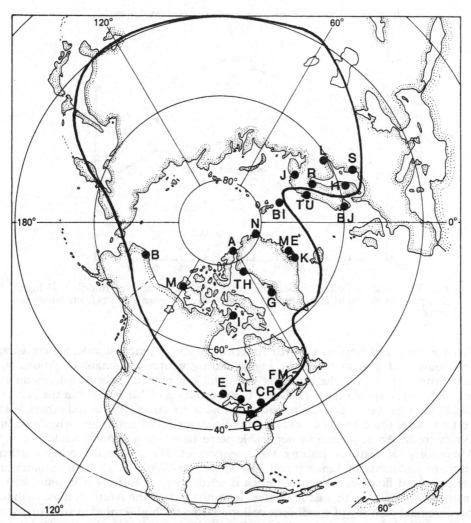

Fig 1. The mean position of the arctic front in January (solid line) from Barrie and Hoff (1984) and the location of air monitoring stations providing observations used in Figure 7. A-Alert; AL-Algoma; B-Barrow; BI-Bear Island; BJ-Birkenes; CR-Chalk River; E-ELA-Kenora; FM-Forêt Montmorency; G-Godhavn; H-Hoburg; I-Igloolik; J-Jergul; K- Kap Tobin; L-Lesogorski; LO-Long Point; M-Mould Bay; ME-Mestersvig; N- Nord; R-Ricklea; S-Suwalik; TH-Thule; TU-Tustervaten.

Pollution sources and transport pathways

Considering the location of pollution sources, the arctic air mass, and the major features of atmospheric circulation in the northern hemisphere, it is apparent that northern Eurasian pollution is far more available to the Arctic than that in North America. An inventory of SO_2 emissions (a major source of arctic haze) in regions that influence arctic air at some time during the winter (Figure 3) reveals that Eurasian SO_2 emissions in winter are more than double those in North America (54×10^6 against 24×10^6 tonnes per year). The difference is even greater in the region north of latitude 60° (6.3×10^6 against 0.013×10^6 tonnes per year). This reflects the relative populations of

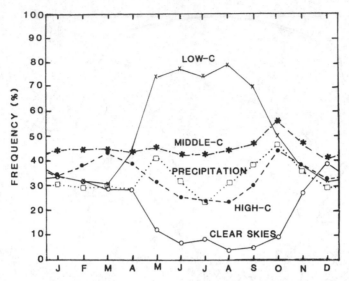

Fig 2. Temporal variation of mean monthly frequencies of low, middle and high cloud, precipitation and clear skies in the western Eurasian Arctic (from Huschke 1969).

northern Eurasia and northern North America. In the absence of volcanic emissions, natural sources of SO_2 in the arctic region during winter are small. Occasionally, a volcano erupts in or near the arctic air mass and can provide substantial amounts of natural SO_2 to the atmosphere. Evidence of the eruption of Mt Katmai in the Aleutian islands in 1912 has been seen in the acidity of glacial ice cores (Barrie and others 1985). Since that time, there has not been a volcanic eruption of sufficient magnitude and proximity to the Arctic to cause a noticeable perturbation to ice core records.

Meteorological features causing the transport of SO_2 and most other northern hemisphere pollutants in January and July appear in Figure 4. In both summer and winter the mean flow over North America is south-east; in Eurasia it is south-east in summer but strongly north-east in winter. In winter, when the Arctic is most polluted, the predominant pathway of windborne pollution from the high-emission areas of eastern North America and south-eastern Asia to the pole is eastward over the stormy Atlantic. In contrast, Eurasian pollution travels north-east over snow covered land-ice areas. There is a marked difference in the precipitation encountered along these respective pathways (Figure 5). The cumulative mean monthly precipitation encountered by North American and south-east Asian pollution on its most probable route to the Arctic is at least four times greater than that encountered by Eurasian sources. Strong transport into the Arctic from Eurasia during winter is caused by the presence of the persistent Siberian high pressure region. In a detailed analysis of weather patterns, Raatz and Shaw (1984) confirmed that the link between the Arctic and North America is weak, while the Eurasian connection is strong. Surges of air northward out of Eurasia during winter were associated with weather patterns that typically involved the tracking of a low pressure cell from Iceland into the Murmansk area or across Europe until it encounters the semi-persistent Siberian high. At this point, a northward flow of air between the low and high pressure zones is maintained for several days. This results in a surge of polluted European air into the Arctic.

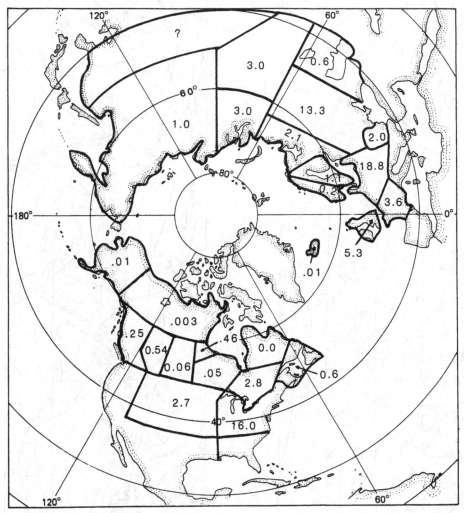

Fig 3. Annual emissions of SO$_2$ (10^6 tonnes) in regions of the northern hemisphere that influence the Arctic.

Observing arctic air pollution

Occurrence of particulate pollution

Temporal and spatial distributions of anthropogenic arosols in the arctic atmosphere has been under routine investigation at ground level for a number of years. Seasonal variation of anthropogenic sulphate (excess SO$_4^-$) concentration is remarkably consistent throughout the Arctic. Weekly mean excess SO$_4^-$ at three locations in the Canadian Arctic for the winters of 1980 to 1983 (Figure 6) demonstrate the year-to-year consistency as well as the spatial coherency of the haze over thousands of kilometers. Observations from the arctic aerosol sampling network, as well as acid rain networks in Europe and eastern Canada, enable one to reconstruct the spatial distribution at ground level of the

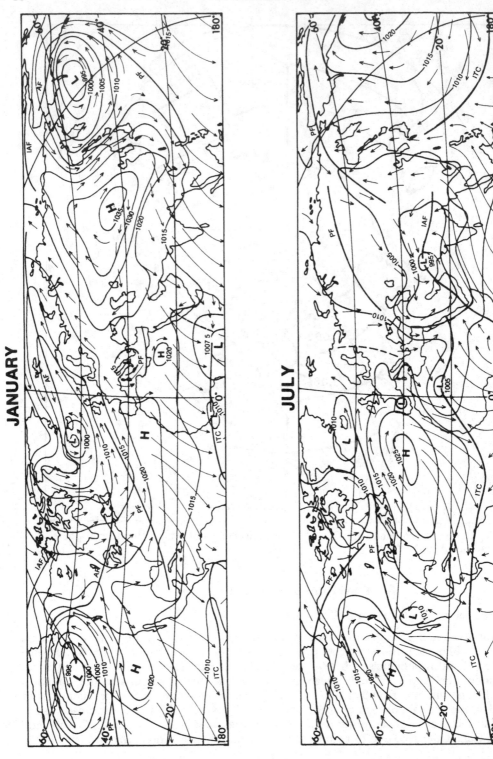

Fig. 4. Average values of surface pressure (thin lines; unit mbar), winds (long arrows: steady winds; short arrows: unsteady winds), frontal zones and convergence zones in the northern hemisphere during January and July (from Liljequist 1970).

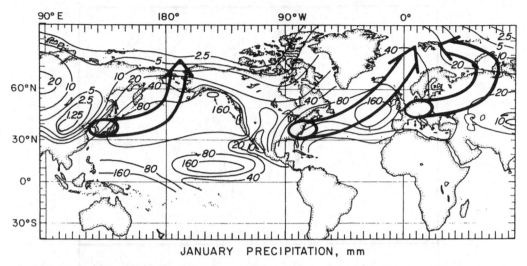

JANUARY   PRECIPITATION, mm

Fig 5. Spatial distribution of mean monthly precipitation in January (mm of water). Arrows indicate the predominant pathway of air out of major pollutant source regions to the Arctic (from Rahn 1982a).

arithmetic mean excess SO_4^- concentration for the months of January-April 1980 (Figure 7). There is a minimum in SO_4^- concentration located in the North America-western Greenland Arctic, between the influence of Eurasian and eastern North American sulphur sources.

The vertical distribution of arctic aerosol is less well-known than the ground-level distribution. Of vertical soundings published to March 1985, 23 are of the aerosol optical scattering parameter b_{scat} (Leaitch and others 1984; Schnell and Raatz 1984; Hoff and Trivett 1984) in dry air, two are of the aerosol optical extinction coefficient b_{scat} (Shaw 1975) and two are of particulate carbon concentration (Hansen and Rosen 1984). They were all taken in the spring months, March and April. No measurements have been reported for the period October to February. Since b_{scat} (and b_{ext} which is approximately equal to b_{scat}) is well correlated with fine particle (accumulaion mode) mass (Waggoner and Weiss 1980), and since most SO_4^- and most other anthropogenic particulate constituents are concentrated in the fine particle mode, vertical profiles of b_{scat}, b_{ext} and particulate carbon were used to estimate the average vertical distribution of pollution in the Arctic. The average profile of particulate pollution and its one-standard-deviation bounds of variability is shown in Figure 8. The height scale is referenced to the lowest level an aircraft measurement can be made, about 300 m. Ground-level concentrations are sometimes lower that at 300 m. The particulate mass concentration decreases with height in the first 2 km of atmosphere to about 50% of the near-surface value. It remains roughly constant up to 5 km. Then between 5 and 7 km, it decreases to about 30% of the near surface value. The large bounds of variability indicate that at any given time a profile can differ greatly from the mean. On average, however, particulate pollution in spring is found throughout the depth of the arctic air mass and tends to be highest in the lowest 2 km.

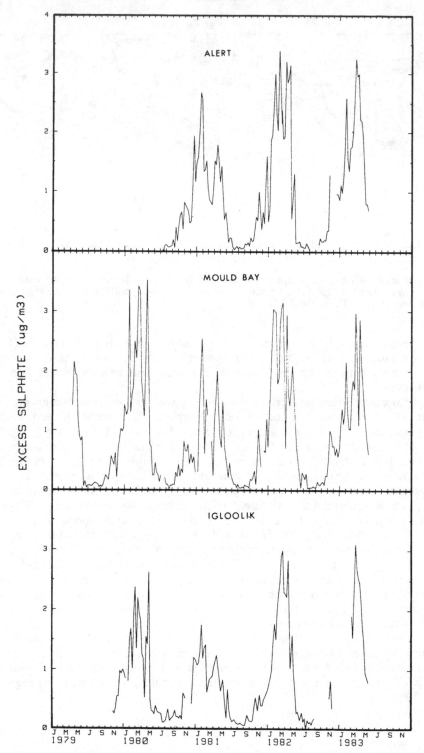

Fig 6. Temporal variation of weekly mean sulphate in the atmosphere at three locations in the Canadian Arctic (A, M and I in Figure 1); from Barrie and Hoff (1985).

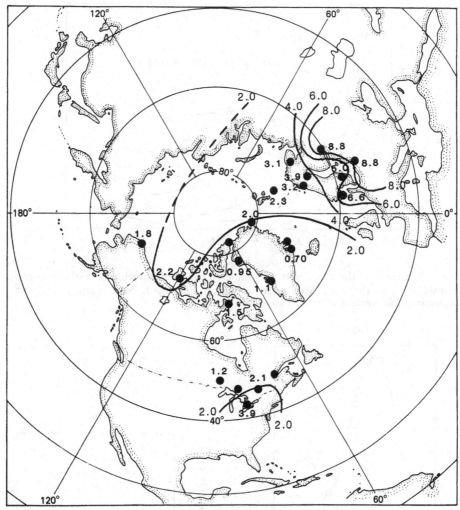

Fig 7. Spatial distribution of the mean ground-level concentration of sulphate in the atmosphere for the period January to April 1980. (See Figure 1 for station identification).

Size distribution of particulates in the Arctic

The number and volume size distributions of aerosols, as well as the mass size distributions of their chemical constituents, are information-rich characteristics that provide insight into mechanisms of haze formation, origin of particulate matter, and removal rates. Typical number and volume size distributions measured in spring haze near the ground and aloft are shown in Figure 9. Particles are most numerous in the diameter range 0.005–0.2 μm. Above 0.2 μm, their number concentration (N) decreases rapidly with increasing diameter (D). In the wide range 0.1–30 μm, N depends roughly on $D^{-3.3}$ (Figure 9a). In the winter arctic air mass, the total particle number concentration (N_T) ranges from approximately 10–4,000 cm^{-3} with a geometric average of 200–350 cm^{-3}.

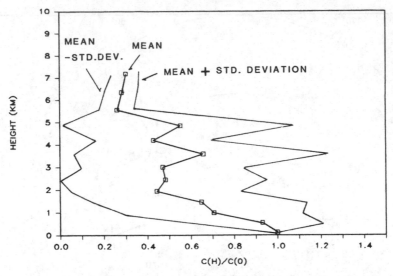

Fig 8. Estimate of the mean vertical profile of the concentration of anthropogenic aerosol mass in the high Arctic during March and April, based mainly on light scattering observations from aircraft.

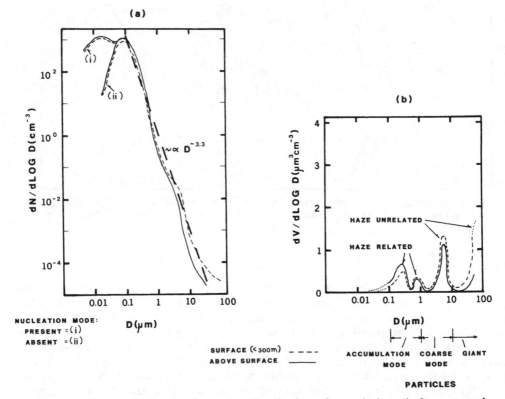

Fig 9. The number and volume size distribution of a typical arctic haze aerosol (adapted from Radke and others 1984; Bigg 1980; Shaw 1984).

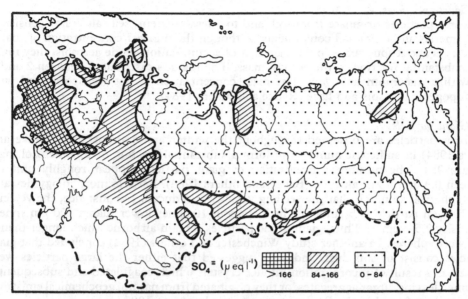

Fig 10. The spatial distribution of snowpack sulphate concentration in the Soviet Union during spring 1982 (from Belikova and others 1984).

The shape of the number size distribution is highly variable below 0.1 μm (Figure 11a). Sometimes it decreases much less with decreasing particle size than at other times, often showing a second maximum at about 0.02 μm (Heintzenberg 1980, Shaw 1984, Radke and others 1984). This is believed to be due to the presence or absence of nucleation-mode aerosols by condensation of unstable gaseous compounds in gas-phase reactions, formed with the involvement of SO_2 and possibly organobromides.

The main volume (or mass) of aerosol in the winter Arctic (Figure 9b) is concentrated in larger particles than aerosol number. Mass is distributed in three aerosol modes whose formation, optical properties, chemical composition, origin and removal rates differ greatly. These include: (1) the accumulation mode (D = 0.1–1.0 μm), (2) the coarse particle mode (D = 1–10 μm) and (3) the giant particle mode (D mm).

(1) *Accumulation mode particles*

Accumulation mode particles efficiently scatter solar radiation and hence reduce visibility. In the absence of precipitation, they are present in the atmosphere longer than other particles because the rate of dry deposition to the earth's surface reaches a minimum between 0.1 and 1 μm diameter (Ibrahim and others 1984). In winter they are generally anthropogenic in origin, forming from matter emitted directly to the atmosphere, from the coagulation of nucleation mode particles and from gas-particle interactions. Several investigations have confirmed that particle mass in this size range is concentrated in a distribution centred at about 0.3 μm. In a detailed study of accumulation mode particle morphology and SO_4^- content at Barrow, Alaska Bigg (1980) noted that: (1) winter size distributions are remarkably constant; (2) sulphuric acid is the dominant winter aerosol; and (3) over half the acid particles contained insoluble single particle inclusions, with shapes ranging from spheres through irregular but compact forms to highly irregular aggregates of small particles.

(2) *Coarse particles*

Coarse particles originate from soil, and to a lesser extent, sea salt. They consist of clay minerals and other soil constituents. Although the number concentration of coarse particles is about four times lower than that of accumulation-mode aerosols, they make up a substantial fraction of the aerosol mass. The mass is distributed between 2 and 8 μm with a maximum at 4 μm. The mass concentration is not well correlated with anthropogenic aerosols and haze (Radke and others 1984).

(3) *Giant particles*

Giant particles are observed throughout the winter arctic atmosphere (Radke and others 1984) in sufficient numbers to form a substantial fraction of total aerosol mass (Figure 9b). The number concentration of giant particles appear roughly uniform between 0.3 and 5 km altitude (Bailey and others 1984). Particle morphology deviates from spherical so much that the settling velocity is reduced. For instance, the velocity of a 100 μm particle is 5 cm s^{-1}, which is about 10 times lower than a 100 μm sphere of density 2.5 g cm^{-3}. Thus, giant arctic particles remain airborne much longer than if they were spheres. In another study Winchester and others (1984) concluded that giant particles consist of soil dust, and they suggested that either the large particles were formed as a result of incorporation into ice crystals which coagulated and subsequently sublimed leaving larger aggregates, or they originated from natural geochemical emissions from Bennett Island in the Soviet eastern Arctic in spring 1983.

The mass size distribution of aerosol trace constituents in the winter arctic aerosol has been measured in a number of studies (Heintzenberg and others 1981, Hoff and others 1983, Pacyna and others 1984, Radke and others 1984). Most of the total mass of particulate NH_4^+, SO_4^-, V, Ni, Cr, Pb, Cu, Br and I is in the accumulation mode range (0.1–1 μm) while NO_3^-, Cl^-, Na^+, Mg^{++}, Al, Mn, Ca, Ti and Fe have not only a sub-micrometer fraction but also a substantial coarse particle component (1–10 μm). The cation-anion budget in aerosol measured by Hoff and others (1983) at Igloolik, Canada in February showed that in particles of diameter less than 1 μm, the dominant soluble ions were H^+, NH_4^+ and SO_4^-, while in particles greater than 3 μ NO_3^-, Na^+ and Mg^{++} and Cl^- were dominant. Between 1 and 3 μm, a mixture of sea salt and acidic sulphates occurred. Thus, the haze-related accumulation mode aerosol is acidic.

Gaseous pollutants

Anthropogenic gases are important as sources of particulate matter, as absorbers of infrared radiation, as agents of biospheric toxification and as tracers of transport pathways to the Arctic. Sulphur dioxide, an important source of SO_4^- particulate matter, is most concentrated in January and February, with levels of 0.5–1.5 ppb(v) in the North American Arctic that increase toward Eurasia. The strong seasonal variation is mainly due to seasonal variations in the oxidation rate of SO_2 in the atmosphere (Barrie and Hoff 1984).

Carbon dioxide gas contributes to the atmospheric greenhouse effect and hence warming. Throughout the world its concentration is steadily increasing through man's activities. Concentrations vary seasonally in the Arctic more than farther south, for instance at Mauno Loa, Hawaii (Peterson and others 1982). The main features of the seasonal variation, which are remarkably similar to that of aerosol, are due to CO_2 uptake during photosynthesis and release during respiration and biological decay. Since winter biospheric uptake is lowest in the arctic air mass, CO_2 levels increase more than over vegetated continents or over oceans. There is direct evidence linking CO_2 with haze aerosol in the Alaskan Arctic. Furthermore, relative maxima in winter CO_2 concentrations at Barrow and Alert are associated with air mass pathways from Eurasia (Peterson and others 1980; Higuchi and Daggupaty 1983; Halter and others 1985).

Kahlil and Rasmussen (1983) and Rasmussen and Khalil (1984) have shown that anthropogenic organic gases with residence times long enough to ensure transport over thousands of km, but short enough for anthropogenic emissions to be readily observed above global background, are useful as transport tracers and possibly as source indicators. Measurements at Point Barrow, Alaska indicate that $CHClF_2$, CCl_3F, CH_3CCl_3, C_2Cl_4, C_2HCl_3, CO, CH_3Cl, $C_2H_4Br_2$, $CBrClF_2$, CHBrCl and $CHBr_2$ fulfill these criteria. These gases have seasonal variation in concentration parallel to that of arctic haze. Variation is due mainly to seasonal variations in transport, rather than to gas phase chemical destruction mechanisms. Thus seasonal variations in arctic haze are caused not only by variations in pollutant removal efficiency, but also in transport between mid-latitudinal sources and the Arctic. This confirms what one would expect from meteorological flow patterns (Figure 4).

Organic gases are important as infra-red absorbers, enhancing the greenhouse effect, and as atmospheric tracers. Kahlil and Rasmussen (1984) investigated the relationship between various atmospheric trace gases and their concentration difference between haze and non-haze layers in spring. Gases with well correlated concentrations, generally of anthropogenic origin, were more abundant within haze layers than outside them; they included high-temperature combustion products CO_2, CO and C_2-C_6 hydrocarbons as well as the industrial and commercial chlorofluorocarbons or chlorocarbons (F-11, F-12, CCl_4, CH_3, CCl_3, perchloroethylene, trichloroethylene, dichloromethane and chloroform).

At Spitsbergen, concentrations of halocarbons and light hydrocarbons were an order of magnitude higher in March 1983 than in July 1982; ethene and propene were exceptions (Hov and others 1984). Compared with samples from Barrow, air reaching Spitsbergen from an easterly direction (ie the central Soviet Union) is much richer in alkanes (presumably from oil and gas combustion), much less abundant in chlorinated ethenes, and comparably abundant in benzene and toluene. Long-range transport of pesticides to the Norwegian Arctic has been documented by Oehme and Ottar (1984). The accumulation of polychlorinated hydrocarbons in arctic ecosystems can be expected. Measurements of the pesticide chlordane at Mould Bay, Canada, in June and July 1984 (Hoff and Chan 1986) confirm the existence of this potentially toxic man-made substance in an area very remote from sources. They indicate that the atmospheric pathway may be important in explaining the levels of such chemicals found in fish and mammals of the North American Arctic (Muir and others 1986). More measurements of airborne organic compounds are needed to establish seasonal variations.

Deposition of pollutants: precipitation and glacier evidence

Pollutants are deposited on arctic ecosystems by precipitation scavenging (wet deposition) and by direct uptake at the earth's surface (dry deposition). The processes involved are discussed in detail by Barrie (1985). No routine, quality-controlled chemical measurements of precipitation are currently being made in the Arctic. Low winter precipitation (typically 5 mm water equivalent per month) together with relatively low pollutant concentrations and the frequent occurrence of blowing snow, makes measurement very difficult. Among summer measurements McNeely (1982) recorded the pH of precipitation on Ellesmere Island, Canada in June and July 1981 as 5.7 ± 0.5. At Poker Flats, Alaska, 16 samples of precipitation collected in December-January 1979 and in

May-December 1980 had a mean pH of 5.0, due mainly to sulphuric acid (Galloway and others 1982). Dayan and others (1985) examined these and other Alaskan summer measurements, confirming the dominance of sulphuric acid; formic and acetic acids contributed at most 33% to free acidity. Few measurements were available for winter when arctic air pollution is at its worst. McNeely and Gummer (1984) surveyed the snowpack on central Ellesmere Island between May and August 1979–81, finding measureable concentrations of several pesticides (Table 1), with lindane and its isomer α-BHC most abundant.

Table 1. A summary of the concentrations (μ g/l) of man-made pesticides in the snowpack of central Ellesmere Island in May-August 1979–81 (McNeely and Gummer 1984).

Pesticide	No. of analyses	No. of samples above detection limit	Concentrations min	max
α-BHC	28	21	<.001	0.018
γ-BHC (Lindane)	28	10	<.001	0.008
α-Chlordane	28	1	<.001	0.002
γ-Chlordane	28	2	<.002	0.002
α-Endosulfan	28	1	<.001	0.002
op' DDT	28	2	<.001	0.002
pp' DDT	28	2	<.001	0.002
HEOD (Dieldrin)	28	5	<.002	0.004

A snowpack chemistry survey in the Soviet Union during spring 1982 (Belikova and others 1984) yielded the spatial distribution of sulphate concentration shown in Figure 10. Results are consistent with the configuration of SO_2 emissions in Eurasia (Figure 3). Concentrations of SO_2 are highest in the western Soviet Union around Norilsk, in eastern Siberia, and in the south central region. A snowpack survey on Spitsbergen in spring 1983 (Semb and others 1984) yielded mean concentrations (μeq/l) in the top 30 cm of H^+, 12; NH_4^+ and NO_3^-, < 1; non-sea salt SO_4^-, 12. Mean snow pH was 4.9, most of the acidity being associated with the SO_4^- ion. A snowpack chemistry survey in the Northwest Territories of Canada in March 1984 (Shewchuk 1985) at sites on the pre-Cambrian shield remote from calcareous soils, yielded H^+, NH_4^+, NO_3^- and SO_4^- concentrations in the range 2–10, 1–2, 5–7, and 7–18 μeq/l respectively. In March 1978 the snowpack in north-eastern Alberta at sites remote from local sources had H^+ concentration of 8–20 μeq/l^{-1} and SO_4^- and NO_3^- as the predominant anions (Barrie 1980). Thus the snowpack is slightly acidic and there is sufficient SO_4^- and NO_3^- to explain the acidity. Weiss and others (1978) and Rahn and McCaffrey (1979b) investigated concentrations of various trace elements in the winter snowpack of northern Alaska.

Glacial ice also provides information on pollutant deposition. Measurements of snow and ice composition have been made in Greenland (Hammer and others 1980; Herron 1982; Davidson and others 1981 and in press; Neftel and others 1985), on northern Ellesmere Island (Koerner and Fisher 1982; Barrie and others 1985) and on Mt Logan, northwestern Canada (Holdsworth and Peake 1984). The elevation and location of these sampling sites are important factors in interpreting the observations. For instance, the elevation of northern Ellesmere Island is 1.6 km,, that of Greenland 2.5–3.0 km and Mt Logan 5.3 km; considering the vertical profile of arctic air pollution (Figure 8), Ellesmere Island glaciers receive polluted lower tropospheric air, while Mt Logan and Greenland glaciers receive snow from the less polluted middle and upper troposphere.

The conductivity and acidity of snow on Agassiz Glacier, northern Ellesmere Island, are well correlated and undergo a strong seasonal variation paralleling that of arctic air

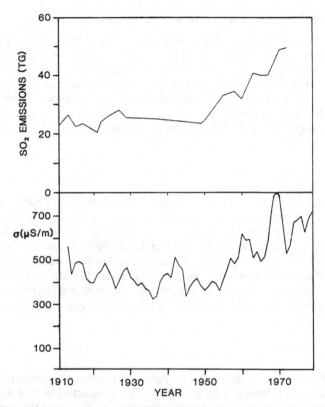

Fig 11. A comparison of the historical record of annual maximum conductivity in meltwater from the Agassiz Glacier, Ellesmere Island, and SO₂ emissions from Europe (including the western Soviet Union): from Barrie and others 1985. The conductivity record is a three-year running mean.

pollution (Barrie and others 1985). The highest annual conductivity (acidity) occurs at the peak of arctic air pollution, and is therefore an indicator of pollution levels. Indeed, comparison of the historical record of maximum ice core conductivity (acidity) with that of SO_2 and NO_X emissions in Europe (including the western Soviet Union) during this century supports this hypothesis (Figure 11). As SO_2 emissions remained roughly constant in the first part of the century, so ice core conductivity remained roughly constant, with the exception of a peak in 1912–13 associated with the eruption of Mt Katmai in the Aleutian Islands. Then between 1956 and 1977, as SO_2 emissions in Europe doubled and NO_X emissions increased remarkably, ice core conductivity increased by 75%. Current levels of maximum winter acidity are 12–14 μeq/l compared with 7–9 μeq/l before 1956. Limited historical records of aerosol optical depth of the arctic atmosphere in March agree with the general trends suggested by the Agassiz ice cap data (Shaw 1982).

Implications and effects of arctic air pollution

Arctic air pollution is widespread and covers large parts of the northern hemisphere during the winter period January–April, but is it really a cause for concern? What are the implications of arctic haze? Are there serious environmental effects? One important implication of haze is that during the winter half of the year the arctic air mass is more easily polluted than any other air mass on the globe. Pollutants are inefficiently dispersed and removed in the cold northern atmosphere, a fact to be taken into account in planning future industrial development. Several real or potential effects of arctic air pollution currently warrant attention.

Visibility reduction

Particles scatter solar radiation, making distant objects obscure and reducing visibility. Using the Koschmieder relationship between visibility and the optical extinction coefficient, and the correlation between aerosol scattering in dry air and SO_4^- aerosol concentration (Barrie and others 1981, Barrie and Hoff 1985), it can be shown that for SO_4^- concentrations of 1–4 μgm^{-3} typically found in the Arctic in winter, one can expect visibilities in situations of low relative humidity to be from 244 km to as low as 78 km. However, visibility is often reduced to 10 km in the winter Arctic, even in the absence of blowing snow. The most likely cause is atmospheric ice crystals, which often accompany haze aerosols.

Climatic effects

Soot-bearing anthropogenic particles in the Arctic can affect the atmospheric energy budget and hence influence climate. The energy budget is affected in two ways. Firstly, as a black aerosol suspended over a white surface, pollution increases the net amount of solar radiation trapped in the troposphere (Valero and others 1984). Secondly, soot-ladened aerosols reduce the albedo of the snowpack when they are deposited, increasing the amount of sunlight trapped. Clarke and Noone (1985) estimated that the albedo is decreased by 1–3% in new snow and by 3–10% in old snow. This effect can last well into May and June, even after airborne anthropogenic pollution has disappeared. During spring melt, the blackness of the snowpack increases as water runs off but soot remains. Resulting changes in the solar radiation budget may modify local climate, and possibly the hemispheric climate. Attempts to model perturbations of the solar radiation budget have been made using radiative transfer models (Shaw and Stamnes 1980; Porch and McCracken 1982, Cess 1983). However, as Cess points out, they suffer from numerous drawbacks one of which is the assumption of clear skies. Furthermore, they are not climatic models that link atmospheric dynamic processes and climate feed-back mechanisms with the perturbation. Arctic aerosols need to be incorporated into global climate models as the next step in assessing the impact of arctic air pollution on climate. A preliminary attempt to examine the effects of aerosols on a global basis (Potter and Cess 1984) led to the conclusion that particulate matter reduces global mean surface temperature by 3–4°C, and that maximum cooling occurs at high latitudes even though particles produce an increase in solar heating at these latitudes. However, this model did not take into account the high absorptivity of black arctic aerosols, reduced albedo of the snowpack discussed above, or seasonal variations in the aerosol.

Pollutant deposition effects

Arctic winter snow is slightly acidic, with pH ranging from 4.9 to 5.2. Analyses of the effects of acidic snow on delicate ecosystems in the Arctic have yet to be published. There is some concern that during spring melt, selective leaching of H^+ ions from the snowpack by the first meltwater can lead to a three-to-five-fold concentration of acidity and an acid shock (Johannessen and Henriksen 1978). Runoff is not likely to be acidic unless the soil is poorly buffered, but large areas of the Canadian and Soviet Arctic have poorly buffered soils; the impact of acidic runoff at the start of spring melt in these areas should be assessed.

Deposition of man-made pesticides transported from mid-latitudes (for example lindane on Ellesmere Island; see above) is potentially more serious. However, any biological effects have yet to be assessed. Few budgets of trace elements in arctic ecosystems have been calculated to assess the importance of the atmosphere as a source. One by Rahn (1981a) estimated inputs of Al, V, Mn, Cd, Pb, SO_4^- and NO_3^- to the Arctic Ocean, concluding that Pb was the only element whose atmospheric source equalled or exceeded riverine and oceanic sources. More such studies are needed for pesticides.

Conclusion

The arctic atmosphere is not a pristine environment untouched by man. In the winter period of December to May, elevated levels of man-made particulate matter, originating mainly from mid-latitudinal sources in Eurasia, are found throughout the Arctic. For the rest of the year, particulate pollution is either absent or present at much lower concentrations. In the last 10 years we have come a long way towards understanding the occurrence, nature, history and origin of arctic pollution. However, much remains to be learned, particularly in describing the occurrence of pollution in the Soviet Arctic, in understanding the interaction between particulate pollution and the water vapor cycle, and in assessing the effects of pollution on visibility, climate and terrestrial/aquatic ecosystems. More measurements of pollutant vertical profiles are needed, and more research toward the development of chemical-transport models for the northern hemisphere north of 30°. These models will be useful in the planning of future industrial development in the north. Models, in conjuction with chemical tracer techniques involving aerosols and gases (see Rahn and Lowenthal, these proceedings) have the potential to define more quantitatively the relative contributions of arctic pollution sources.

Acknowledgements

I thank R. Hoff, K. Rahn, N. Neidam and A. Semb for providing data and helpful comments, and A. Smith and B. Martin for technical assistance.

References

AGASP. 1984. *Geophysical Research Letters*, 11(5).
ARCTIC SYMPOSIUM. 1981. *Atmospheric Environment*, 15(8).
ARCTIC SYMPOSIUM. 1985. *Atmospheric Environment*, 19(12).
BAILEY, I. H. AND OTHERS. 1984. Airborne observations of arctic aerosols, 2: giant particles. *Geophysical Research Letters*, 11: 397–400.
BARRIE, L. A. 1980. The fate of particulate emissions from an isolated power plant in the Oil Sands area of western Canada. In KNEIP, T. AND LIOY (editors). *Aerosols: anthropogenic and natural, sources and transport. Annals of the New York Academy of Science* 338: 434–52.

BARRIE, L. A. 1985. Atmospheric particles; their physical/chemical characteristics and deposition processes relevant to the chemical composition of glaciers. *Annals of Glaciology*, 7: 100–108.

BARRIE, L. A. AND HOFF, R. M. 1984. The oxidation rate and residence time of sulphur dioxide in the arctic atmosphere. *Atmospheric Environment*, 18: 2711–22.

BARRIE, L. A. AND HOFF, R. M. 1985. Five years of air chemistry observations in the Canadian Arctic. *Atmospheric Environment*,

BARRIE, L. A. AND OTHERS. 1981. The influence of mid-latitudinal pollution sources on haze in the Canadian Arctic. *Atmospheric Environment* 15: 1407– 19.

BARRIE, L. A. AND OTHERS. 1985. Twentieth century trends in arctic air pollution revealed by conductivity and acidity observations in snow and ice in the Canadian high Arctic. *Atmospheric Environment*, 19(12): 1995–2010.

BELIKOVA, T. V. AND OTHERS. 1984. Characteristics of snow-cover background pollution by sulphates in the territory of the USSR. *Meteorology and Hydrology*, 9: 47–56.

BIGG, E. K. 1980. Comparison of aerosol at four baseline monitoring stations. *Journal of Applied Meteorology*, 19: 521–23.

CESS, R. D. 1983. Arctic aerosols: model estimates of interactive influences upon the surface-atmosphere clear-sky radiation budget. *Atmospheric Environment*, 17: 2555–64.

CLARKE, A. D. AND NOONE, J. 1985. Measurements of soot arosol in arctic snow. *Atmospheric Environment*, 19(12): 2045–54.

DAVIDSON, C. I. AND OTHERS. 1981. Wet and dry deposition of trace elements onto the Greenland ice sheet. *Atmospheric Environment*, 15: 1429–38.

DAVIDSON, C. I. AND OTHERS. 1985. Atmospheric transport and deposition of trace elements onto the Greenland Ice Sheet. *Atmospheric Environment* 19(12): 2065–82.

DAYAN, U. AND OTHERS. 1985. An analysis of precipitation chemistry data from Alaska. *Atmospheric Environment*, 19(4): 651–58.

GALLOWAY, J. N. AND OTHERS. 1982. The composition of precipitation in remote areas of the world. *Journal of Geophysical Research*, 87: 8771–86.

HALTER, B. C. AND OTHERS. 1985. A study of winter variability in carbon dioxide and Arctic haze aerosols at Barrow, Alaska. *Atmospheric Environment*, 19(12): 2033–38.

HAMMER, C. U. AND OTHERS. 1980. Greenland ice sheet evidence of past glacial volcanism and its climatic nature. *Nature*, 288: 230–35.

HANSEN, A. D. A. AND ROSEN, H. 1984. Vertical distribution of particulate carbon, sulphur and bromine in the arctic haze and comparison with ground- level measurements at Barrow, Alaska. *Geophysical Research Letters*, 11: 381–84.

HEINTZENBERG, J. 1980. Particle size distribution and optical properties of arctic haze. *Tellus*, 32: 251–60.

HEINTZENBERG, J. AND OTHERS. 1981. The chemical composition of arctic haze at Ny-Ålesund, Spitsbergen. *Tellus*, 33: 162–71.

HERRON, M. M. 1982. Impurity sources of F^-, Cl^-, NO_3^- and $SO_4^=$ in Greenland and antarctic precipitation. *Journal of Geophysical Research*, 87: 3052–60.

HIGUCHI, K. AND DAGGUPATY, S. M. 1985. Variability of CO_2 at station Alert. *Atmospheric Environment* 19(12): 2039–44.

HOFF, R. M. AND CHAN, K. W. 1986. Atmospheric concentrations of chlordane at Mould Bay, NWT, Canada. *Chemosphere*, in press.

HOFF, R. M. AND TRIVETT, N. B. A. 1984. Ground based measurements of arctic haze made at Alert NWT, Canada, during the arctic gas and aerosol sampling project (AGASP). *Geophysical Research Letters*, 11: 389–92.

HOFF, R. M. AND OTHERS. 1983. Mass-size distributions of chemical constituents of the winter arctic aerosol. *Journal of Geophysical Research*, 88: 10947–56.

HOLDSWORTH, G. AND PEAKE, E. 1985. Acid content of snow from a mid- troposphere sampling site on Mt Logan, Yukon Territory. *Annals of Glaciology*, 7: 153–60.

HOLMGREN, B. AND OTHERS. 1974. Turbidity in the arctic atmosphere. *AIDJEX Bulletin*, 27: 135–48.

HOV, O. AND OTHERS. 1984. Organic gases in the Norwegian Arctic. *Geophsyical Research Letters*, 11: 425–28.

HUSCKE, R. E. 1969. *Arctic cloud statistics from 'air calibrated' surface weather observations*. US Air Force Project Rand Contract No F44620–67–C– 0015.

IBRAHIM, M. AND OTHERS. 1984. An experimental and theoretical investigation of the dry deposition of particles to snow, pine trees and artificial collections. *Atmospheric Environment*, 17: 781–88.

JOHANNESSEN, M. AND HENRIKSEN, A. 1978. Chemistry of snowlmelt: changes in concentration during melting. *Water Resources Research*, 14(4): 615–19.

KHALIL, M. A. K. AND RASMUSSEN, R. A. 1983. Gaseous tracers of arctic haze. *Environmental Science Technology*, 17: 157–64.

KHALIL, M. A. K. AND RASMUSSEN, R. A. 1984. Statistical analysis of trace gases in arctic haze. *Geophysical Research Letters*, 11: 437–40.

KOERNER, R. M. AND FISHER, D. 1982. Acid snow in the Canadian high Arctic. *Nature*, 295: 137–40.

LEAITCH, W. R. AND OTHERS. 1984. Some physical and chemical properties of the arctic winter aerosol in northeastern Canada. *Journal of Climatology and Applied Meteorology*, 23: 916–28.

LILJEQUIST, G. H. 1970. *Klimatologi* (in Swedish). Stockholm, General stabens Litografiska Anstalt.

McNEELY, R. 1982. Ambient pH levels in environmental samples from the high Arctic. *Scientific and Technical Notes in Current Research* Part C, Geological Survey of Canada Paper 82–1C.

McNEELY R. AND GUMMER, W. D. 1984. A reconnaissance survey of the environmental chemistry in east-central Ellesmere Island, NWT. *Arctic*, 37: 210–23.
MITCHELL, M. 1956. Visual range in the polar regions with particular reference to the Alaskan Arctic. *Journal of Atmospheric and Terrestrial Physics*, Special Supplement: 195–211.
MUIR, D. AND OTHERS. 1986. Chlorinated hydrocarbons and heavy metals in arctic marine mammals, fish and polar bears; preliminary results in 1983/84 samples. (Proceedings of International Conference on arctic water pollution research: applications of science and technology, Yellowknife, 28 April–1 May 1985). *Water Science and Technology*, in press.
NEFTEL, A. AND OTHERS. 1985. Sulphate and nitrate concentrations in snow from south Greenland, 1895–1978. *Nature*, 314: 611–14.
OEHME, M. AND OTTAR, B. 1984. The long-range transport of polychlorinated hydrocarbons to the Arctic. *Geophysical Research Letters*, 11: 1134–36.
PACYNA, J. M. AND OTHERS. 1984. Size-differentiated composition of the arctic aerosol at Ny-Alesund Spitsbergen. *Atmospheric Environment*, 18: 2447–59.
PETERSON, J. T. AND OTHERS. 1980. Dependence of CO_2, aerosol and ozone concentrations on wind direction at Barrow, Alaska during winter. *Geophysical Research Letters*, 7: 349–52.
PETERSON, J. T. AND OTHERS. 1982. Atmospheric carbon dioxide measurements at Barrow, Alaska 1973–1979. *Tellus*, 34: 166–75.
PORCH, W. M. AND MacCRACKEN, M. C. 1982. Parametric study of the effects of arctic soot on solar radiation. *Atmospheric Environment*, 16: 1365–71.
POTTER, G. L. AND CESS, R. D. 1984. Background tropospheric aerosols: incorporation within a statistical-dynamical climate model. *Journal of Geophysical Research*, 89(D6): 9521–26.
RAATZ, W. E. 1981. Trends in cloudiness in the Arctic since 1920. *Atmospheric Environment*, 15: 1503–06.
RAATZ, W. E. AND SHAW, G. E. 1984. Long-range transport of pollution aerosols into the Alaskan Arctic. *Journal of Climatology and Applied Meteorology*, 23: 1052–64.
RADKE, L. F. AND OTHERS. 1984. Airborne observations of arctic aerosols, I. Characteristics of arctic haze. *Geophysical Research Letters*, 11: 393–96.
RAHN, K. A. 1981a. Atmospheric riverine and oceanic sources of seven trace constituents to the Arctic Ocean. *Atmospheric Environment*, 15: 1507–16.
RAHN, K. A. 1981b. The arctic air-sampling network in 1980. *Atmospheric Environment*, 15: 1349–52.
RAHN, K. A. 1982a. On the causes, characteristics and potential environmental effects of aerosol in the arctic atmosphere. In REY, L. (editor). *The Arctic Ocean: the hydrographic environment and the fate of pollutants.* London, Macmillan: 163–95.
RAHN, K. A. 1982b. Elemental tracers for source regions of arctic pollution aerosol. *Journal of the Hungarian Meterological Service*, 86: 1–14.
RAHN, K. A. AND McCAFFREY, R. J. 1979a. Long range transport of pollution aerosol to the Arctic: a problem without borders. *Proceedings of the WMO Symposium on the long range transport of pollutants and its relation to general circulation including stratospheric/tropospheric exchange processes* Sofia, 1–5 October. WMO No 538: 25–35.
RAHN, K. A. AND McCAFFREY, R. J. 1979b. Compositional differences between arctic aerosol and snow. *Nature*, 280: 479–80.
RAHN, K. A. AND McCAFFREY, R. J. 1980. On the origin and transport of the winter arctic aerosol. *Annals of the New York Academy of Sciences*, 338: 486–503.
RASMUSSEN, R. A. AND KHALIL, M. A. K. 1984. Gaseous bromine in the Arctic and arctic haze. *Geophysical Research Letters*, 11: 433–36.
SCHNELL, R. C. AND RAATZ, W. E. 1984. Vertical and horizontal characteristics of arctic haze during AGASP; Alaskan Arctic. *Geophysical Research Letters*, 11: 369–76.
SEMB, A. AND OTHERS. 1984. Major ions in Spitsbergen snow samples. *Geophysical Research Letters*, 11: 445–48.
SHAW, G. E. 1975. The vertical distribution of tropospheric aerosols at Barrow, Alaska. *Tellus*, 27: 39–49.
SHAW, G. E. 1982. Atmospheric turbidity in the polar regions. *Journal of Applied Meteorology*, 21: 1080–88.
SHAW, G. E. 1984. Microparticle size spectrum of arctic haze. *Geophysical Research Letters*, 11: 409–12.
SHAW, G. E. AND STAMNES, K. 1980. Arctic haze: perturbations of the polar radiation budget. *Annals of the New York Academy of Science*, 338: 533–39.
SHAW, G. E. AND WENDLER, G. 1972. Atmospheric turbidity measurements at McCall Glacier in northeast Alaska. *Conference Proceedings on Atmospheric Radiation.* Fort Collins, Colorado. Boston, American Meteorological Society: 181–87.
SHEWCHUK, S. R. 1986. Acid deposition sensitivities within the Northwest Territories and current depositions to the snowpack and small lakes chemistry of a selected area of the Mackenzie District. (Proceedings of the International Conference on Arctic Water Pollution Research: applications of science and technology, Yellowknife, 28 April–1 May 1985). *Water Science and Technology*, in press.
VALERO, F. P. AND OTHERS. 1984. The absorption of solar radiation by the arctic atmosphere during the haze season and its effects on the radiation balance. *Geophysical Research Letters*, 11: 465–68.
WAGGONER, A. P. AND WEISS, R. E. 1980. Comparison of fine particle mass concentration and light scattering extinction in ambient arosol. *Atmospheric Environment*, 14: 623–26.
WEISS, H. V. AND OTHERS. 1978. Natural enrichment of elements in snow. *Nature*, 274: 352–53.
WINCHESTER, J. W. AND OTHERS. 1984. Coarse particle soil dust in arctic aerosols, spring 1983. *Geophysical Research Letters*, 11: 995–98.

PHYSICAL AND CHEMICAL PROPERTIES OF ARCTIC AEROSOLS AND CLOUDS

J. HEINTZENBERG, H-C. HANSSON, J. A. OGREN, D. S. COVERT
AND J-P. BLANCHET

ABSTRACT. This paper reviews aspects of Swedish arctic aerosol research, including aerosol particle size distribution and composition variability, and how the aerosol influences arctic climate. Possible reasons are discussed for large variations in winter and summer measurements of physical and chemical aerosol properties. Particle size distribution studies on formation, transformation and removal processes of the arctic aerosol indicate that the accumulation mode of the aerosol is narrow, with small median diameter compared to midlatitude aerosols; sulfate is the mass-dominant component. Nuclei modes appear in both winter and summer aerosol. Model calculations evaluate different transformation pathways leading to sulfate particles. To assess meteorological effects of anthropogenic aerosols, local heating was calculated using measured aerosol properties and radiative transfer models; results indicated heating rates of 0.1 to 1.0 K/day. Our model studies showed the need of knowing where in the earth/atmosphere system the light absorbing particles are located–in snow, under clouds, in clouds or above the cloudy layer. A discussion of unresolved issues and further work concludes the paper.

Contents

Introduction

A full understanding of the arctic aerosol and its effects on the ecosystem requires a large body of physical, chemical and geographical knowledge. Local and long distance source areas need to be identified and quantified. Primary and secondary source processes need to be known to understand its chemical composition, size and shape. Atmospheric transport and physico-chemical transformation processes determine its evolution with time, locally and with distance from remote sources (eg changes of chemical signatures which were established in certain source regions). Scavenging processes during the transport to the polar region and in the Arctic need to be assessed to determine atmospheric lifetimes and aerosol deposition. Finally, the three-dimensional distribution of arctic aerosol particles in their environment of gases, clouds and ice or snow determines the climatic effect of the arctic aerosol on the arctic radiation budget. Besides their direct radiative effect, their influence as cloud condensation nuclei has to be considered.

In the present paper we assess the extent to which our studies in the European Arctic sector contributed to the understanding of the arctic aerosol. Results derived by other groups have been used to strengthen arguments based on our own data. In summarizing the results of the past six years of arctic studies we identify gaps in knowledge, and outline our present research and plans for future work to fill in missing pieces of information.

State of knowledge about the arctic aerosol

Concentrations

The most basic aerosol properties, such as total atmospheric burdens or total number and mass concentrations, should be known in the atmospheric layer exhibiting greatest aerosol effects. The first steps in chemical and particle size resolution are chemical analyses of two size fractions that separate aerosol particles with strongly different generating processes. 1–3m diameter is a generally accepted cut-size between fine and coarse particles.

Because of drastic seasonal variations in the arctic aerosol (discussed below), the presentation of mean annual results is not very meaningful. Instead, late winter (or spring) results from our experiments at Ny-Ålesund, Spitsbergen are compared to summer results from the Swedish icebreaker expedition Ymer-80 in Table 1. The contrast between winter and summer levels is clearly visible. Depending on the measured aerosol property, winter to summer ratios between 5 and 50 were recorded. Except for the cleanest summer months, inter-annual variability of the aerosol concentrations is large. As shown in Figure 1 for the years 1978–83, during the period of peak pollution

Table 1. Integral aerosol properties and chemical composition in the European Arctic in winter and summer. CNC = total number of aerosol particles; σsp = particle light scattering coefficient at 550 nm wavelength; TSP = total particulate mass; Cl_t = total particulate chlorine < 1 μm radius; S_f = particulate sulfur <1 μm particle radius; S_{fNM} = as S_f except for non-marine sulfur; NH_4^+ = total particulate ammonium; CE_t = total elemental carbon; K_{fNM} = non-marine particulate potassium <1 μm radius; Ca_{fNM} = non-marine particulate calcium; Fe_f = non-marine particulate iron; Ti_f = particulate titanium <1 μm radius; Ni_f = same for nickel; Zn_f = same for zinc; Pb_f = same for lead; Cu_f = same for copper. [1]Söderlund 1982. [2]Heintzenberg and others 1981.

Parameter	Unit	Spring Spitsbergen[2]	Ymer-80 Grand average	July	Aug	Sept
CNC	cm^{-3}	380	130	100	240	76
σsp	$10^{-6}m^{-1}$	15	1.5	1.1	1.6	1.9
TSP	$ng\ m^{-3}$	4700	1900	390	3100	1800
Cl^t	$ng\ m^{-3}$	96	910	45	1200	1200
S^f	$ng\ m^{-3}$	620	99	42	1200	1200
S^fNM	$ng\ m^{-3}$	620	82	41	100	130
NH_4^+	$ng\ m^{-3}$	150	—	—	76	110
EC_t	$ng\ m^{-3}$	70	4.5	1.9	32^1	—
K_{fNM}	$ng\ m^{-3}$	15	2.6	1.2	5.7	4.5
Ca_{fNM}	$ng\ m^{-3}$	21	2.7	3.4	1.7	4.0
Fe_f	$ng\ m^{-3}$	26	1.1	0.54	2.1	2.8
Ti_f	$ng\ m^{-3}$	< 2	0.16	< 0.063	1.4	1.3
Ni_f	$ng\ m^{-3}$	< 0.36	< 0.58	< 0.027	0.22	0.17
Zn_f	$ng\ m^{-3}$	2.6	0.14	0.11	< 0.25	0.10
Pb_f	$ng\ m^{-3}$	3.0	0.14	< 0.0	0.23	0.084
Cu_f	$ng\ m^{-3}$	1.6	0.041	0.033	0.30	0.056
					0.054	0.036

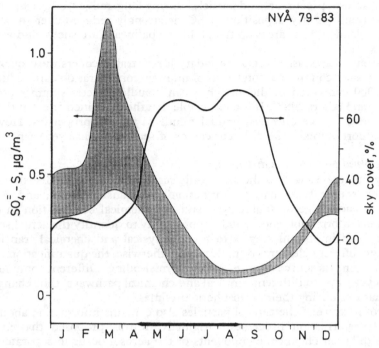

Fig 1. Annual variation of sulfate as $SO_4^- - S$ on Spitsbergen and sky-cover in % in the European Arctic. The length of the arctic day (midnight sun) is marked as an arrow on the abscissa.

Table 2. Reactive and climatically important trace gases in the European Arctic. [1]Söderlund 1982. [2]Ockelmann and Georgii 1984. [3]P. Winkler, Meteorological Observatory, Hamburg. [4]Heintzenberg and others 1983; Heintzenberg and Larssen 1983.

Gas	Unit	Spring Spitsbergen	July	Aug	Sept
			Ymer-80		
NH_3[1]	ng m^{-3}	—		15	
SO_2-S	ng m^{-3}	600[4]		7 – 140[2]	
O_3[3]	ppbv			16 – 25	
CO_2[4]	ppmv	342	333	326	328

this variation is as large as mean seasonal variation. As a complement to the discussion of particle formation processes and climatic effects below, Table 2 presents surface level concentrations of the reactive gases SO_2, O_3 and NH_3 together with the climatically important CO_2 results. The CO_2 concentrations can be updated by the commonly used annual increase of 0.4%.

Causes of seasonal variations

There are a number of reasons for the large seasonal variations in the arctic aerosol (and varying seasonality for different aerosol parameters). Midlatitude industrialized regions are suspected to be major contributors to the arctic aerosol (Rahn and McCaffrey 1980). Emission of aerosol precursors varies about a factor of two between winter and summer in these regions. Atmospheric transport from these source regions to the Arctic is 2–5 times more frequent in spring than in summer (Heintzenberg and Larssen 1983).

Scavenging of aerosol precursors and particles during long-range transport to the Arctic has a seasonal trend (eg dry deposition of SO_2 is strongly reduced over snow and ice). In winter, both land and sea areas on the pollution pathways are snow- and ice-covered to a greater extent.

During the summer season, arctic boundary layer stratus covers most of the central Arctic. During such conditions, total aerosol number concentrations are reduced up to a factor of 1,000 compared to the already clean cloudless arctic summer atmosphere (Heintzenberg and others 1983). It is conceivable that the combined effect of the seasonal dependencies leads to the observed annual trends in aerosol properties. However, no quantitative apportioning of the different causes of seasonality has yet been achieved.

Chemically resolved particle size distribution

Measuring the total as well as the chemically resolved aerosol size distribution yields information on source type, particle formation and transformation, and scavenging processes. The knowledge of source type and the chemical composition in each size fraction contains information about, and the possiblity to quantify the actual source. On the other hand, an important step is to reveal physical and chemical transformation processes of the different elements or molecules. Otherwise, the quantification of sources in a receptor region, via source signatures, can be misleading. Different compounds have different life cycles due to different physical and chemical pathways, thus changing their chemical signatures during their atmospheric lifetimes.

The state of mixture of the aerosol particles also contains information about sources and aerosol transformation processes. Originally, it was proposed that in external mixtures, the different chemical components of the aerosol occur in separate particles. In internal mixtures, various chemical components occur in the same particle. Considering the range of microcomponents which is analytically accessible today, all sources produce internally mixed particles to some extent. Even though microcomponents comprise insignificant mass fractions in the aerosol, they can determine important macroscopic properties of the aerosol (eg light absorption). Thus, it is of interest to investigate the degree of internal mixture, ie to study how the different components which determine physico-chemical aerosol properties, are related to each other during the different phases of long-range transport.

Hygroscopic particle properties are closely connected to their chemical composition and source. The best-known hygrophobic aerosol component is soot or elemental carbon. Sulfate particles on the other hand are hygroscopic. Hence, the state of mixture of the aerosol can be inferred to some extent by separately sampling and anlyzing hygroscopic and non-hygroscopic particles. Determining hygroscopic properties of the aerosol has its own value because of the importance of water uptake in radiative transfer processes and particle deposition.

In recent years several arctic experiments have been performed to obtain information concerning the size distribution of the arctic aerosol, mostly in connection with the arctic haze phenomenon in winter. Measurements concerned total (Heintzenberg 1980; Jaenicke and Schuetz 1981) and chemically resolved (Heintzenberg and others 1981; Pacyna and others 1984) size distributions. However, sampling and analytical facilities were rather limted in each of the experiments. With the data at hand today, some general observations can be made about the arctic aerosol.

— the accumulation mode (0.05–0.5 m diameter) is narrow and exhibits smaller diameters than mid-latitude continental aerosols (Heintzenberg 1980, Shaw 1983);
— nuclei modes (m diameter) have been found in winter and summer aerosols,

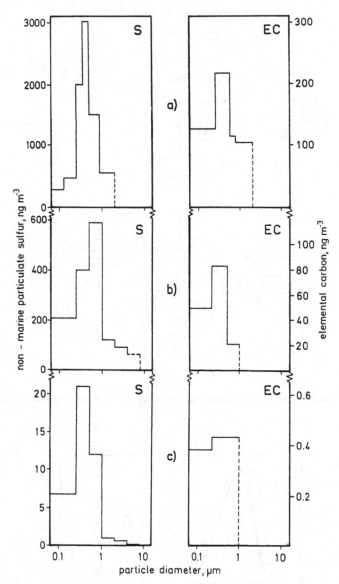

Fig 2. Mass size distributions of sulfur (S) and elemental carbon (EC): a) late winter, Spitsbergen; b) summer pollution episode, Svalbard area; c) summer Arctic background, north of Spitsbergen. Dashed horizontal lines indicate results at the detection limit. Dashed vertical lines mark incomplete size distributions because of instrumental size cuts.

indicating particle formation through homogeneous nucleation from gas phase precursors (Flyger and others 1980, Jaenicke and Schuetz 1981, Shaw 1983);

— the main components are acid sulfates and sea salt. Microcomponents seem to follow roughly the sulfur distribution (see Figure 2 and Pacyna and others 1984);

— winter and summer size distributions do not differ strongly from each other (see Figure 2).

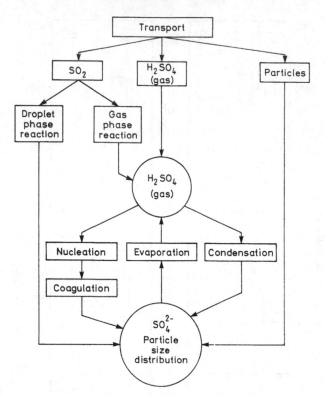

Fig 3. Physical and chemical pathways leading to sulfate particles in the arctic aerosol.

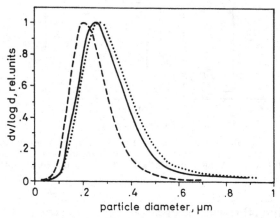

Fig 4. Volume size distributions resulting after a volume doubling of an initial arctic aerosol (—) (Heintzenberg 1980) through condensation growth (– – –) and through volume growth (.....).

Hering and Friedlander (1982) have shown that different types of size distributions can be produced by different transformation processes. In Figure 3 we present a schematic overview of the different transformation processes leading to a sulfate size distribution as in the arctic aerosol. There are three different pathways from SO_2 to SO_4^{2-}. We applied Hering and Friedlander's aerosol growth model to the case of arctic haze by letting the processes in Figure 3 act on an initial (imported background) size distribution which is typical for the winter Arctic.

Because of its negligible contribution to the mass growth (Shaw 1983), homogeneous nucleation was excluded. In Figure 4 we see that after a doubling of the particulate volume, neither gas phase nor droplet phase reactions significantly change the original distribution. This is due to the fact that surface and colume distribution initially had coinciding peaks at about 0.25 m diameter.

From our model calcultions, we conclude that the shape of the initial size distribution decides if the size distribution will reveal which transformation process dominated. Consequently, during clean summer situations with coarse particle sea salt dominating the initial aerosol, the different formation processes should reveal themselves more clearly in the resulting size distribution.

The observational base for chemically resolved size distributions in the different seasons is still too small to allow conclusions about the state of mixing and its change during transport and ageing processes. The same holds for the hygroscopic aerosol properties.

In summary, the information content of the size distribution of arctic aerosol particles has not yet been exhausted. Measurements of trace gases related to particle formation need to be made together with high resolution aerosol measurements in well-defined air masses.

Climatic effects

The main climatic effects of atmospheric aerosols are caused by their scattering and absorption of radiation plus their involvement in cloud formation. The overall aerosol effect is very hard to assess because of a multitude of feedback loops between the various parts of the earth-atmosphere system. Figure 5 depicts some of these loops schematically. Aerosol absorbers (eg soot) and aerosol scatterers (eg sulfate) will influence air and surface temperatures through their interaction with radiation. They will influence cloud albedo either directly as absorbers in the cloudy air or through their effect as condensation nuclei on cloud droplet size distribution.

A reduction of the planetary albedo due to light absorbing particles is one of the most clear-cut aerosol effects. However, even in the case of only black particles, interactions between surface, clouds and particles make the total aerosol effect dependent on where in the system the particles are located. We illustrate this problem with one-dimensinal radiation calculations utilizing a delta-Eddington model described in Blanchet and List (1983). The change of planetary albedo was calculated for particles suspended either in a surface snow layer, under a stratus cloud in 1 to 1.5 km height, in that cloud as external mixture (ie outside the cloud drops), and spread over the entire above-cloud troposphere.

Figure 6 gives the results for varying amounts of EC (expressed as relative optical density τ_{EC}). A mid-latitude winter atmosphere at the equinox over a surface with an albedo characterized the system. Figure 6 shows the importance of knowing where in the system the absorbing particles are acting on the solar radiation. Under a cloud, large amounts of absorbing particles can be hidden without strong radiative effects. In a cloud, (or in snow which can be seen as a geometrically thin, optically thick cloud), the absorbing effect is enhanced due to multiple scattering processes. Finally, over a cloud (or over any

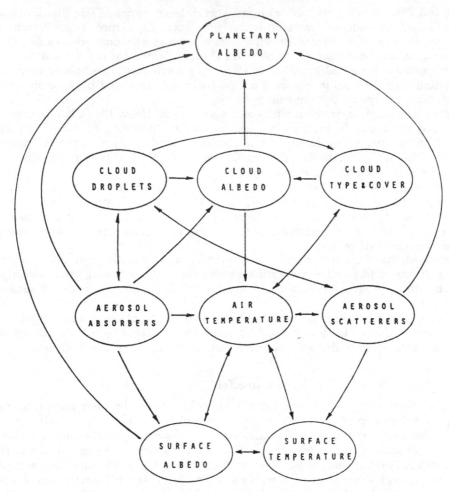

Fig 5. Aerosol related feedback loops in the earth-atmosphere system.

strongly reflecting surface) the greatest heating effect is achieved, emphasizing the need to know the distribution of the particles in the climate system.

The situation in the Arctic is complicated by the fact that the essential climate-affecting factors show the strong seasonality, as depicted in Figure 1. Aerosol concentrations exhibit a strong maximum in February-April while the sun is hardly yet over the horizon. On the other hand, when the sun is highest and stratus is most widespread, surface aerosol concentrations are at their minimum, reducing the possiblities of cloud-amplified aerosol effects.

So far, investigations of the climatic effect of the arctic aerosol consist mostly of model calculations based upon measured dry aerosol properties or measured radiative fluxes in the hazy spring Arctic. Blanchet and List (1985) evaluated our aerosol results from the spring-1979 campaign at Ny-Ålesund. Combining atmospheric heating in the solar spectral range and thermal cooling in the long-wave region they found about 0.1 K/day heating in the arctic haze layer. Our spring-1983 results include airborne aerosol

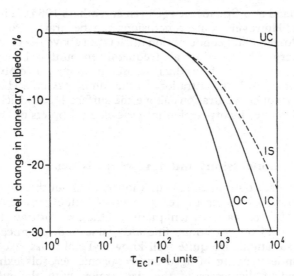

Fig 6. Change in planetary albedo with varying burdens of elemental carbon, expressed as optical density τ_{EC}, in the earth- atmosphere system. The same amounts of EC are located in a surface snow layer (IS), in the boundary layer under a stratus between 1 and 1.5 km (UC), as individual particles inside the stratus (IC), and evenly distributed in the troposphere above the cloud (OC).

absorption data and vertical profiles of meteorological and aerosol parameters. EC-concentrations were a factor 2–4 higher than during the first campaign. Wendling and others (1985) evaluated the measured data with their radiation model in the solar spectral range. They derived heating rates between 0.5 and 1.2 K/day which seems consistent with the previous results by Blanchet and List (1985) when considering the higher particle concentrations in 1983 and the omittance of thermal cooling by Wendling and others (1985). The long-term consequences of these heating rates have not been determined yet.

For the energy budget of the Arctic, aerosol effects during the long arctic day (Figure 1) are more interesting than the winter haze which occurs with nil or weak illumination. Here our knowledge is largely confined to the cloud-free surface air. The low concentrations of aerosol scatterers and the extremely low concentrations of aerosol absorbers do not cause significant heating or cooling effects. On the other hand, pollution particles affecting the arctic stratus as condensation nuclei, in or above clouds as intensified absorbers cannot be ruled out a priori.

Our first aerosol/cloud studies in arctic air were conducted in summer 1984. At the cloud chemistry station on Åreskuta, Sweden, cloud samples were taken with back-trajectories leading to the Greenland/Spitsbergen sector. Chemical analyses of these samples yielded extremely small amounts of dissolved material. Median sulfate concentrations were 6 μeq/l, (about 100 times lower than in cloudy air from the European continent). From the liquid water content and cloud water compositions, clear-air aerosol composition was inferred yielding sulfate concentrations of about 15 ng m^{-3}. This value is lower than the arctic clear-air background concentrations determined during the Ymer-80 expedition (Lannefors and others 1983).

During the same period, we participated in the Marginal Ice Zone Experiment (MIZEX-84) with airborne sampling and real-time measurements in and above arctic stratus between Greenland and Spitsbergen (Ogren and others 1985). The samples have not been analyzed yet. However, some conclusions can be drawn from the real-time measurements. Nephelometer and condensation nuclei data above the cloud layer did not show the pronounced aerosol layers which are frequently encountered in the arctic winter troposphere. Our measurements of residual aerosol mass in the cloud droplets are consistent with clear-air aerosol concentrations in the summer Arctic. Hence, we saw no indication of higher aerosol concentrations above the surface layer which might lead to significant aerosol effects after incorporation into the cloud droplets or above the highly reflecting cloud layer.

Conclusions and unresolved issues

In the previous sections, we summarized our knowledge about the arctic aerosol based upon Swedish research during the past 6 years, together with extensive results gathered by other groups. We feel there is a consistent picture of the composition and the radiative effects of the arctic aerosol. Concentrations, fine and coarse mode chemical compositions, and average seasonal variations are quite well known. Total mass size distribution is consistent with our understanding of aged anthropogenic aerosols exhibiting narrow distributions with median diameters in the size region with the weakest removal processes. Systematic chemical data covering gaseous precursors and high resolution size distributions are still lacking. Our model calculations showed that the shape of the initial size distribution determines whether certain chemical transformations will be reflected in the arctic haze size distribution.

So far, the investigation of climatic effects of anthropogenic arosols in the Arctic yielded heating rates between 0.1 and 1 K/day. To interpret these results in terms of climatic consequences would require firstly the establishments of pre-industrial arctic aerosol burdens (eg from ice-core data) as references, and secondly the incorporation of anthropogenic heating rates into a general circulation model which covers ocean, sea-ice and atmosphere.

Despite substantial existing knowledge, several unresolved issues are worth further study. These include: variation of concentrations between years, especially during the period of peak aerosol particulate mass concentration; chemical processes of aerosol formation and transformation in both dark and illuminated Arctic (an issue that includes the state of mixture and the molecular composition of acidic aerosol components and their involvement in haze formation); identification of regional and remote sources.

Swedish arctic research is exploiting two approaches to further our knowledge of the arctic aerosol–long-term monitoring and short-term experiments. An air pollution monitoring station is being assembled at Ny-Ålesund, Spitsbergen (Heintzenberg and others 1985). Aerosol absorption and scattering, total number and mass concentration, and elemental composition are the monitoring parameters besides CO_2. A similar station in the Swedish mountains will focus on aerosol and cloud water composition. Long-range transport via Scandinavia to the Arctic is being monitored with a series of 4 automatic sampling stations for multi-element analyses. On a short-term basis, arctic aerosol and cloud water composition will be studied in the Spitsbergen area in 1986–88.

Acknowledgements

We gratefully acknowledge the support we received during our Spitsbergen experiments, and through measurements in Norway by the Norwegian Institute for Air Research. Hans Lannefors and Peter Winkler provided part of the summer data during the Ymer expedition. Ulla Jonsson drafted the figures.

References

BLANCHET, J.–P. and LIST, R. 1983. Estimation of optical properties of Arctic haze using a numerical model. *Atmosphere Ocean*, 21: 444–64.

BLANCHET, J.–P. and LIST, R. 1985. Radiative effects of the anthropogenic aerosol components of the Arctic aerosol in a surface-atmosphere model: I. Arctic haze. *Tellus*, in press.

FLYGER, H. AND OTHERS. 1980. The background levels of the summer tropospheric aerosol and trace gases in Greenland. *Journal of aerosol Science*, 11: 95–110.

HEINTZENBERG, J. 1980. Particle size distribution and optical properties of Arctic haze. *Tellus*, 32: 251–60.

HEINTZENBERG, J. and LARSSEN, S. 1983. SO_2 AND SO_4^- in the Arctic: interpretations of observations at three Norwegian Arctic-Subarctic stations. *Tellus*, 35B: 255–65.

HEINTZENBERG, J. AND OTHERS. 1981. The chemical composition of Arctic haze at Ny-lesund, Spitsbergen. *Tellus*, 33: 162–71.

HEINTZENBERG, J. AND OTHERS. 1983. *An investigation of possible sites for a background monitoring station in the European Arctic*. Stockholm, University of Stockholm. (Report AP-22, Department of Meteorology).

HEINTZENBERG, J. AND OTHERS. 1985. Concept and realization of an air pollution monitoring station in the European Arctic. *Ambio*, 14: 152–57.

HERING, S. V. AND FRIEDLANDER, S. K. 1982. Origins of aerosol sulfur size distributions in the Los Angeles basin. *Atmospheric Environment*, 16: 2647– 56.

JAENICKE, R. AND SCHÜTZ, L. 1981. Arctic aerosols in surface air. *Időjaras*, 86: 235–41.

LANNEFORS, H. AND OTHERS. 1983. A comprehensive study of physical and chemical parameters of the Arctic summer aerosol. *Tellus*, 35B: 40–54.

OCKELMANN, G. AND GEORGII, H. – W. 1984. The distribution of sulfur dioxide over the Norwegian Arctic ocean in summer. *Tellus*, 36B: 179–85.

OGREN, J. A. AND OTHERS. 1985. *In situ* sampling of clouds with a droplet to aerosol converter. *Geophysical Research Letters*, 12: 121–24.

PACYNA, J. M. AND OTHERS. 1984. Size-differentiated composition of the Arctic aerosol at Ny-(AA)lesund, Spitsbergen. *Atmospheric Environment*, 18: 2447–59.

RAHN, K. A. AND McCAFFREY, R. J. 1980. On the origin and transport of the winter Arctic aerosol. *Annals of the New York Academy of Science*, 338: 486–503.

SHAW, G. E. 1983. On the aerosol particle size distribution spectrum in Alaskan air mass systems: Arctic haze and non-haze episodes. *Journal of Atmospheric Sciences*, 40: 1313–20.

SÖDERLUND, R. 1982. Ammonia in surface air. PhD thesis, University of Stockholm.

WENDLING, P. AND OTHERS. 1985. Calculated radiative effects of Arctic haze during a pollution episode in spring 1983 based on ground-based and airborne measurements. *Atmospheric Environment*,

THE GREENLAND AEROSOL: ELEMENTAL COMPOSITION, SEASONAL VARIATIONS AND LIKELY SOURCES

NIELS Z. HEIDAM

ABSTRACT. Danish research into arctic aerosols began in the early 1970s with short-term airborne measurements of particles and gases over most of Greenland. Pollutant levels were low, but often thin layers with high concentrations of submicron particles extended for several hundred km over the ice cap. In 1979 began a long-term ground-based study of chemical composition, seasonal variations and possible origins of the aerosol. From four stations on the Greenland perimeter north of 70°, particulate samples were collected twice weekly for three years. Analyzed by PIXE, these yielded results for about 15 elements. Factor analysis indicates that three to five source-related components, seemingly of distant origin, make up the aerosol. Two are natural, one of marine origin and the other a crustal component originating from erosion in dry mid-latitude regions. The remaining components are anthropogenic, the most important being a combustion component originating from fossil fuel combustion in mid-latitudes. Elements of this component, particularly sulphur, vary seasonally, rising in winter 100-fold over summer levels. This is presumably the dominating Arctic haze component. Metallic composition of other components indicates origins in industrial processes, notably from the Ural region of the USSR. They peak sharply in midwinter, coincident with clearly discernible transport episodes from this area.

Contents

Introduction

During the 1970s widespread concern over increasing air pollution led to internationally coordinated projects to assess regional effects and study pollutant transport in the atmosphere. The best known European project was probably the OECD Long Range Transport of Air Pollution (LRTAP) Project, 1972–75 (OECD 1979), of which the final report established that air pollution can travel several hundred kilometers, and that the pristine state of the atmosphere had been disturbed over large areas. Parallel 'clean air' or 'global background aerosol' studies were carried out in remote regions to assess the degree of disturbance. In Greenland the Aerosol Laboratory, now the Air Pollution Laboratory, undertook tropospheric aerosol studies, taking airborne measurements in the summers of 1971, 1973, and 1976 (Flyger and others 1973, 1976, 1980), and carrying out

37

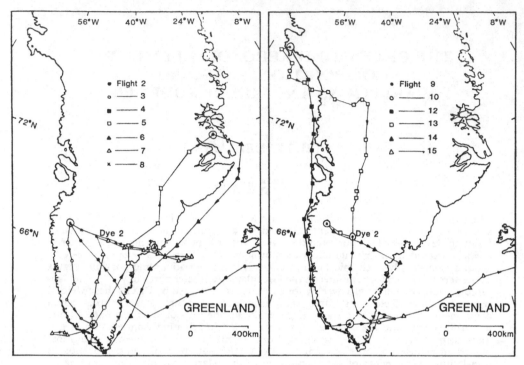

Fig 1. Flights over Greenland, August 1973.

ground-based fieldwork in northern Greenland in summer 1974 (Flyger and Heidam 1978).

Flights of 1973 appear in Figure 1 and resulting measurements of particles in Figure 2. Measurements of particle concentration yielded the most interesting results, in particular of submicron Aitken particles: whereas concentrations of Aitken nuclei around 200 cm^{-3} were expected and observed, rapid excursions to sizes close to 10^4 cm^{-3} often occurred (Figure 2), in close relation with a distinct yellow-brownish haze layer observed throughout the flights. Similar haze layers, albeit not so extensive, with large variations in particle concentrations were observed on several flights over sea and land. But even when haze layers were absent, large variations in Aitken nuclei were observed. These observations provided evidence of a local particle production by gas-phase reactions. In view of later research it now seems reasonable to suppose that the particles or their gaseous precursor, presumably SO_2, originated in mid-latitudes and had been transported to the Arctic in the atmosphere.

Other arctic aerosol projects, notably at Point Barrow, Alaska (Rahn and others 1977, Rahn and McCaffrey 1980), in the Canadian Arctic (Barrie and others 1981), and at Spitsbergen in the Norwegian Arctic (Heintzenberg 1980; Heintzenberg and others 1981; Rahn and others 1980) confirmed that such concepts as 'the global background aerosol' and 'naturally clean air' had limited validity. Their results showed that, although the air in these remote places was extremely clean, it nevertheless contained minute amounts of anthropogenic pollutants. It gradually became clearer that there is probably no place on the globe—including the Arctic—totally unaffected by industrial air pollution.

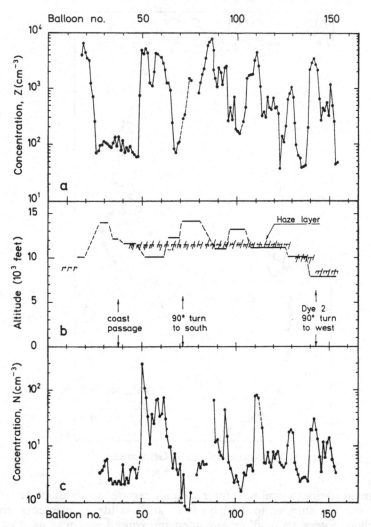

Fig 2. Results of particle measurements from Flight 13 in 1973. Abscissae show time intervals of about 2.5 min, corresponding to a flight distance of about 12.5 km. (a) Aitken nuclei concentrations, Z cm^{-3}; (b) flight altitude in thousands of feet; (c) concentrations of large nuclei (0.3–3 μm), N cm^{-3}.

The short-term Danish campaigns in the Greenland Arctic had produced only 'snapshots', inadequate for a comprehensive understanding of the origin, level and composition of the tropospheric aerosol. Several fixed, long-term measuring stations were needed. In autumn 1979 the Air Pollution Laboratory therefore started SAGA (Studies of the Aerosol in the Greenland Atmosphere) Project (Heidam 1983), operating four stations—THUL, NORD, GOVN and KATO/MEST—for more than three years (Figure 3: note that THUL is at the settlement of Qanaq and should not be confused with Thule Air Base at Dundas 100 km to the south). Observations were also recorded at PCS, near the southern tip of Greenland.

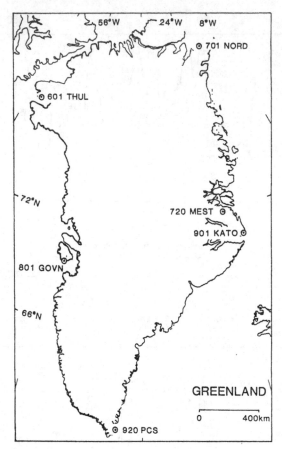

Fig 3. Locations of the SAGA stations.

At the four main stations membrane filter samples exposed to about 200 m³ of air were taken twice a week and shipped to the laboratory at Roskilde for elemental analysis by the PIXE method. Every tenth filter remained unexposed and was used as a blank. Heidam's (1983) report on SAGA includes site descriptions, wind statistics, sampling and analytical techniques and detailed results from the first year of operation.

The Greenland aerosol: elemental composition

Determination of elements in a filter sample by PIXE is 10–15% accurate. Up to 25 elements heavier than Al can be detected if present in sufficient quantity. Only those found in more than 70–75% of the samples were retained. Concentrations of these constituents were to a good approximation log-normally distributed, and therefore geometric means and standard deviations are reported. Table 1 shows results of the August 1979 to August 1980 study (Heidam 1984): THUL and PCS data are for limited periods only. Concentration levels on the west coast fall off towards the north; GOVN shows the highest and THUL the lowest annual levels. On the east coast levels at NORD and KATO are similar, lying between those of THUL and GOVN; notable exceptions

Table 1. Geometric mean concentrations in ngm^{-3}, and standard deviations. THUL: Feb-Aug 1980 only. PCS: Sept 1979 to Jan 1980 only. SAGA 1979–80.

Element	THUL conc	THUL dev	NORD conc	NORD dev	GOVN conc	GOVN dev	KATO conc	KATO dev	PCS conc	PCS dev
13 Al	1.91	2.42	8.39	3.18	18.20	2.97			6.39	2.42
14 Si	19.1	2.64	161	2.80	260	2.61	69.2	3.04		
16 S	93.8	3.00	179	3.27	206	1.76	111	2.41	155	1.78
17 Cl	21.5	3.22			504	3.80	91.6	5.94	1170	2.12
19 K	4.15	2.31	15.9	2.83	33.6	2.14	11.0	2.79	31.3	1.84
20 Ca	2.71	2.54	19.2	3.10	51.0	2.36	11.3	2.81	35.2	1.89
22 Ti	0.24	2.46	1.29	3.51	3.56	3.46	0.78	3.07	0.57	3.15
24 Cr	0.03	1.95	0.16	2.52	0.14	2.30			0.06	1.81
25 Mn	0.13	2.33	0.50	2.99	0.77	2.92	0.21	2.83	0.19	2.86
26 Fe	4.16	2.74	24.2	3.36	53.1	3.18	12.5	3.09	14.0	2.68
28 Ni					0.17	2.12				
29 Cu	0.07	2.63	0.30	3.91	0.21	2.27	0.12	2.48	0.09	2.58
30 Zn	0.68	3.08	1.97	3.69	1.52	2.81	0.59	3.10	1.41	2.17
35 Br	2.02	3.20	3.73	3.23	4.05	2.26	1.02	3.05	4.95	1.98
38 Sr	0.06	2.66	0.25	2.95	1.00	2.17	0.20	3.04	0.85	1.68
82 Pb	1.82	2.57	4.98	3.50	1.28	2.48	0.80	2.44	0.92	3.87
100 TSP	1177	2.18							4840	1.68

are metals Cr, Cu, Zn, and Pb, which are most abundant at NORD. PCS, the southernmost station, has some of the lowest levels for these metals; its very high Cl is due to strong marine influences. Geometric standard deviations, range 2.0–3.5, indicate that annual mean concentrations are not very representative of the levels as they occur throughout the year. Maxima for most elements, occurring in winter or late spring, may be 10 to 20 times larger than summer minima.

These results agree well with concentrations observed elsewhere in the Arctic. Concurrent measurements in the Canadian Arctic archipelago (Barrie and others 1981) showed very similar maxima and minima for S, Mn, Cu, and Zn; summer values from the Greenland ice cap (Davidson and others 1981) and central North Greenland (Flyger and Heidam 1978) agreed within a factor of 2–3. Pb is occasionally high at NORD, but in better agreement at THUL and GOVN.

Additional information on the composition and origin of the Arctic aerosol can be obtained by studying crustal enrichment factors (Heidam 1985). These are measures of excess concentrations of constituents relative to a crustal source. Constituents of a sample are assumed to be present in the same proportion as they occur in the source. To a good approximation, some constituents are specific to the source and, if sufficiently abundant, can be used as a source reference. For elemental aerosol concentrations c_k the crustal enrichment factor of element k is defined as:

$$E_k = \frac{c_k/c_0}{e_k/e_0}$$

where e_k is the abundance of element k in the globally averaged crust (Mason 1966), and subscript zero designates the reference element. Constituents derived from sources other than the earth's crust will have high crustal enrichments much greater than one. Elements with low enrichments are assumed to be of crustal origin only; enrichments of less than one signify depletion with respect to the crustal source.

As demonstrated by Schütz and Rahn (1982), the basic assumption for enrichment factors—that ratios of crustal abundances are preserved from source to sampling site—is well-founded for particles less than 10μm in diameter. At remote locations such as the Greenland sites, long-range transported particles will most likely be in this size range.

Table 2. Annual geometric means and standard deviations for Ti-enrichments.
[1]Feb 1980–May 1981. SAGA 1979–80.

Element	THUL[1] enrich	dev	NORD enrich	dev	GOVN enrich	dev	KATO enrich	dev
13 Al	0.65	1.73	0.35	1.34	0.28	1.78		
14 Si	0.79	1.83	1.99	1.73	1.16	2.40	1.41	2.24
16 S	5410	2.64	2350	5.96	979	4.30	2420	2.80
17 Cl					4800	4.33	3990	3.92
19 K	2.48	1.95	2.10	2.01	1.60	2.37	2.40	1.87
20 Ca	1.39	2.08	1.81	1.88	1.74	1.79	1.76	1.81
24 Cr			5.58	2.62	1.77	1.92		
25 Mn	1.94	1.78	1.80	2.07	1.00	1.43	1.22	1.60
26 Fe	1.39	1.35	1.66	1.28	1.31	1.25	1.41	1.25
28 Nl					2.74	2.31		
29 Cu	18.0	2.90	19.0	4.96	4.68	2.50	12.0	2.91
30 Zn	97.2	3.43	96.4	5.33	27.0	4.31	47.6	3.79
35 Br	7200	3.36	5100	5.58	2010	4.21	2320	2.64
38 Sn	2.64	2.22	2.27	3.12	3.30	2.17	3.07	2.06
82 Pb	1330	3.16	1310	5.70	122	3.77	348	3.60

CRUSTAL ENRICHMENTS

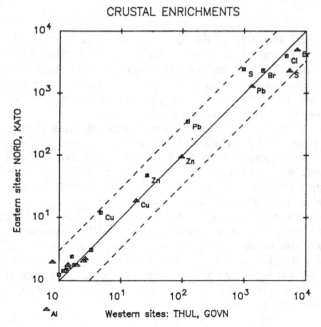

Fig 4. Geometric mean enrichments at the western sites THUL and GOVN plotted
against the values at eastern sites NORD and KATO. Annual values, 1979–80. Dotted
lines represent factors of three. Squares: southern sites GOVN and KATO. Triangles:
northern sites THUL and NORD.

Titanium, an excellent but rarely-used reference element, was selected for the
calculation of crustal enrichment factors. The PIXE method precisely detects very low
concentrations of Ti, and Ti correlates well with crustal aerosol components. To avoid
confusion with the statistical factors introduced later, enrichment factors are referred to
simply as 'enrichments'. Total geometric means and standard deviations for Ti-
enrichments relative to the average global crust are shown in Table 2 and Figure 4.

In general, over Greenland there is a pronounced homogeneity in the geographical distribution of the excess material in the atmosphere, in particular the non-enriched elements. Annual east-west deviations are within a factor of three (dotted lines in Figure 4). Enriched elements are, however, more enriched at the northern sites, except for excess S which exhibits a large latitudinal gradient on the west coast and no gradient on the east coast. The other characteristic feature of the distribution of non-crustal material is the large difference in enrichments among groups of aerosol constituents.

Crustal elements Al, Si, K, Ca, Mn and Fe are characterized by low enrichment factors from 1.0 to 3.0 at all stations, except that Al appears to be somewhat depleted. That may, however, have been due to fairly high detection limits. Metals Ni and Sr also fall into this group of non-enrichment elements. For metals Cr, Cu and Zn there are considerable enrichments of 5–100, so these do not appear to be crustally derived. Finally a group of highly enriched elements S, Cl, Br and Pb show enrichments ranging from 120 to 7,200, and their geographic spread is considerable.

Reports on crustal enrichments in the Arctic aerosol are scarce. However, those found by Flyger and Heidam (1978) in the summer aerosol of central North Greenland showed similar groupings; so did enrichments observed in rural Sweden by Lannefors and others (1983), who incidentally also used Ti as a crustal reference. Annual mean Ti-enrichments in Greenland are in general considerably less than enrichments in rural Sweden. Exceptions are S and Br, which are similarly enriched in these two areas some 4,000 km apart.

Arctic aerosol components

To investigate in more detail the occurrence of high and low concentrations or enrichments, their mutual relationship and causes, logarithms of the data have been subjected to factor analysis. Factor analysis is a statistical technique that allows a linear model to relate a large number of interdependent variables to a smaller number of independent variables or 'factors' (Harman 1976). Factors are fictitious stochastic variables that describe or measure common features, or even causes of variation in the variables. In effect, factor analysis analyses the covariance structure of the data. The starting point is, therefore, the matrix of correlation coefficients, shown for NORD in Table 3. The table contains two sets of correlation coefficients, those for log E are shown in the lower triangle and those for log c in the upper one.

Large differences between these two sets of correlation coefficients can be illustrated

Table 3. Correlation coefficients at NORD 1979–80. Upper triangle: r(log c); lower triangle: r(log E).

	13 Al	14 Si	16 S	19 K	20 Ca	24 Cr	25 Mn	26 Fe	29 Cu	30 Zn	35 Br	38 Sr	82 Pb
13 Al		0.89	−0.05	0.84	0.87	0.65	0.79	0.96	0.23	0.13	0.03	0.55	0.04
14 Si	0.33		−0.11	0.78	0.83	0.62	0.73	0.90	0.29	0.22	−0.05	0.50	0.06
16 S	0.38	0.39		0.22	0.11	0.45	0.37	0.02	0.46	0.58	0.78	0.41	0.42
19 K	0.39	0.42	0.69		0.92	0.79	0.83	0.88	0.44	0.45	0.27	0.77	0.16
20 Ca	0.36	0.44	0.54	0.78		0.74	0.77	0.88	0.42	0.39	0.22	0.74	0.18
24 Cr	0.36	0.46	0.82	0.74	0.60		0.87	0.75	0.65	0.67	0.41	0.74	0.35
25 Mn	0.21	0.26	0.80	0.62	0.39	0.83		0.88	0.53	0.52	0.31	0.71	0.22
26 Fe	0.17	0.33	0.50	0.58	0.44	0.71	0.72		0.35	0.29	0.05	0.61	0.09
29 Cu	0.18	0.43	0.69	0.61	0.55	0.79	0.69	0.57		0.84	0.39	0.54	0.42
30 Zn	0.24	0.50	0.78	0.74	0.65	0.87	0.79	0.72	0.90		0.47	0.61	0.40
35 Br	0.41	0.38	0.90	0.69	0.59	0.77	0.68	0.38	0.64	0.72		0.41	0.56
38 Sr	0.36	0.38	0.73	0.79	0.74	0.77	0.68	0.57	0.66	0.77	0.70		0.24
82 Pb	0.33	0.45	0.72	0.53	0.50	0.70	0.55	0.39	0.64	0.66	0.78	0.56	

by noting how many correlations are numerically larger than some critical value, say r_0 = 0.5. In the following list, numbers in parentheses refer to the (log c)–correlations:

Al: 0 (7), Si: 1 (7);
K: 10 (7), Ca: 8 (7), Cr: 10 (9), Mn: 9 (9), Fe: 7 (7), Sr: 10 (9);
S: 10 (2), Cu: 10 (4), Zn: 11 (5), Br: 9 (2), Pb: 9 (1).

Thus definitely crustal elements Al and Si are effectively decoupled from the rest of the aerosol constituents when log E is considered. As expected, the non-crustal parts of these two constituents are random residuals; they are therefore of crustal origin, do not carry additional information, and are consequently excluded from the factor analysis. The other extreme is represented by the definitely non-crustal elements S, Cu, Zn, Br, and Pb. The non-crustal parts of these element are considerably better correlated than their totals. The remaining elements in the middle row exhibit both increases and decreases in their correlations. The non-crustal parts do, however, tend to be the better correlated ones.

Analysis of crustal enrichments, with the exclusion of definitely crustal elements, effectively removes 'crustal noise' from the data. In view of the findings just described we expect to arrive at factor models with fewer and perhaps better-defined factors than in the analysis of the total concentrations. A simpler and more firmly knitted pattern may emerge when 'the crust is broken off'.

A factor model is described by a (m x p) matrix A of factor loadings. Each row of the matrix refers to one of the m elements included in the analysis, and each column refers to one of the p independent factors or components of the aerosol that contribute significantly to the variations observed. The matrix-elements or loadings are in fact the correlation coefficients between the elements and the factors.

Using the results of analysis of the concentrations measured at NORD in 1979–80 (Table 4), the correlation matrix shows that variations over the year in composition of the aerosol, as represented by the log-concentrations of the 14 elements and their intercorrelations, can be explained in terms of variations of only four distinct and statistically independent factors. This four-dimensional model of the deviates of the logarithmized concentrations was rotated to fulfil the varimax criterion of either maximum or minimum loadings on the factors (Harman 1976). Matrix A of loadings of this rotated model is shown in Table 4, which also includes communalities, absolute and relative factor variance contributions and standard deviations of factor loadings. For clarity, loadings of less than three standard deviations have been deleted as insignificant (Heidam 1982).

The first factor, with prominent members such as Al, Si, K, Ca, Ti, and Fe, clearly represents a crustal component of the aerosol. The second factor is anthropogenic; the most prominent elements are S and Br, but there is also some coupling to Cr, Zn, Sr, and Pb. This aerosol component may arise from various combustion processes in power plants, incinerators, and motor vehicles in distant source areas. The third factor couples mainly to Cu and Zn, but Cr, Mn, Sr and S also have significant loadings. This factor is also considered to constitute an anthropogenic aerosol component, largely of a metallic composition and statistically independent of Factor 2. Factor 4 represents Pb. There is a considerably weaker but no less significant coupling to Br, pointing to an aerosol component that originates from motor vehicle exhausts. Decoupling from the long-range transported combustion component indicates that this aerosol component to some extent may have a local origin.

Table 4 shows that at NORD the model accounts for a variance of 12.3, corresponding

Table 4. Varimax solution of four-factor solution. Log-concentrations at NORD, 1979–80.

Variable	Communality	Factor 1	Factor 2	Factor 3	Factor 4	Std dev S(A)
13 Al	0.9563	0.977				0.021
14 Si	0.8623	0.905	−0.160	0.123		0.038
19 K	0.8848	0.876	0.215	0.267		0.035
20 Ca	0.8674	0.894		0.230		0.037
22 Ti	0.9705	0.982	−0.072			0.018
24 Cr	0.8388	0.676	0.325	0.514		0.041
25 Mn	0.8558	0.812	0.273	0.349		0.039
26 Fe	0.9775	0.976		0.154		0.015
38 Sr	0.6940	0.594	0.378	0.444		0.057
16 S	0.8439		0.846	0.347		0.041
35 Br	0.8493		0.854	0.175	0.295	0.040
29 Cu	0.8525	0.222	0.172	0.857	0.196	0.039
30 Zn	0.9144	0.158	0.332	0.877	0.097	0.030
82 Pb	0.9550		0.318	0.227	0.895	0.022
14	12.3225	6.805	2.100	2.454	0.963	Variances
100	88.02	48.6	15.0	17.5	6.9	Percentages

Table 5. Concentration components of the Arctic aerosol. Factor model survey, SAGA 1979–80. MMC: minor member collecting factor. [1] Feb 1980–May 1981.

Station	Sums	Crust	Marine	Combustion	Metal	Eng Exh
601 THUL[1]						
Prominent members		Al, Si, Ti, Fe		Br, Pb	Cu, Zn	
Weak members		Mn, S, K, Ca, Sr		S, Mn, Zn	K, Ca, Sr, S, Mn, Pb	
Variance 13	11.1	5.0		2.6	3.5	
Percentage 100	85.1	38.2		20.3	26.6	
701 NORD					MMC	
Prominent members		Al, Si, K, Ca, Ti, Mn, Fe		S, Br	Cu, Zn	Pb
Weak members		Cr, Sr		Cr, Sr	S, Cr, Mn, Sr	
Variance 14	12.3	6.8		2.1	2.5	1.0
Percentage 100	88.0	48.6		15.0	17.5	6.9
801 GOVN					MMC	MMC
Prominent members		Al, Si, K, Ca, Ti, Cr, Mn, Fe, Ni, Sr	Cl	S, Br	Zn	Pb
Weak members		Cu	K, Ca, Br, Sr	Pb, Zn	Cu, Ni	Si, Cu
Variance 16	14.5	8.1	1.8	2.0	1.4	1.1
Percentage 100	90.4	50.7	11.1	12.6	9.0	7.0
901 KATO				MMC	MMC	
Prominent members		Si, Ti, Mn, Fe	Cl, K, Ca, Br, Sr	Pb	Cu, Zn	
Weak members		K, Ca, Cu, Sr	S, Ti, Mn, Fe	S, Cu, Br	S	
Variance 13	11.4	3.4	4.7	1.4	1.8	
Percentage 100	87.4	26.2	36.4	10.8	14.0	

to 88% of the total variance of 14. Almost 50% comes from the crustal factor, whereas the combustion and metallic factors contribute 15% and 17.5% respectively. Finally, 7% of the variance of the log concentrations can be ascribed to aerosol components related to engine exhausts. The four-factor solution is also a very good model of the individual variables, and the column of communalities shows that the model explains well over 80% of the variance of each variable with the exception of Sr with a communality of 0.69.

Similar factor models for aerosol composition at other SAGA stations (except PCS, which operated only for a short winter period) are summarized in Table 5. For each station and each factor, the more important elements loaded on that factor are listed. Elements have been divided into three categories. Prominent members (defined as those with a loading higher than 0.7) contribute at least $a^2 = 0.5$ to the communality ($h^2 < 1$) and can therefore be prominent in one factor only. Weak members of the factor have loadings numerically larger than 0.3, so they contribute at least 0.1 to the communality. Minor members—elements in the third category—still have statistically significant loadings, but the values are very small ($a^2 < 0.1$); they are not listed in Table 5. However, if they are collectively important the factor is labelled minor member collector (MMC). Table 5 shows that the atmospheric aerosols in Greenland can be described by the same four or five statistically independent components. In view of the distance between the sampling locations (up to 1,600 km) that is a remarkable result. It means that, at least on an annual basis, the tropospheric aerosol in Greenland has a uniform composition except for minor local variations. With the general similarity of results for aerosol composition in the Arctic, this concept of one aerosol may possibly be extended to the whole Arctic region.

The Greenland aerosol is thus composed of at most two naturally occurring components, one crustal and one marine, and at most three anthropogenic components, a combustion, a metallic and an engine exhaust component. The components vary considerably in their composition of prominent and weak members; however, similarities in aerosol composition suggest that the components observed over various parts of Greenland originate from common or similar sources, and are subject to common or similar large-scale meteorological transport mechanisms.

Table 6. Components of the non-crustal Arctic aerosol, the Ti- enrichments. Factor model survey SAGA 1979–80. [1] Feb 1980–May 1981; 124 samples. [2] Communality $h^2 < 0.7$; [3] Many data below d.l.; reconstructed.

Station	Sums	General Industry	Marine	Ferro-metal
601 THUL[1]				
Prominent members		S, Br, Pb	K, Ca, Sr	
Weak members		Mn, Zn, K, Cu, Sr	Cu, S, Mn, Zn, Pb	
Variance 9	7.2	3.7	3.9	—
Percentage 100	80	41	39	—
701 NORD				
Prominent members		S, Br, Pb	K, Ca	Mn, Fe
Weak members		K, Cr, Mn, Cu[2], Sr	Sr, S, Cr, Cu[2], Zn, Br, Sr	Cr, Cu, Zn, S, K, Sr
Variance 11	9.1	3.6	2.5	3.0
Percentage 100	83	33	23	27
801 GOVN				
Prominent members		S, Cu, Zn, Br, Pb	Cl, Sr	Mn, Fe
Weak members		Cr, Ni, K, Ca, Mn, Sr	K, Ca, S, Cr, Ni, Zn, Br	S, K, Ca, Cr, Ni, Br, Sr
Variance 13	10.9	4.8	3.2	2.9
Percentage 100	84	37	25	22
901 KATO				MMC
Prominent members		S, Cu, Zn, Pb	Cl[3], K, Ca, Sr	Fe
Weak members		Mn[2], K, Ca, Br[2]	Br[2], S	Mn[2]
Variance 11	8.7	3.6	3.8	1.3
Percentage 100	79	33	34	12

Similar conclusions are reached from analysis of the non-crustal aerosol (Table 6), which can also be described by a few independent and source-related components that are the same throughout Greenland, and presumably over the entire Arctic. The first factor is an industrial component derived from combustion (power production, motor vehicles, waste incineration) and metal smelting. The variance contribution is 30–40%. The second factor, almost certainly a marine component of predominantly Atlantic origin, explains 25–40% of the total variance. The third factor (not observed at THUL) has prominent non-crustal Mn and Fe, but is also strongly influenced by Cr, Ni, Cu and Zn, presumably from ferro-metal processing sources. It explains 10–30% of the variance. At all sites these components contain about 80% of all variance in the data.

The descriptive names used above were chosen partly because of the elemental compositions and partly on the basis of their seasonality. A comparison with the factor models in Table 5 shows that the crustal component has disappeared. The enriched elements show much the same behaviour, but anthropogenic factors of the total aerosol tend to merge when the non-crustal aerosol is considered. A reduction in factor number was anticipated, and indeed the non-crustal aerosol can be described by fewer, but different, components. The factors were also expected to be better defined and easier to identify. The compositions of components in Table 6 are actually more similar between sites than was the case for the components of the total aerosol in Table 5. On the other hand components of the non-crustal aerosol have a relatively smaller number of prominent members and a correspondingly larger number of elements distributed with weak couplings over several factors; they may thus be more representative for conditions over the whole Arctic, but they are also more difficult to identify.

Seasonal variations, transport and sources

The temporal behaviour of source-related components of the aerosols reflects the importance of various atmospheric transport mechanisms over the year. Seasonal variation may thus be an important clue to the identification of a factor. In addition, a study of the individual concentration or enrichment data of elements dominant in a given factor, in conjunction with synoptic weather maps, may disclose transport paths and likely origins. Such a study was carried out for the total aerosol models at NORD in 1979–80 (Heidam 1984).

Monthly mean concentrations or enrichments for selected elements at THUL and NORD are shown in Figures 5–10, covering the whole experimental period 1979–83. Concentrations of S, representing the combustion or industrial component of the aerosol (Figure 5) show a characteristic annual cycle, with month-by-month synchronization; concentrations are high in January–April, and summer values are lower by a factor of 10–100. In winter, high arctic troposphere constitutes a mixed aerosol reservoir subject to injections of pollutants from lower latitudes. In spring the polar front migrates northward from latitudes south of the major source areas, its scavenging progressively isolating the Arctic from mid-latitude influences. In autumn the polar front returns southward, allowing more southern pollutants to penetrate. This pattern is amplified by the very low precipitation rate in the Arctic winter. Combustion products may arrive episodically in northern Greenland from both North American and European sources, though episodes were rare and the enhanced winter levels are essentially due to an Arctic-wide increase in pollution levels by injections all along the Arctic perimeter.

Concentrations of Zn, representing the metallic component (Figure 5) show the clearest annual cycle at NORD, with peaks in January and February. Weather maps for this time of year show NORD subject to a strong, well-defined atmospheric flow from

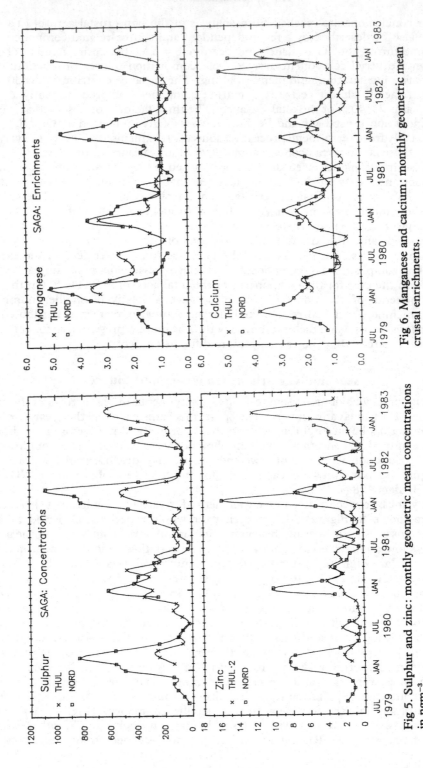

Fig 5. Sulphur and zinc: monthly geometric mean concentrations in ngm⁻³.

Fig 6. Manganese and calcium: monthly geometric mean crustal enrichments.

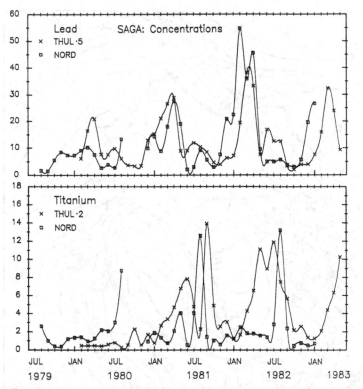

Fig 7. Lead and titanium: monthly geometric mean concentrations in ngm^{-3}.

central USSR, which is deflected westward over the Arctic ocean by an Atlantic cyclone residing over north Scandinavia. The average synoptic situation in the first quarter of the year is shown in Figure 8 (Liljequist 1970). The major source for the metallic component is therefore most likely to be the Urals region, with its vast complex of mines, smelters and other metal processing plants. Figure 8 also shows that THUL, in northwestern Greenland, is less subject to this flow. For the rest of the year the metallic component is more erratic, but short autumn episodes of transport from both North America and Europe are identified. At NORD a minor peak appears consistently in August. It has not been analyzed meteorologically, but may be connected to the Pb factor described later.

The crustal enrichment of Mn, representative of the ferro-metal component of the non-crustal aerosol (Figure 6) behaves very much like Zn. It may be another manifestation of the metallic component from the Urals, since episodes essentially coincided. Enrichments of Mn may better represent this metal component, since the secondary summer values noted for Zn concentrations are almost completely absent. It is noteworthy that these low enrichments which, if considered separately, might have been discarded as insignificant, collectively show such a systematic recursive behaviour. The good signal-to-noise ratio results from removing the crustal noise. It is also worth noting that, on this monthly basis, the station-to-station correlation appears to have improved compared to the bare concentrations (Figure 5) of Zn.

Concentrations of lead (Figure 7) at THUL formed part of the long-range combustion

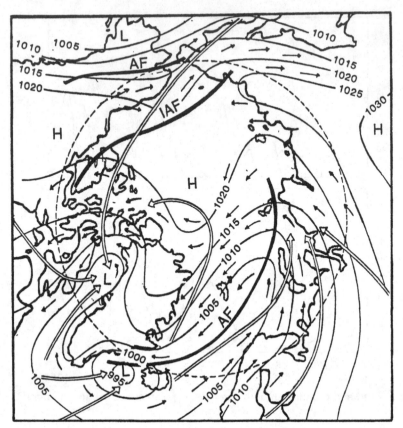

Fig 8. Mean distributions of surface pressures, winds, fronts and cyclone trajectories in January. Winds are marked by short arrows, cyclone trajectories by long double arrows and fronts by heavy lines. AF — arctic front, IAF — inner arctic front. Data from 1931–60 or 1954–60 (after Liljequist 1970).

or industrial components of the aerosol (Tables 5 and 6); at NORD only the non-crustal part, as expressed by enrichment factors, can be so described. Figure 7 shows that total concentrations behave somewhat differently, with large values in August. At NORD, therefore, the Pb-containing component was partly of local origin and related to engine exhaust of local vehicles. Zn showed the same behaviour. As there is a gravel airstrip with slight summer traffic at NORD, and Zn is a product of tyre wear, the coincidence of enhanced Zn and Pb levels in August was probably due to resuspension of road and airstrip dust.

Concentrations of titanium (Figure 7) represent the natural crustal component of the Arctic aerosol. High summer values fall to minima in September-November, and moderate winter values. This component in summer is presumably indigeneous, from dried-up soil; in winter it may originate in distant dry continental regions, governed by the same reservoir mechanism as described for the anthropogenic industrial component. Autumn minima may then result from a quenching of indigeneous sources by increasing snow-cover, and absence of a long-range transport mechanism that reaches sufficiently far south to pass over dry continental regions. High summer Ti values coincide with the August maxima for zinc and lead at NORD, already attributed to resuspension of ground dust. As resuspension of dust and wind erosion of soils are governed by the same forces, the coincidence of peak values in summer strengthens the interpretation that the Zn and Pb maxima in August originate locally from road and airstrip dust.

Crustal enrichments of calcium (ie non-crustal Ca) are shown in Figure 6. The corresponding aerosol component element is potassium (Table 6) and at the more southerly stations GOVN and KATO chlorine is prominent, indicating a marine origin. The most abundant elements in sea-water sequentially are Cl, Na, Mg, and S followed by K and Ca with concentrations of about 400 ppm, about half of the concentration of S (Brewer 1975). Removing the crustal part of Ca may thus leave us with a marine component. Episode studies showed that high individual values at NORD in October, November and December 1979 coincided with short transport episodes of both combustion and metal-processing products from North America and Europe, registered at both NORD and KATO. These transport paths passed over the North Atlantic ocean to encompass the whole of eastern Greenland. Figure 6 therefore shows the influence of a long-range transported marine component, originating in the North Atlantic and characterized by high levels in autumn and sometimes in winter, with lower summer values. Total concentrations of marine components may well reach maxima in summer, but that will then be an influence caused by mesoscale transport from Greenland waters.

Conclusions

Application of factor analysis to data on element concentrations or enrichment factors have shown that the arctic aerosol may be composed of three to five statistically independent components. Elemental composition and seasonal variations identify the components as source-related, and generally of distant origin. Despite minor differences in composition, essentially the same components appear everywhere; their appearance is a result of large-scale atmospheric transport mechanisms that affect the whole Arctic in much the same way. Variations over the years reveal, however, that the components vary in relative strength in a complex way, due presumably to variations in relative strengths of the otherwise recursive air mass movements.

It should be stressed finally that the factor analyses apply to the logarithmized data. The resulting models are not physical in the sense that source-apportioning receptor models are physical. Rather they explain data, providing a conceptual framework that has aided our understanding of the arctic aerosol.

References

BARRIE, L. A. AND OTHERS. 1981. The influence of mid-latitudinal pollution sources on haze in the Canadian Arctic. *Atmospheric Environment*, 15: 1407– 19.
BREWER, P. G. 1975. Minor elements in sea water. Chapter 7 in RILEY, J. P. AND SKIRROW, G. (editors). *Chemical oceanography*. Academic Press, London, New York, San Francisco.
DAVIDSON, C. I. AND OTHERS. 1981. Wet and dry deposition of trace elements onto the Greenland ice sheet. *Atmospheric Environment*, 15: 1429–37.
FLYGER, H. AND HEIDAM, N. Z. 1978. Ground level measurements of the summer tropospheric aerosol in northern Greenland. *Journal of Aerosol Science*, 9: 157–68.
FLYGER, H. AND OTHERS. 1973. The background level of the summer tropospheric aerosol over Greenland and the North Atlantic Ocean. *Journal of Applied Meteorology*, 12: 161–74.
FLYGER, H. AND OTHERS. 1976. The background level of the summer tropospheric aerosol, sulphur dioxide and ozone over Greenland and the North Atlantic Ocean. *Journal of Aerosol Science*, 7: 107–40.
FLYGER, H. AND OTHERS. 1980. The background levels of the summer tropospheric aerosol and trace gases in Greenland. *Journal of Aerosol Science*, 11: 95–110.
HARMAN, H. H. 1976. *Modern factor analysis*. (3rd edn.) Chicago, The University of Chicago Press.
HEIDAM, N. Z. 1982. Atmospheric aerosol factor models, mass and missing data. *Atmospheric Environment*, 16: 1923–31.

HEIDAM, N. Z. 1983. *Studies of the aerosol in the Greenland atmosphere*. Air Risø, NAEP Air Pollution Laboratory. (Report MST LUFT-A73).
HEIDAM, N. Z. 1984. The components of the Arctic aerosol. *Atmospheric Environment*, 18: 329–43.
HEIDAM, N. Z. 1985. Crustal enrichments in the Arctic aerosol. *Atmospheric Environment*, 19: 2083–97.
HEINTZENBERG, J. 1980. Particle size distribution and optical properties of arctic haze. *Tellus*, 32: 251–60.
HEINTZENBERG, J. AND OTHERS. 1981. The chemical composition of arctic haze at Ny-Ålesund, Spitsbergen. *Tellus*, 33, 162–71.
LANNEFORS, H. AND OTHERS. 1983. Background aerosol composition in southern Sweden. *Atmospheric Environment*, 17: 87–101.
LILJEQUIST, G. M. 1970. *Klimatologi*. Stockholm, Generalstabens Litografiska Anstalt.
MASON, B. 1966. *Principles of geochemistry*. (3rd edn.) New York, Wiley.
OECD. 1979. *The OECD programme on long range transport of air pollutants. Measurements and findings*. Paris, OECD.
RAHN, K. A. AND OTHERS. 1977. Particulate air pollution in the Arctic: large-scale occurrence and meteorological controls. *Proceedings of 9th International Conference on Atmospheric Aerosols and Nuclei*. Galway, Galway University Press: 223–27.
RAHN, K. A. AND McCAFFREY, R. J. 1980. On the origin and transport of the winter Arctic aerosol. *Annals of the New York Academy of Science*, 338: 486–503.
RAHN, K. A. AND OTHERS. 1980. High winter concentrations of SO_2 in the Norwegian Arctic and transport from Eurasia. *Nature*, 287: 824–26.
SCHÜTZ, L. AND RAHN, K. A. 1982. Trace element concentrations in erodible soils. *Atmospheric Environment*, 16: 171–76.

ORIGIN AND CHARACTERISTICS OF AEROSOLS IN THE NORWEGIAN ARCTIC

BRYNJULF OTTAR AND JOZEF M. PACYNA

ABSTRACT. Movements of air masses over several thousand kilometres affect the quality of arctic air in both summer and winter. This paper summarizes the results of a programme by the Norwegian Institute for Air Research (NILU) to identify and characterize sources of pollutants measured in the Norwegian Arctic, and to model the movement of pollutants within the air masses. Aircraft-based data on scattering coefficient, particle size distribution and chemical composition of aerosols were used to trace the origins of polluted layers of arctic air. Polluted air masses were observed at altitudes above 2,000 m in summer and 2,500 m in winter. Polluted layers below 1,500 m were related to episodic movements of air masses from areas where temperatures were similar to those of the Arctic, for example the northern Soviet Union: these layers also gave chemical indications of their origins. Profiles derived from airborne data were extended downward by measurements from the ground. From information on (1) total and size-differentiated chemical composition of surface aerosol, (2) enrichment factors of chemical constituents, (3) emission inventories for potential sources of air pollution in the Norwegian Arctic, and (4) wind trajectories at 850 mb, we conclude that emissions from anthropogenic sources mainly in Eurasia, and occasionally in North America, contribute to the pollution of arctic air over the whole year. Good agreement between these measurements and estimates from Lagrangian transport model estimates supports this conclusion.

Contents

Introduction

During the last decade interest has increased in air pollution studies in the Norwegian sector of the Arctic. Since 1977 regular measurements have shown the frequent occurrence of highly polluted air masses at Ny-Ålesund on Spitsbergen (Svalbard) and Bjørnøya, particularly in winter (Larssen and Hanssen 1979). Pollutants measured include sulphur dioxide, sulphates and trace elements, mostly from anthropogenic sources in Europe and the USSR. This study, together with measurements from other parts of the Arctic (Rahn and McCaffrey 1979, 1980) revealed a need to assess sources and transport paths of arctic aerosols. In 1981 the Norwegian Institute for Air Research (NILU) started a programme to identify and characterize sources of pollutants measured in the Norwegian Arctic, and to model the transport of pollutants within the air masses. The programme is financed by British Petroleum Ltd. A first summary of its results is presented here.

Characteristics of aerosols aloft

Arctic haze was first described by Mitchell (1956). Later both the chemical composition and the origin of turbid layers of polluted air were extensively studied (Shaw and Wendler 1972; Rahn and others 1977). Ground-level measurements., however, did not explain to what extent the pollution was due to a general influx of air pollutants into the Arctic, or whether episodic long-range movement of pollutants were involved. To clarify the point 40 flights were made over Svalbard during four measurement campaigns in 1983–84. The NILU aircraft, a twin-engined Piper Navajo, was equipped with a nephelometer constructed by Heintzenberg and Bäclin at the Meteorological Institute of Stockholm University, a Knollenberg particle mass spectrometer (PMS) for measuring

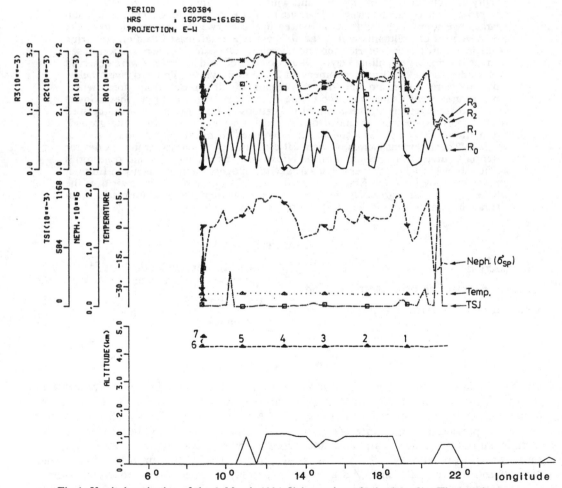

Fig 1. Vertical projection of the 2 March 1984 flight track vs the land profile. The numbers indicate reference points.

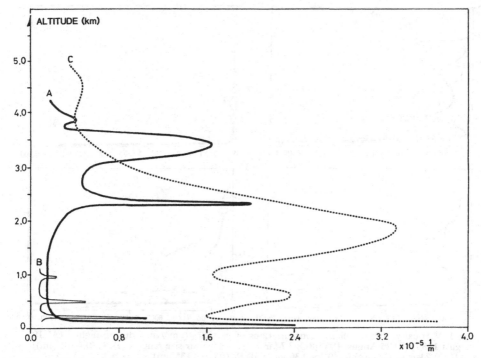

Fig 2. The σ_{sp} values vs flight levels during flights: (A) 18 August 1983; (B) 25 August 1983; (C) 3 March 1984.

aerosol volume-size distributions (0.09 − 3.0 μm), a TSI condensation nuclei counter, and low and high volume aerosol samplers, as well as instruments for measuring height, temperature, and dew point. The measured data were reduced and stored in a specially-designed computer aboard the aircraft. An Omega navigation system referred all data automatically to time and position in space.

Data displays were produced shortly after each flight. The example in Figure 1 shows the nephelometer measurements of σ_{sp}, volume concentrations of particles in four size ranges: 0.09 − 0.135μm (R_3), 0.15 − 0.30 μm (R_2), 0.24 − 0.84 μm (R_1) and 0.60 − 3.0 μm (R_0), temperature, and concentration of condensation nuclei plotted as a function of time along a vertical projection of the flight track against the land profile. The position of the aircraft at 10–minute intervals is indicated by numbered reference points. These data, supported by information on the chemical composition of the particles and wind trajectory analysis, were used to assess the origin of the measured aerosols.

Flights on 18 and 25 August 1983 and 3 March 1984 were selected for further evaluation (Ottar and others 1985). Vertical profiles of σ_{sp} from these flights are given in Figure 2, and particle volume concentrations in four size ranges in Figure 3. High concentrations of R_3 and R_2 particles (mm) and high σ_{sp} values were observed in the air below 2,000 m during the 3 March flight (Curve C in Figures 2 and 3). During winter flights in 1983 and 1984 enhanced σ_{sp} and R_2 and R_3 concentrations were measured up to 2,500 m. Generally, pollution air up to 2,000–2,500 m seems to be well mixed in a vertical projection, though peaks of σ_{sp} and R_2 and R_3 concentrations can be identified at about 2,000 m.

The typical size distribution of particles collected below 2,500 m is represented

ALTITUDE (km)

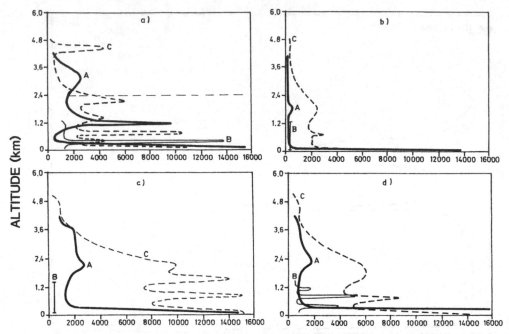

Fig 3. Vertical profile of volume concentrations of particles ($\mu m^3/cm^3$) during flights: (A) 18 August 1983; (B) 25 August 1983; (C) 3 March 1984, for four size ranges: (a) R_0 (0.6–3.0 μm); (b) R_1 (0.24–0.84 μm); (c) R_2 (0.15–0.30 μm); (d) R_3 (0.09–0.135 μm).

by Curve C in Figure 4. Peaks in the range 0.15 to 0.50 μm may be explained by long-range transport of air pollutants from sources a few thousand kilometres away with a similar potential temperature. Particles of 0.15 to 0.50 μm are commonly emitted from high temperature sources, eg coal combustion, roasting of ores, steelmaking and waste incineration. For sources located 2,000–3,000 km from the Arctic, the primary particle size distribution may not change significantly during transport in winter. Thus, the size distribution of the particles measured in the Arctic, at the likely level of air pollution transport, may be given by the size distribution of particles emitted from high-temperature sources.

Isobaric and isentropic wind trajectories at 850 mb for 3 March 1984 indicated movement of air across Kola Peninsula [Kol'skiy Poluostrov] to Svalbard archipelago. Volume concentration peaks in the 0.15–0.5 μm range shown in Figure 4 fit well with the emission from fuel combustion and copper-nickel smelters in Kola Peninsula.

Analysis of aerosols measured at 800 m during the flight of 3 March showed the presence of V, As, Cd, Sb, Au, Cu and partly Mn, Co and Zn. This group of elements is typical for emissions from the Kola Peninsula sources. During episodes of air pollution from other sources, the chemical composition of aerosols measured below 2,500 m was different, and related to emisions in the regions of origin (Pacyna and others 1985). The lateral variations of the aerosol concentration during episodes showed the presence of isolated areas with high σ_{sp} and R concentrations, like wide pollution plumes. During intervals between episodes σ_{sp} and R concentrations (except R_0) below 2,500 m are smaller. An interesting point in Figure 4 is the absence of particles of less than 0.1 μm; there is little or no production of new particles in the winter aerosol (Joranger and Ottar 1984).

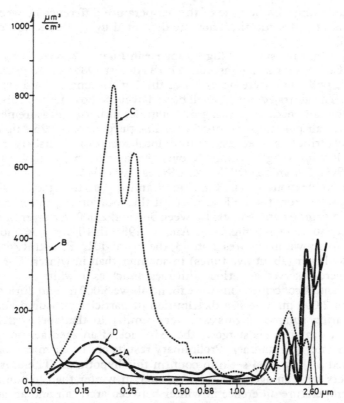

Fig 4. Volume concentration of particles vs particle size: (A) at 3,300 m on 18 August 1983; (B) at 900 m on 25 August 1983; (C) at 1,800 m on 3 March 1984; (D) at 3,500 m on 3 March 1984.

On 3 March a layer of clean air was observed between 3,000 and 4,000 m. The σ_{sp}, R_3 and R_2 values increased slightly above 4,000 m (see Figures 2 and 3). Results from other flights in March 1983 show that the increase of σ_{sp} and volume concentrations may start at 2,500 m and continue with increasing altitudes. A size distribution of particles collected during the 3 March flight in this second layer of pollution is presented as Curve D in Figure 4. The peak observed in the range 0.15–0.50 μm at altitudes below 2,000 m is less pronounced at altitudes above 3,500 m. On the other hand, the concentration of particles larger than 2 μm has increased. This is also evident from R_0 profiles in Figure 3a. Furthermore, the lateral distribution of the R_0 values show a patchy structure (Figure 1). Observed at various altitudes all over the Arctic, this was not related to the episodic transport below 2,500 m. This difference in the particle size distribution indicates different origins of the aerosols measured below and above 2,500 m in winter. Differences in origin were confirmed by differences in chemical composition. The ubiquitous R_0 particles (0.6–3.0 μm), particularly the 2.0–3.0 μm fraction, consist mainly of Al, Si, Cr, Ti, Sc and Fe, which are typical of natural sources such as windblown dust. In addition, traces of anthropogenic elements (V, As, Sb, Sm, and Au) were measured sporadically.

Size distribution and chemical composition of these larger particles indicate that the layers of polluted arctic air above 2,500 m during winter may be due to pollution from distant sources in North America, Europe, and USSR. These aged aerosols must enter the Arctic at higher altitudes because of the temperature difference between air masses from lower latitudes and arctic air. They are deposited in the Arctic and can be found at various altitudes.

Results obtained for the summer flights appear in Figures 2, 3 and 4 by Curve A for 18 August, and Curve B for 25 August. On 17–18 August 1983 very high concentrations of anthropogenic pollutants were observed at the Ny-Ålesund ground station (Pacyna and Ottar 1985). The air trajectories had all passed over the Soviet Union about four days earlier. It was the only occasion during our summer airborne measurements when an episode of elevated air pollution was observed. The purpose of the 25 August flight was to examine the contribution of emissions from local sources in Spitsbergen to arctic air pollution. This low level flight was made over the four towns, Longyearbyen (LYR), Barentsburg (BBG), Pyramiden (PYR), and Sveagruva (SVA).

Enhanced concentrations of all R particles and σ_{sp} at the start of all flights are due to local emissions from the LYR airport. At this location σ_{sp} and R concentrations are comparable in summer and winter. Between 500 and 2,000 m a layer of clean air was observed. During some of the flights in August 1983 this layer was wider, reaching 3,000–3,500 m. As shown in Figures 2 and 3, the σ_{sp} and R_3, R_2 and R_1 concentrations were significantly lower (about five times) in summer than in winter. For R_0, elevated concentrations were observed at various altitudes, much as in winter.

Enhanced σ_{sp} and R concentrations were found above 3,000 m, and during some of the flights even from 2,000 m. The size distribution of particles measured above 3,000 m (or 2,000 m) during summer flights was very similar to that for particles collected above 4,000 m in winter. This suggests the existence of an aged aerosol in the arctic air during both winter and summer. Preliminary results on the chemical composition of aerosols measured in July 1984 seem to confirm this hypothesis. The altitude of these polluted air layers with an aged aerosol is greater in winter than in summer because of the greater temperature difference in winter between arctic air and air arriving from lower latitudes.

During the 18 August 1983 flight, an additional layer of polluted air was observed at 2,400 m. The particle size distribution indicated a significant contribution of particles within the range 0.15–0.50 μm not found during the other flights in August. This, and the high concentrations of anthropogenic constituents measured at the Ny-Ålesund ground station suggest that an episode of long-range transport of pollutants occurred and penetrated into the Arctic, mainly at a level of 2,400 m.

Sources of local air pollution on Spitsbergen can sometimes generate enough particulate air pollutants to disturb air sampling. Several of the σ_{sp} and R concentration peaks on Curve B (Figures 2 to 4) representing the 25 August flight, are due to emissions from the four settlements. Particle size distribution shows a very high contribution of the finest particles (<0.1 μm) in Curve B (Figure 4). This is a result of fresh local discharges of pollutants (see also Figure 3d, Curve B). For particles larger than 0.1 μm the size distribution Curve B follows Curve A for particles collected during summer.

It is difficult from our data to assess the contribution of sea-salt particles to overall air pollution; these particles are probably larger (>3.0 μm) than our equipment measures.

Characteristics of surface aerosols

Surface-based measurements are an important supplement to the lower part of the vertical profiles derived from airborne measurements. Surface data are collected continuously. Much more equipment can be installed at a ground station than aboard an aircraft, resulting in a wider variety of information available for surface aerosols.

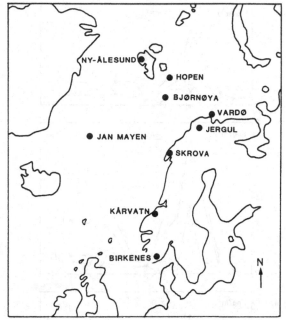

Fig 5. Norwegian Institute of Air Research (NILU) sampling stations.

The main measuring station was at Ny-Ålesund. In addition to the routine equipment for collection of meteorological data and samples of sulphur dioxide and sulphate aerosol, the station was equipped for this study with high-volume samplers for organic and inorganic components, a Battelle cascade impactor, a Winkler impactor, and diffusion batteries, in addition to an integrating nephelometer and a condensation nuclei counter. Other measurement stations were on Hopen, Jan Mayen and Bjørnøya, and there were five on mainland Norway (Figure 5) where additional equipment was limited to high-volume samplers. Additional measurements were carried out over five intensive measurement periods, each of five to six weeks. Chemical analysis of collected samples included instrumental neutron activation analysis (INAA), atomic absorption spectroscopy (AAS), and particle induced x-ray emission (PIXE).

The following information was collected: (1) the total and size-differentiated chemical composition of surface aerosols, (2) enrichment factors of chemical constituents, (3) emission inventories for potential sources of air pollution in the Norwegian Arctic, and (4) wind trajectories at 850 mb during the episodes of long range transport of pollutants.

A few episodes of long-range movement of pollutants to the Norwegian Arctic occurred during the winter measurements of March-April 1983 and February-April 1984; peaks of Pb, Zn and Ni concentrations at Ny-Ålesund, Jan Mayen and Bjørnøya observed during 6–13 March 1983 (Ottar and Pacyna 1984) are shown in Figure 6. On this occasion the peak concentrations at Vardø, 100 km north of the large copper-nickel smelter at Nikel in the USSR, just beyond the Norwegian border, appeared two days earlier than the peaks at Bjørnøya and four days earlier than the peaks at Ny-Ålesund. Apparently it takes two days for the pollution to move from one station to the other. No corresponding concentration peaks were found in the samples collected at mainland stations farther south along the coast. The 850 mb isobaric trajectories for this period indicated air mass movement across Kola Peninsula (where the Nikel smelter is located) to the Norwegian Arctic.

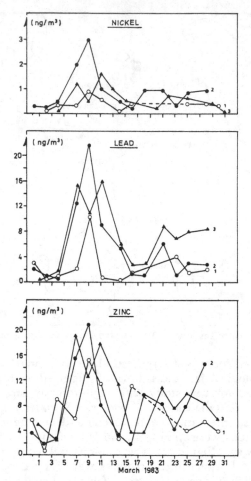

Fig 6. Observed aerosol concentrations: (o) Jan Mayen; (●) Bjørnøya; (▲) Ny-Ålesund.

A further clue to the origin of these pollutants came from the size-differentiated composition of the aerosol at Ny-Ålesund (Pacyna and others 1984). The 6–13 March episode showed increased concentrations of several anthropogenic pollutants in the fractions below 0.5 μm equivalent aerodynamic diameter (EAD), and in the fraction 1–2 μm EAD. The results for Pb, Zn, and Ni, given in Figure 7, fit well with the chemical composition and size distribution of particles emitted during the high temperature processes in copper-nickel smelters and oil combustion. Enrichment factors (based on Ti in bulk crustal rock) were calculated for the chemical constituents measured at the different impactor stages (Pacyna and others 1984): those for Pb, Zn, Ni, and Cu appear in Figure 8 for the 6–13 March episode (Curves 3 and 5 representing two different samples) and during non-episodic conditions (Curve A representing the average values). During the episode enrichment factors increased significantly in particle fractions below 0.5 μm EAD, confirming that additional pollutants at Ny-Ålesund were due to emissions from high-temperature sources, such as non-ferrous metal smelters.

Similar episodes were detected in the winters of 1983 and 1984, when emission sources

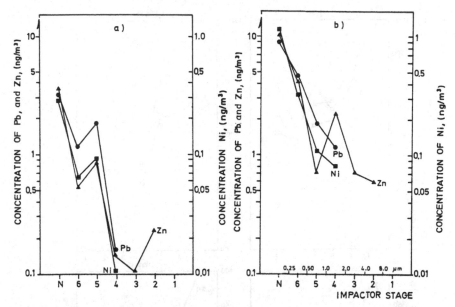

Fig 7. Observed aerosol concentrations vs particle size in March 1983: (a) average values for all samples except those collected during the 6–13 March 1983 episode; (b) average values for samples collected during the 6–13 March 1983 episode.

in the north-western Soviet Union, particularly on Kola Peninsula, contributed to high concentration of pollutants in the Norwegian Arctic. Aircraft measurements showed that these were transported at altitudes below 2,500 m, and resulted in increased Pb, Zn, Ni, Cd, Sb, Se, V, As, Cr, Mn, Au and Sm concentrations among particles less than 0.5 μm EAD. Thus the chemical composition observed at ground level fits well with size distributions of particles obtained from the aircraft measurements during episodic transport. Evidently air in the boundary layer is well-mixed during winter episodes, and 2,500 m seems to be the maximum height of this layer.

Other Eurasian and North American sources may also contribute to high concentrations of anthropogenic pollutants in the surface aerosol at Ny-Ålesund. However, the impact of these sources is more evident at altitudes above 2,500 m.

The three other NILU measurement campaigns in August-September 1982, August-September 1983, and June-July 1984 confirmed that Eurasian and North American anthropogenic emissions contribute to arctic air pollution in both winter and summer. For example there were three periods of elevated concentrations of several pollutants during August-September 1983 in the Norwegian Arctic (Figure 9) (Pacyna and Ottar 1985). Pollutants measured earlier at Ny-Ålesund and Hopen probably originated in North America and Greenland, as shown by an examination of the wind trajectories and confirmed by the low Ni concentrations (only lead-zinc smelters). A few days later, very high concentrations of Cd and Zn at Ny-Ålesund suggested air mass movement from parts of the northern Soviet Union other than Kola Peninsula (low Ni and Pb concentrations). The largest episode of long-range transport to Ny-Ålesund and Hopen was observed in September; air mass trajectories and composition of anthropogenic pollutants indicated a Western European origin.

Maximum concentrations of anthropogenic pollutants are lower during summer episodes than in winter. In summer movements are more strongly influenced by the

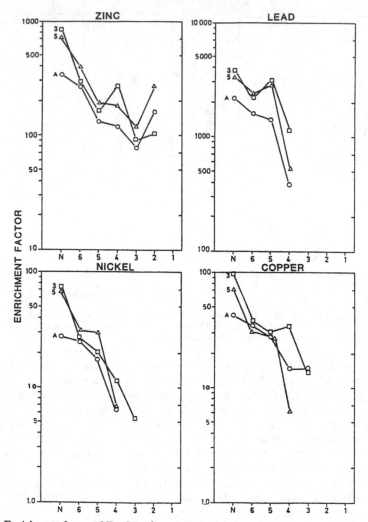

Fig 8. Enrichment factor of Zn, Pb, Ni, and Cu in particular size fraction. The impactor stages from 6 to 1 represent the particle size ranges from 0.25 to μm EAD. Stage N represents particles μm EAD collected on a Nuclepore filter. Curve A represents the average values of enrichment factors for samples other than 3 and 5. Samples 3 and 5 were collected during the 6–13 March 1983 episode.

location of the polar front, more pollutants are deposited en route, and emissions from natural sources, particularly the open sea, are more significant. Figure 10 shows summer measurements of episodic and non-episodic size-differentiated composition of the arctic aerosol. Generally, more large particles are transported in summer than in winter, possibly due to coagulation with new particles formed by conversion of gases, especially those emitted from natural sources. These processes are more efficient in summer than in winter. Very high values of S, Cd, Pb, and Zn enrichment factors (Al, crustal rock) at Ny-Ålesund during the September 1983 episode confirm the anthropogenic origin of aerosols measured at that time (Pacyna and Ottar 1985).

Of particular interest are the relatively high concentrations of S, Fe, and Cl in the

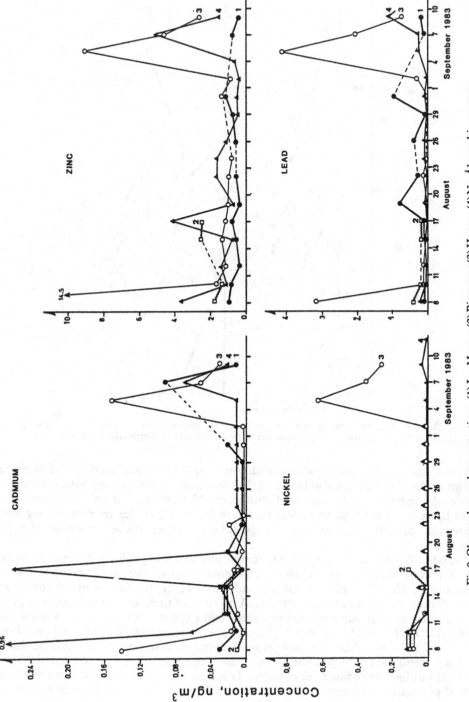

Fig 9. Observed aerosol concentrations: (1) Jan Mayen; (2) Bjørnøya; (3) Hopen; (4) Ny-Ålesund in summer.

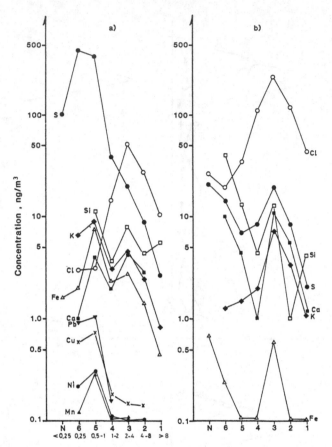

Fig 10. Observed aerosol concentrations vs particle size in summer 1983: (a) concentrations during episode of long range transport of pollutants from Europe; (b) non-episodic concentrations.

N-fraction (mm) under non-episodic conditions: high Fe in this fraction was measured also during the September episode, when the disappearance of Cl may have been due to excess acid sulphates. As Si was not found in the N-fraction, it is unlikely that the Fe content bounces off the large particles; alternatively these particles are part of the general background, possibly due to some degradation process taking place in the free troposphere.

Organic pollutants also reach the Arctic by wet and dry deposition (Oehme and Ottar 1984). The content of different organic compounds in a given sample can be used to assess the origin of polluted air masses. Of particular interest is the ratio between the α- and γ- hexachlorocyclohexanes (α-to γ-HCH). A mixture of these constitutes the insecticide Lindane, used in large amounts during spring and summer. After re-emission the ratio of α-to γ-HCH changes, because in daylight the γ-isomer is transformed into the α-form, with a half life of less than one year. However, the ratio observed in arctic air during episodes depends mainly upon the ratio of the same isomers in the insecticides used; those ratios differ from country to country. This may also allow one to suggest the likely origin of pollutants. In a similar way the ratio of Freon 11 to Freon 12, and the ratio of carbon tetrachloride to trichloroethylene, can be used to differentiate between air pollutants from eastern and western Europe.

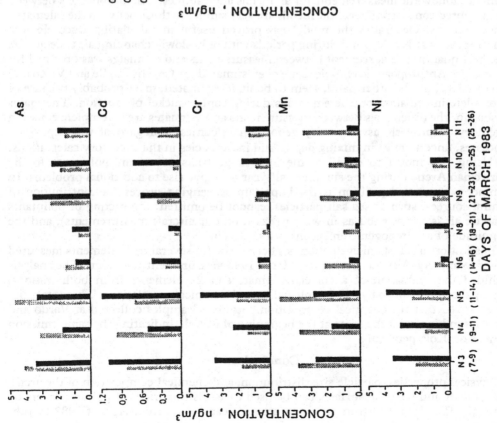

Fig 11. Measured (open bars) and calculated concentrations (full bars) of trace elements at Ny-Ålesund, March 1983.

Source receptor relationships for air pollutants on Spitsbergen

Origin of the aerosol at Ny-Ålesund during episodes of long-range transport has been verified also by use of a simple receptor-oriented Lagrangian transport model and estimated emissions of air pollutants from various source regions in Europe, the USSR and North America. In this way, the high concentrations of several air pollutants observed at Ny-Ålesund during the 6–13 March 1983 episode were related to emissions from sources in the USSR (Pacyna and others 1985). The emission data were taken from a more detailed report (NILU 1984) based on production and consumption statistics for the Soviet Union, available information on manufacturing techniques, and the efficiency of cleaning installations. Then, with the help of a simple trajectory model described earlier (Eliassen and Saltbones 1982; Pacyna and others 1985), the emissions were used to estimate the concentrations. The model is based on the mass-balance equation:

$$\frac{dq}{dt} = (1-\alpha)\,\frac{Q}{h} - kq$$

where q = concentration in ng/m^3; t = time in s; Q = emission per unit area and time, ng/m^2 . s; L = height of mixing layer in m; k = decay rate 1/s (only dry deposition was considered); α = local deposition (a fraction of emission deposited in the same grid element as it is emitted).

The most interesting feature of the comparison presented in Figure 11 is that the estimates follow the measurements for all metals and samples. The proportions between the measured concentrations of the metals are very similar to those between the calculated concentrations. Generally, the model has proved useful in calculating trace element concentrations at Ny-Ålesund during periods with only slowly changing air trajectories. The best quantitative agreement between measurements and estimates was obtained for Sb and Cd. Anthropogenic emissions are overestimated for Cr, Ni, Pb, Se and V. Zn and As emissions, on the other hand, seem to be underestimated, most probably because of incomplete information on the emissions during copper-nickel production. The major reasons for the differences between measurements and estimates are incomplete emission inventory, improperly assumed dry deposition velocities, disregard of wet deposition processes, uncertainties in mixing depth, and inaccuracies in the trajectory calculations.

Using this model to calculate the long-range transport of air pollutants to the Norwegian Arctic during the summer half-year may give rise to additional problems. In summer when wet deposition is the important scavenging process, a contribution of natural sources, such as sea salt particles, cannot be omitted; the mixing of pollutants in arctic air is not as good as in winter (based on our aircraft measurements), and the magnitude of anthropogenic emissions is less certain.

In addition, a statistical evaluation has been made for several trace elements measured at Vardø and Ny-Ålesund during the March 1983 measurement campaign. The analysis included the evaluation of a correlation matrix of 29 elements from both stations, principal component analysis of 10 elements, and canonical correlations. These statistics have proved that the chemical composition pattern of samples collected at Vardø and Ny-Ålesund is almost the same at the beginning of March, indicating the same emission sources on Kola peninsula.

Conclusions

Physical properties, particle size distribution, and chemical composition of the arctic aerosol aloft and near the surface, can be used to assess the origin of polluted layers of arctic air. Based on data from five measurement campaigns from August 1982 to July

1984, movement of air masses over thousands of kilometers seems to affect the quality of arctic air throughout the year. Aircraft measurements showed polluted air masses over the Arctic above 2,000 m in summer and 2,500 m in winter. Lower layers of polluted air may be due to episodes of air mass transport from emission areas with a potential temperature similar to that in the Arctic, and from local anthropogenic and (in summer) natural sources. Industry in the northern Soviet Union is the main contributor to air pollution of the Norwegian Arctic in winter: in summer these and other sources in Eurasia and North America contribute to this pollution in the boundary layer, but observed concentrations are much lower.

Prevailing conditions for long-range transport of air pollutants to the Norwegian Arctic during winter (low precipitation, favourable wind patterns, etc) make it possible to relate measured pollutant concentrations to their emissions. Good agreement between measurements at Ny-Ålesund and estimated emissions for various regions in the Soviet Union has been obtained for a number of air pollutants.

References

ELIASSEN, A. AND SALTBONES, J. 1982. *Modelling of long range transport of sulphur over Europe: a two-year model run and some model experiments.* EMEP/MSC-W Report 1/82.

JORANGER, E. AND OTTAR, B. 1984. Air pollution studies in the Norwegian Arctic. *Geophysical Research Letters*, 11: 365–68.

LARSSEN, S. AND HANSSEN, J. E. 1979. *Annual variations and origin of aerosol components in the Norwegian Arctic-Subarctic region.* Geneva, WMO. (WMO Special Environment Report 14.)

MITCHELL, M. 1956. Visual range in the polar regions with particular reference to the Alaskan Arctic. *Journal of Atmospheric and Terrestrial Physics*, Special Supplement: 195–211.

NORWEGIAN INSTITUTE FOR AIR RESEARCH (NILU). 1984. *Emission sources in the Soviet Union.* Norway, Lillestrøm. (Technical Report 4/84, NILU.)

OEHME, M. AND OTTAR, B. 1984. The long-range transport of polychlorinated hydrocarbons to the Arctic. *Geophysical Research Letters*, 11: 1133–36.

OTTAR, B. AND PACYNA, J. M. 1984. Source of Ni, Pb, and Zn during the Arctic episode in March 1983. *Geophysical Research Letters*, 11: 441–44.

OTTAR, B. AND OTHERS. 1985. Aircraft measurements of air pollution in the Norwegian Arctic. Norway, LillestrPACYNA, J. M. AND OTTAR, B. 1985. Transport and chemical composition of summer aerosol in the Norwegian Arctic. *Atmospheric Environment*, 19: 2109– 20.

PACYNA, J. M. AND OTHERS. 1984. Size-differentiated composition of the Arctic aerosol at Ny-Ålesund, Spitsbergen. *Atmospheric Environment*, 18: 2447–59.

PACYNA, J. M. AND OTHERS. 1985. Long-range transport of trace elements to Ny-Ålesund, Spitsbergen. *Atmospheric Environment*, 19: 857–65.

RAHN, K. A. AND OTHERS. 1977. The Asian source of Arctic haze bands. *Nature*, 268: 713–15.

RAHN, K. A. AND McCAFFREY, R. J. 1979. *Long-range transport of pollution aerosol to the Arctic. A problem without borders.* Proceedings of the WMO Symposium on the Long Range Transport of Pollutants and its Relation to General Circulation including Stratospheric/Tropospheric Exchange Processes, Sofia, 1–5 Oct. WMO No.j 538: 25–35.

RAHN, K. A. AND McCAFFREY, R. J. 1980. On the origin and transport of the winter Arctic aerosol. *Annals of the New York Academy of Science*, 338: 486–503.

SHAW, G. E. AND WENDLER, G. 1972. *Atmospheric turbidity measurements at McCall Glacier in north-east Alaska.* Conference proceedings on atmospheric radiation, Fort Collings, Colorado. Boston, American Meteorological Society: 181–87.

PROBLEMS OF AIR QUALITY IN LOCAL ARCTIC AND SUB-ARCTIC AREAS AND REGIONAL PROBLEMS OF ARCTIC HAZE

CARL BENSON

ABSTRACT. A strong and persistent air temperature inversion at ground level, resulting from outgoing radiation, is a feature of Arctic and sub-Arctic regions. In sheltered valleys the lowest air layers become extremely stable and prone to pollution. Temperature gradients of 30°C per 100 m are common in the lowest 50 m. The stable air structure extends 1 or 2 km upward from the surface and permits transport of pollutants from mid-latitudinal sources in sheet-like layers across the Arctic with negligible vertical dispersion. This produces widespread Arctic haze, most effectively in winter and early spring during cold, dry periods when the air is stable and well-stratified, and ending in April when conditions become turbulent and moist. Considering the distances involved, the concentration of pollutants in the haze seems surprisingly high, but it is a natural consequence of exceptionally stable air. Urban haze is most intense during December and January, when the inversions are undisturbed even at midday: pollution levels in a small community like Fairbanks during winter can equal or exceed those of large industrial, urban centres with populations two orders of magnitude larger, such as Detroit, Los Angeles or New York. Local winter problems become especially acute at latitudes north of 60° in continental regions such as Siberia, interior Alaska and Canada.

Contents

Introduction

Air pollution in the Arctic and sub-Arctic is still not taken seriously: the popular image of snow-covered mountain ranges, vast spaces with clean air, and a sparse population persists. However, in winter when these regions are snow-covered and incoming solar radiation is low, sheltered urban areas become especially prone to air pollution. In Alaska both Fairbanks and Anchorage frequently violate the US Environmental Protection Agency (EPA) standards for carbon monoxide. In the Fairbanks atmosphere concentrations of lead, hydrocarbons, carbon monoxide and nitric oxide match those of Detroit and Los Angeles (Winchester and others 1967; Holty 1973; Jenkins and others 1975; Coutts 1979; Benson and others 1983) though the Fairbanks population is 250 times less than that of Los Angeles area and the city has no significant industry. Similar problems are apparent in Whitehorse (McCandless 1982) and Anchorage (Bowling 1984). Winter pollution levels in the Arctic sometimes compare with those of US mid-western rural areas (Shaw 1985). What conditions permit such high concentration of pollutants in places where populations are relatively small? This paper shows that these and other large-scale Arctic atmospheric phenomena stem mainly from the lack of incoming solar radiation.

Types of air pollution

Chambers (1976) distinguishes three kinds of air pollution:

(1) *Coal smoke and gases*: a major source of air pollution in the industrial world;

(2) *Specific toxicants*: usually related to the effluent from a specific industrial source such as metal smelters;

(3) *The Los Angeles type of air pollution*: identified by Haagen-Smit (1952) as resulting from complex photo-chemical reactions involving ultraviolet radiation on unburned hydrocarbons and NO_2, primarily from vehicle exhausts. It is assocated with high temperatures and an oxidizing atmosphere. This kind of haze does not involve smoke and fog, but it is popularly known as 'smog'. It occurs with varying intensity in large metropolitan areas worldwide.

This paper focusses on a fourth kind of air pollution:

(4) *Fairbanks winter-type air pollution*: depends on scanty or absent solar radiation and the formation of inversions by the radiation of heat from snow surfaces to space. It is associated with low temperature and a reducing atmosphere.

A SPECTRUM OF AIR POLLUTION SETTINGS

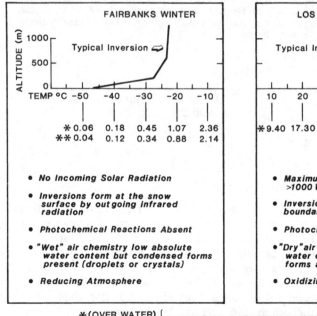
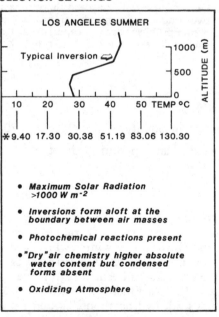

Fig 1. A spectrum of air pollution types with opposite extremes represented by winter in Fairbanks, Alaska and summer in Los Angeles, California.

Figure 1 illustrates the full spectrum of air pollution with extreme cases represented by winter values in Fairbanks and summer values in Los Angeles. High or low air temperatures, and the presence or absence of photochemical reactions, result directly from solar radiation; the significance of air temperature and photochemistry is discussed in the section on urban air pollution. Most air pollution research has been concerned with the high solar radiation, high temperature end of the spectrum, where dry air chemistry and strongly oxidizing conditions prevail. The other end of this spectrum prevails in high latitudes for prolonged periods, demanding regional and local attention.

Air temperature usually decreases with increasing altitude. In an inversion cold, high-density air is near the ground with warm air above: the lower atmosphere is stratified and stable so that vertical mixing is inhibited. Stability of an inversion increases with temperature gradient , which is expressed in °C per 100 m to facilitate comparison with the dry adiabatic gradient of $-1°C$ per 100 m. Inversions permit the concentration of pollutants which could otherwise mix and become diluted. When the base of an inversion lies above ground level, mixing is possible in the layer, called the mixing layer. Mixing height is the height to the base of the inversion.

Wexler (1936) presented the mechanism responsible for surface inversions over snow. He showed how, when skies are clear and winds slight, fresh polar maritime air transforms into polar continental air by heat losses from the snow surface, which radiates in all wavelengths as a black body. The overlying warmer air radiates selectively in wavelength bands governed primarily by CO_2 and H_2O molecules. Surface inversions do not form at the high-temperature end of the air pollution spectrum (Figures 1 and 2). Inversions occur instead at the boundary where a warm air mass flows over a cooler one.

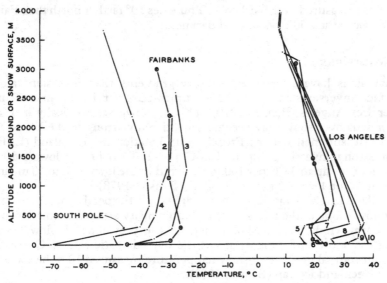

Fig 2. Comparison of inversions over the South Pole (2,792 m), Fairbanks (135 m) and Los Angeles (38 m). The stations are plotted at a common altitude to allow direct comparison of temperature profiles; times GMT. Data from radiosonde records. Curves: South Pole (1) 2315, 21 Apr 1958: Fairbanks (2) 1200, 26 Jan 1962; (3) 0000, 22 Dec 1961; (4) 1200, 26 Dec 1961: Los Angeles (Santa Monica) (5) 1200, 13 Aug 1963; (6) 0000, 23 Sept 1963; (7) 0000, 25 Sept 1963; (8) 0000, 26 Sept 1963; (9) 1200, 26 Sept 1963; (10) 0000, 27 Sept 1963.

Above Los Angeles this frequently happens when warm air, sometimes from the Mojave Desert, overrides cool air from the Pacific Ocean. The base of such inversions is some distance above the surface, the famous Los Angeles inversion having a mixing height of about 500 m.

Inversions and pollution in the Arctic

Radiation inversions are widespread throughout the Arctic and sub-Arctic. In polar continental air they often extend 1–2 km above the earth's surface and are overlain by a nearly isothermal layer; the normal negative lapse rate is generally not found below 2 km in winter (Figure 2). This temperature distribution ensures an extremely stable stratification of air over a very large area. Because the ice-covered Arctic Ocean has a dry snow surface in winter, the boundary between land and sea becomes blurred, enlarging the range of polar continental air. The presence of these persistent inversions has multiple consequences, some of which will be briefly considered.

Frequency of occurrence

Billelo (1966) summarized statistics on inversion at 11 Arctic and sub-Arctic stations in Alaska, Canada and Greenland, all coastal except for Fairbanks. He averaged the strength of the inversions up to the height at which normal lapse rates began; thus the strengths of extreme surface inversions (Figures 1 and 2) were not considered. All stations showed frequent surface inversions. From 1952–61 seven of the stations, including Fairbanks, had surface inversions more than 50% of the time between November and April. Only Thule, Greenland, had a higher frequency of average inversion gradients stronger than those measured over Fairbanks. Thule lies 10° farther north than Fairbanks and experiences more than 50 days of total darkness.

Strength of inversions

Surface inversions have steeper temperature gradients than inversions aloft; the Fairbanks surface inversions are up to three times stronger in the first 50 m than inversions over Los Angeles (Benson 1970). This is shown schematically in Figure 1. Radiation inversions over snow may become exceptionally strong in the lowest few cm (Nyberg 1938). Measurements at the French central station in Greenland (EPF 1954) show frequent strong inversions (up to 100°C per 100 m) in the lowest 10 m. In Fairbanks, gradients of up to 30°C per 100 m occurred in the lowest 30 to 50 m, and over 200°C per 100 m in the lowest 2 m (Bowling and Benson 1978).

On the Greenland ice sheet katabatic winds originate in the persistent inversions (EPF 1954), flowing downslope in the lowest, stable, high density air layer, some 500 m thick. Except when storm winds blow onshore, they are the prevailing air flow. As the air descends along the snow surface it warms at the dry adiabatic rate (1°C per 100 m), controlling the altitudinal gradient of snow temperature measured 10 m below the snow surface on the Greenland ice cap (Benson 1962).

Bowling (1986) showed that in satellite images of Alaska and Canada (Figure 3) dendritic valley bottoms appear persistently colder than uplands, with temperatures matched only on the highest mountains above 3,000 m.

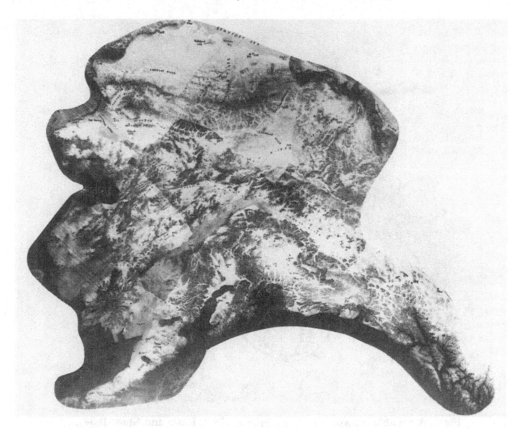

Fig 3. Mosaic of thermal infrared imagery from NOAA satellites over Alaska: courtesy E. Wendler.

Inversions and transportation

Winter in Alaska provides smooth flying conditions through cold, stable air masses. Wintertime flights from Fairbanks to Barrow are often done without a ripple of turbulence. In northern Alaska these conditions predominate from September to April (R. Wien, personal communication), the period of greatest snowcover. Stable air stratification is also critical in the long-distance transport of air pollutants. Raatz and Shaw (1984) showed how air pollutants can be transported from central and western Eurasian sources across the polar regions into Alaska (see Figure 4). The presence of well-developed anticyclones is of special importance in blocking the eastward movement of cyclonic disturbances. Zones of northward flowing air occur in steep pressure gradients between cyclones and anticyclones; when such flow patterns exist over a source of pollution, pollutants escape along the western side of the anticyclone into the Arctic.

Within well-developed anticyclones, inversions persist for weeks at a time, producing stable air masses in the lower 1–2 km. These allow cold dry air to flow in sheet-like layers along the border of the anticyclones, with little vertical mixing, only slight condensation and no precipitation. During winter the most likely source areas in central Eurasia are also snow-covered so the process is intensified, and the resulting haze flows into northern

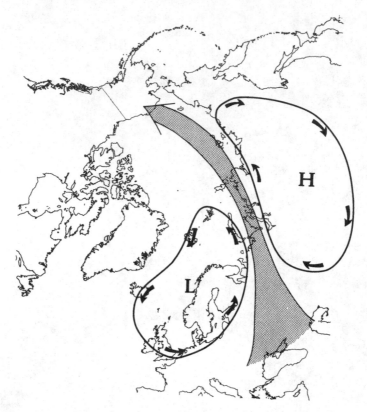

Fig 4. A possible transport path (generalized from Raatz and Shaw 1984).

Alaska with significant pollutant content. The sheet-like flow of air does not occur in summer because of turbulence and precipitation associated with movements of cyclonic systems (Figure 5). Raatz and Shaw (1984) show that the pollution transport by northerly winds into Alaska is maximal in winter and early spring, but does not occur during summer. Heidam (this symposium) reported that air pollutants are similarly transported into northern Greenland only in winter.

Urban air pollution in the Arctic and sub-Arctic
Fairbanks as a type locality

Billelo (1966) showed that surface inversions are present at Fairbanks in more than 60% of all night-time soundings the year round. During the dark months of December and January, when there are few diurnal changes in weather, surface inversions are present on more than 80% of day and night soundings: frequency of night-time inversions exceeds 70% for eight months of the year and 80% for five months.

From mid-November to mid-February incoming solar radiation is very small (Figure 6); from 1 December to 12 January the sun rises less than 3.5°, there is no measurable rise in midday temperature, and conditions for surface inversions persist day and night. Ball (1960) referred to this as a 'semi-permanent nocturnal condition' for Greenland and Antarctica. The effects of even short-lived nocturnal radiation inversions in concentrating

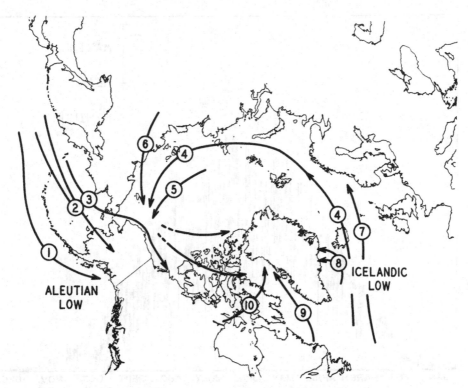

Fig 5. Generalized Arctic and sub-Arctic cyclone paths, summarized by Benson (1967) from Klein (1957), Keegan (1958) and Kunkel (1959).

air pollutants were well-displayed by Remsberg and others (1979) in Hampton, Virginia. They found the highest concentrations of CO occurred at night, despite a significant reduction in the principle source (vehicular traffic). Fairbanks and other communities in sheltered, snow-covered regions, with prolonged winter darkness and strong surface inversions lasting weeks at a time, are prime candidates for air pollution.

The lowest air layers in the Fairbanks area are both stable and complex. Using an acoustic sounder Holmgren and others (1975) showed that 10 to 20 separate, quasi-horizontal backscatter bands persisted within the first 500 m above the snow surface (Figure 7): at the same time wind and temperature sensors on a tethered balloon showed very strong temperature gradients, and winds of less than 0.5 m per second in the first 10 to 20 m. Above this extremely stable lowest layer there was often a step-like build-up of the inversion; between layers of small temperature gradients, positive or negative, there were thin layers of sharp inversions. Typically, there were two or more shear winds in the lowest 500 m with sheet-like layers moving in different directions (Figure 8). The volume of air available to dilute air pollutants becomes increasingly restricted during the same time that the needs for heat and power are increasing; thus the prerequisites for air pollution (pollutant input and restricted air volume) reinforce one another.

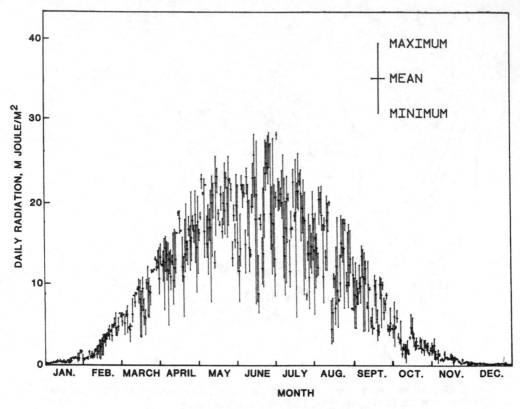

Fig 6. Solar radiation measured at Fairbanks (courtesy of G. Wendler).

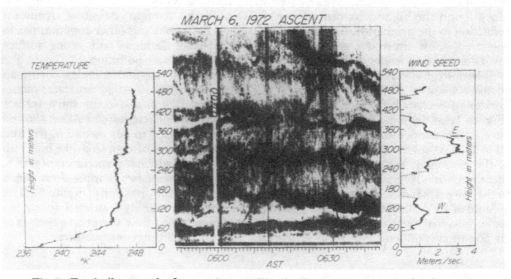

Fig 7. Facsimile record of acoustic sounding in Fairbanks air with simultaneous temperature and wind profiles. The latter were measured from a tethered balloon. The trace of the balloon ascent can be seen on the acoustic sounding (from Holmgren and others 1975).

Fig 8. Aerial photograph of shallow ice fog over Fairbanks, 3 January 1975 at the start of a severe ice fog episode. Later during this episode the two lower exhaust plumes from power plants merged with the ice fog. (Photo: Carl Benson)

Urban heat island

Areas of enhanced urban temperature have been observed since 1833, and studied over the past 50 years because of their interactions with air pollution (Duckworth and Sandberg 1954). They are produced by a combination of factors which are difficult to separate, and some of which oppose each other (Myrup 1969). Important factors include (1) heat released from sources within the city (anthropogenic or self-induced heat); (2) lack of transpiring plants (reduced evaporation); (3) high heat capacity and conductivity of building materials, and the generally increased roughness of cities over their surroundings, reducing the amplitude of the daily temperature wave (Nappo 1972); and (4) radiative effects of dirty city air (Atwater 1971): if the city albedo differs from that of its surroundings, this will also have an effect. Fairbanks in winter provides ideal conditions for urban heat island study, because its heating depends almost entirely on anthropogenic sources relatively easy to measure in an isolated town.

There are several empirical relationships between heat island intensity, population, meteorological conditions and geographical setting (Landsberg 1974; Oke 1972; Ludwig 1970). When applied to Fairbanks, these relationships yield heat island values of 4–8°C. The Fairbanks heat island observed by Benson and Bowling (1975) and Bowling and Benson (1978) with clear night skies, snow cover and light winds was very rarely less than

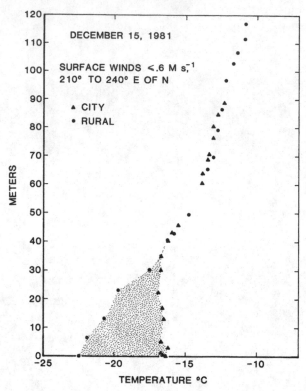

Fig 9. Air temperature profiles measured in and over Fairbanks, 15 December 1981 (from Bowling 1986).

10°C, and on one occasion was 14°C. This discrepancy between measured and computed values may be because the combination of lapse rate, wind speed and per capita heat production at Fairbanks is completely outside the range for which the empirical models were developed.

During a study of the heat island and ice fog, an inventory was made of fuel combustion in the area. The anthropogenic heating was 10 kw per person during winter—a high value comparable to that for the urbanized strip along the east coast of the US. Yet anthropogenic heating does not break the inversions in the Fairbanks area unlike in most cities where heat input is adequate to prevent inversions in their lowest 100 m (Pack 1964). Soundings by Bowling (1986) through the lowest air layers in Fairbanks and its surroundings show the effect of the heat island on the inversion (Figure 9).

The pollutants

Pollutants in the Fairbanks winter air mass, like those of other cities, are mostly exhaust products of combustion. Stable air structure permits high concentrations to build up from relatively small sources. Low radiation and low temperatures affect relative concentrations of some components and govern the air chemistry.

Benson (1970) gave combustion equations for gasoline, fuel oil and coal. A typical equation for combustion of gasoline is:

$$C_8H_{18} \xrightarrow[\text{excess } O_2]{\text{burned with}} 8CO_2 + 9H_2O.$$

The molecular weight of the gasoline is 114; the molecular weight ratio of H_2O:fuel is 1.42 and for CO_2:fuel it is 3.08. This equation is too simple because it assumes complete combustion which does not occur. However, when combustion is efficient the equations may be used for overall estimates of the major components. This particular one shows that for every kilogram of gasoline burned, more than 4 kg of exhaust products are produced, namely 1.4 kg of H_2O and 3 kg of CO_2. Levels of CO are notoriously high in Fairbanks and Anchorage. In 1980 Fairbanks exceeded the EPA standard of 9 ppm for an 8 hour average value during 30 days (Coutts 1983). Between 1979–83 it averaged four days per year with more than the EPA alert level of 15 ppm CO on an 8 hour average (Bowling 1986). In Anchorage the 9 ppm, 8 hour limit was exceeded 45 times during the 1980–81 winter (Coutts 1983).

Holty (1973), Jenkins and others (1975), Leonard (1975), and Coutts (1979, 1981, 1983) have all noted high levels of CO, winter values approaching annual values in residential sections of Los Angeles and exceeding those of New York and Detroit (Jenkins and others 1975). Leonard (1975) showed that CO concentrations in exhaust fumes are highest during cold starts; short trips combined with stop-and-go driving also lead to higher CO output. Maximum values of CO occur not at the lowest temperatures but at about $-20°C$. Possible reasons for this are discussed below.

Water is of special importance as an air pollutant because:

(1) its solubility as vapour in air is strongly temperature dependent, for example air at $+40°C$ can dissolve about 430 times more water vapour than air at $-40°C$ (Figure 1);

(2) when water is present in the condensed phases, liquid or crystal, (wet air rather than dry air chemistry), the condensed water droplets or crystals interact with other gaseous and particulate components. This is especially important when sulphur compounds are present. The most serious air pollution disasters (Meusse Valley, Belgium, London and Donora, Pennsylvania) occurred under conditions of wet air chemistry;

(3) suspended droplets or crystals radiate as black bodies and modify the thermal structure of fog layers; and

(4) it is visible and makes the public aware of a degredation in air quality.

Low air temperature creates dense, low-lying air layers and severely restricts the amount of water vapour which can be dissolved in the air. Water vapour condenses into droplets as cooling proceeds. When the temperature is low enough (–30 to $-35°C$) the droplets freeze, and the freezing has a dramatic effect on the amount of water that condenses as cooling proceeds (Benson 1970). For example, if the air is in equilibrium with water, 0.027 g m^{-3} of water vapour is condensed per degree of cooling at $-35°C$. If freezing occurs at $-35°C$ with cooling, air saturation changes from its value over water to its value over ice (Figure 1). This results in the condensation of a further 0.08 g m^{-3}, or a total of over 0.1 g m^{-3}. About 10^9 m^3 of air are involved in the Fairbanks air mass; as this saturated air cools through the freezing boundary, 83 tons of water are condensed as crystals. This is about twice the daily cold-weather output of the university power plant, yet it occurs in only a few hours and is widespread; it explains why the ice fog seems to appear so suddenly everywhere as cooling proceeds.

The presence of ice fog establishes a new radiative equilibrium which reduces or eliminates the lowest inversion layer (Figure 9). This decreases the concentration of some pollutants inserted at street level (eg CO) while it increases the ground level concentration of components that are introduced at high levels (eg SO_2).

Benson (1970) and Bowling and Benson (1978) estimated that over Fairbanks, 2,200 tons of water were put into the air daily as combustion exhaust during November and December 1975. Evaporation from power plant cooling ponds and open water areas of the Chena River contributed another 3,000 tons per day. Including condensation from the saturated air as it cooled, total wintertime input was approximately 6,000 tons of water vapour per day. This would be insignificant at air temperatures of $+30°$ or $+40°C$, but at $-30°$ or $-40°C$ it produces a dense ice fog of tiny ice crystals.

Ice fog has been recognized as an air pollution problem for many years (Oliver and Oliver 1949, Robinson and others 1957). Thuman and Robinson (1954) and Robinson and Bell (1956) studied the physical properties of ice fog and its micrometeorological conditions at air bases near Fairbanks. Benson (1965, 1969, 1970) studied the genesis and mass balance of several aspects of Fairbanks ice fog, while Kumai (1964) carried out electron microscope studies of the nuclei upon which the ice crystals form; these studies were continued in greater detail by Ohtake (1967, 1968, 1970). The subject is summarized by Weller (1969).

Ice fog consists of irregularly shaped, often rounded crystals 3–10 μm in diameter (Ohtake 1970), which resemble a fog of water droplets. When formed from hot exhaust gases ejected into cold air, their cooling rates are 10^6 times greater than cooling rates of the free atmosphere (Benson 1970); hence their small size and poorly developed shapes. Robinson and Bell (1956) observed that ice fog modifies strong surface inversions by shifting the radiating surface to the upper part of the fog (Figure 9); Bowling (1970) is a quantitative contribution. The net result is a weakened inversion, or even a negative lapse rate within the fog, with a capping inversion near the fog top (Figure 2, Curve 4). This phenomenon permits some upward mixing in the ice fog and is a major reason for the decrease in CO concentrations during ice fog. It also contributes to weakening the heat island, especially if the ice fog is widespread into the surroundings (Bowling and Benson 1978).

A mass balance equation can be written for ice fog (Benson 1970):

$$I = PA_B + EA_T + \rho \frac{dv}{dt},$$

where I is the input rate in tons H_2O per day, P is the precipitation rate, A_B is the area of the bottom of the fog, E is the evaporation rate, A_T is the area of the top, ρ is the density of the fog, and dv/dt is the rate of volume change which vanishes at equilibrium. The input rate is approximately 6,000 tons per day. Precipitation rate (ie ice fog fallout) has been measured by placing polyethelene sheets on plywood mounts at various points in the city. (This excludes natural precipitation.) Values up to 80 g m^{-2} day^{-1}, measured near the city centre, included 1.5% of matter other than ice. The residence time of the ice crystals is less than 10 hours. Their role in cleansing the air of particals is not clear (Ohtake and Eaton 1982), though they probably adsorb gases (Bowling 1986; Benson 1970). Total precipitation rate of 5,500 tons H_2O per day was estimated, which is close to the estimated input rate. No data are available for evaporation from the top of the fog but it should not be large because the overlying air is most likely about 80 to 90% saturated with ice.

The ratios of stable isotopes O^{18}/O^{16} and H^2/H^1 have been measured in ice fog in an attempt to obtain an independent estimate of the amount of man-made water contributed to the ice fog by combustion processes. Results so far indicate that about 20% of the water which makes up ice fog is created in the combustion process, when hydrogen from fossil fuel combines with oxygen from the air (Benson, Friedman and Gleason, in preparation). This figure is in the same order of magnitude as that derived from the direct source estimates (Benson 1970; Bowling and Benson 1978).

Chemical properties of ice fog residue are being investigated; pH values of the derived liquid generally exceed 8, with a maximum value of 9.9 (Gosink and Benson 1982). Surprising at first, these basic values are consistent with a reducing atmosphere.

Most nitrogen oxides from combustion sources are emitted as nitric oxide (NO) which is subsequently oxidized in the atmosphere to nitrogen dioxide (NO_2). Unlike CO and unburned hydrocarbons, more nitrogen oxides are formed at higher combustion efficiencies (National Academy of Sciences 1977). The ratio of NO to NO_2 in background non-urban air measured at Pikes Peak and in rural North Carolina averaged between 0.6 and 0.7 (ibid 1977: 57). For average urban air the ratio was 1.45 at six Continuous Air Monitoring Project (CAMP) cities (ibid 1977: 65–66). In contrast, the Fairbanks midwinter ratio on a monthyly mean basis was 10 (Coutts 1979). NO values were 0.2 to 0.3 ppm, and NO_2 values were 0.02 to 0.03 ppm.

The NO:NO_2 ratio is an excellent indicator of oxidation rate in the atmosphere. In non-urban background air, NO is oxidized to NO_2 and the ratio is less than one. In cities, the rate of input of NO is high enough for the ratio just to exceed one. Variation in NO:NO_2 ratio with solar radiation was demonstrated in Los Angeles during October 1974. The NO:NO_2 ratio reached values slightly above two between 0600 and 0800 hrs when traffic was heavy but before solar radiation was strong. At 0800 hrs the values plummeted to reach 0.2 by 1000 hrs, remaining in the range 0.2–0.25 until 2000 hrs (National Acadamy of Sciences 1977: 60–61). This range of values measured in Los Angeles, with and without strong solar radiation, helps explain the extremely high value of 10 for the NO:NO_2 ratio measured in Fairbanks in the prolonged absence of solar radiation.

Ozone values in Fairbanks averaged between 0.01 and 0.002 ppm, reaching 0.037 ppm in rural Alaska (Coutts 1979). The 'alert' level for O_3 in Los Angeles is 0.1 ppm. The effect on O_3 of a single exhaust plume containing NO was dramatic; simultaneous recording of the two gases showed a precipitous drop in O_3 as NO reached the sensors (Coutts 1979: 42).

Sulphur has been measured as sulphur dioxide (SO_2) and sulphate (SO_4^-) by Holty (1973) and Coutts (1979); their values in the air mass are not high. However, Holty found them to increase 50 fold (from 0.6–33.8 $\mu g\ m^{-3}$) when thick ice fog was present. This increase indicates that exhaust gases from power plant plumes, which are SO_2 sources, become incorporated into the ice fog and mix downward. This mixing occurs when the ice fog becomes thick enough to blend with the plume (Figures 8 and 9); the same mixing decreases the CO values at street level. Sulphur and water have a complex equilibrium phase diagram, with stable liquid forms existing at $-50°C$ (Mellor 1930).

Lead and halogens had the same ratios in the Fairbanks air as in tetraethyl lead (Winchester and others 1967), indicating the prime source. Pb:Cl ratios remain constant while Pb:Br ratios do not: thus lead bromine compounds break down to become a source of bromine. Holty (1973) observed a three-fold increase in lead concentrations during ice fog. Maximum values of 6 $\mu g\ m^{-3}$ measured by Winchester and others (1967) and Coutts (1979) exceed the EPA standard of 1.5 $\mu g\ m^{-3}$. The average atmospheric Pb conentration for the Northern Hemisphere is 0.005 $\mu g\ m^{-3}$. Urban values are in the range of 2 or 3 $\mu g\ m^{-3}$ and Patterson (personal communication 1964) has measured 10 to 20 $\mu g\ m^{-3}$ in Pasadena, California.   Jenkins and others (1975) showed that unburned hydrocarbons

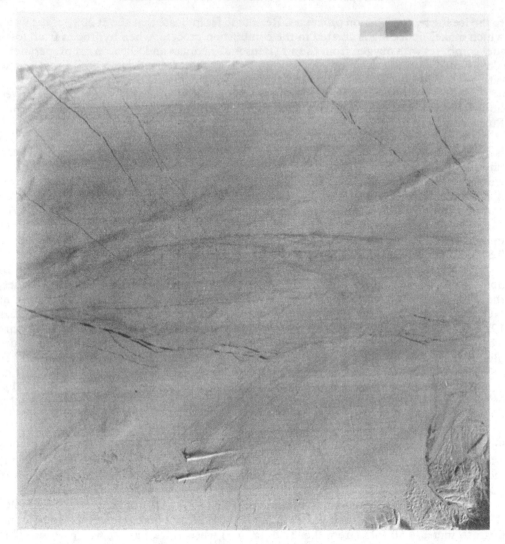

Fig 10. LANDSAT satellite photo (185 x 185 km) of the North Slope, Alaska near Prudhoe Bay; note long pollution plumes from the oil-producing area.

were present in Fairbanks and Los Angeles areas in the same concentrations. A study by Reichardt and Reidy (1980) showed polycyclic aromatic carbon (PAH) values to be comparable to those measured in other major cities. Organic compounds in cold weather air pollution have been studied less than gases or inorganic conpounds, and the need for more studies is emphasized.

The impact of man-made air pollution in the pristine Arctic is dramatically apparent in Figure 10, which shows exhaust plumes over Prudhoe Bay, Alaska as seen by satellite. The longest (highest altitude) plumes extend over 20 km. Lower plumes can be seen going in different directions. The components of these plumes are unknown, but could be sampled in detail because they are deposited on the snow surface of the Arctic North Slope and Arctic Ocean for about nine months each year. This snow cover can be though of as a giant filter, which catches and retains all the pollutants dumped on it. A programme to sample the impact of Prudhoe Bay on its land and sea environment could make good use of this snow filter.

References

ATWATER, M. A. 1971. The radiation budget for polluted layers of the urban environment. *Journal of Applied Meteorology* 10: 205–14.

BALL, F. K. 1960. Winds on the ice slopes of Antarctica. *Antarctic Meteorology: proceedings of the symposium held in Melbourne*. New York, Pergamon Press: 6–16.

BENSON, C. S. 1962. Stratigraphic studies in the snow and firn of the Greenland ice sheet. SIPRE (CRREL) Research Report 70: 1–93; summarized in *Folia Geogr. Danica* IX, 1961: 13–35.

BENSON, C. S. 1965. *Ice fog, low temperature air pollution*. UAGR 63. Geophysical Institute, University of Alaska.

BENSON, C. S. 1969. The role of air pollution in Arctic planning and development. *Polar Record* 14(93): 783–90.

BENSON, C. S. 1970. Ice fog;—low temperature air pollution defined with Fairbanks, Alaska as type locality. Cold Regions Research and Engineering Laboratory (CRREL) Research Report 121.

BENSON, C. S. AND BOWLING, S. A. 1975. The sub-arctic heat island as studied at Fairbanks, Alaska. In WELLER, G. and BOWLING, S. A. (editors) *Climate of the Arctic*: 309–11.

BENSON, C. S. AND OTHERS. 1983. Urban climates in Alaska. *Environments* 15(2): 23–26.

BILLELO, M. A. 1966. Survey of Arctic and subarctic temperature inversions. CRREL Technical Report TR 161. Hanover, New Hampshire.

BOWLING, S. A. 1970. *Radiative cooling rates in the presence of ice crystal aerosols*. PhD Dissertation, University of Alaska.

BOWLING, S. A. 1984. Meteorological factors responsible for high CO levels in Alaskan cities. EPA Report; NTIS PB 85–115 137.

BOWLING, S. A. 1986. Climatology of high latitude air pollution as illustrated by Fairbanks and Anchorage, Alaska. *Journal of Climate and Applied Meteorology*, 25: 22–34.

BOWLING, S. A. AND BENSON, C. S. 1978. Study of the subarctic heat island at Fairbanks, Alaska. EPA Report; 600/4–78–027.

CHAMBERS, L. A. 1976. Classification and extent of air pollution problems. Chapter 1 in STERN, A. C. (editor) *Air pollution* vol 1, 3rd edition. Academic Press.

COUTTS, H. J. 1979. A study of winter air pollutants at Fairbanks, Alaska. EPA Report; EPA-600/3–79–100, NTIS PB 80–111495.

COUTTS, H. J. 1981. Automotive cold-start carbon monoxide emissions and pre- heater evaluation. US Environmental Protection Agency Special Report 81– 32.

COUTTS, H. J. 1983. Low temperature automotive emissions. Final Report prepared for State of Alaska Department of Environmental Conservation.

DUCKWORTH, F. S. AND SANDBERG, J. S. 1954. Th effect of cities upon horizontal and vertical temperature gradients. *Bulletin of the American Meteorological Society* 35: 198–207.

EPF. 1954. *Les observations météorologiques de la station Française du Groenland. Conditions atmospheriques en surface du 5 septembre 1949 au 20 juin 1950. Fascicule II: Études et commentaires sur les observations*. Paris, Expéditions Polaires Françaises, Mission Paul-Emile Victor.

GOSINK, T. A. and BENSON, C. S. 1982. Aspects of far-northern air pollution with particular reference to Fairbanks, Alaska. Geophysical Institute report UAG R 291, Fairbanks.

HAAGEN-SMIT, A. J. 1952. Chemistry and physiology of Los Angeles smog. *Industrial and Engineering Chemistry* 44: 1342–46.

HOLMGREN, B. AND OTHERS. 1975. Acoustic sounding of the Fairbanks temperature inversion. In WELLER, G. AND BOWLING, S. (editors). Climate of the Arctic. Fairbanks, Geophysical Institute, University of Alaska at Fairbanks: 293–306.

HOLTY, J. G. 1973. Air quality in a subarctic community, Fairbanks, Alaska. *Arctic* 25: 292–302.

JENKINS, T. R. AND OTHERS. 1975. Accumulation of atmospheric pollutants near Fairbanks, Alaska during winter. special Report 225, Cold Regions Research and Engineering Laboratory, Hanover, New Hampshire.

KEEGAN, T. J. 1958. Arctic synoptic activity in winter. *Journal of Meteorology* 15: 513–21.

KLEIN, W. H. 1957. Principal tracks and mean frequencies of cyclones and anticyclones in the northern hemisphere. US Weather Bureau Research Paper no 40.

KUMAI, M. 1964. A study of ice fog and ice fog nuclei at Fairbanks, Alaska. Cold Regions Research and Engineering Laboratory (CRREL) Research Rerpot 150.

KUNKEL, B. A. 1959. A synoptic-climatological study of the Arctic circulation in summer. *Scientific Reports* 7, *Department of Meteorology and Climatology, University of Washington*. (AF Contract 19 (604)–3603).

LANDSBERG, H. 1974. Inadvertent atmospheric modifications through urbanization. In HESS, W. N. Jr. (editor) *Weather and climate modification*. New York, Wiley and Sons.

LEONARD, L. E. 1975. *Cold start automotive emissions in Fairbanks, Alaska*. Fairbanks, Geophysical Institute, University of Alaska at Fairbanks. (UAG R-239).

LUDWIG, F. 1970. Urban air temperatures and their relation to extra-urban meteorological measurements. *Symposium on Survival Shelter Problems, American society of Heating, Refrigerating and Air-Conditioning Engineers, January 19–22, San Francisco, California*: 4–45.

McCANDLESS, R. 1982. Wood smoke and air pollution at white Horse, Yukon Territory. Regional Program Report 82–16, Department of Environment Canada (Pacific Region, Yukon Branch).

MELLOR, J. W. 1930. *Comprehensive treatise on inorganic and theoretical chemistry*: 10, *Sulfur and selenium*. New York, Longmans Green.

MYRUP, L. O. 1969. A numerical model of the urban heat island. *Journal of Applied Meteorology* 8:4–45.

NAPPO, C. J. Jr. 1972. A numerical model of the urban heat island. *Conference on the Urban Environment and Second Conference on Biometeorology, American Meteorological Society Philadelphia*, 1–4, Oct 31– Nov 2, 1972.

NATIONAL ACADEMY OF SCIENCES. 1977. *Nitrogen oxides*. EPA-600/1–77–013, US Environmental Protection Agency, Research Triangle Park, North Carolina.

NYBERG, A. 1938. Temperature measurements in an air layer very close to a snow surface. *Geografiska Annaler*, 20: 234–75.

OKE, T. R. 1972. City size and urban heat island. *Conference on the Urban Environment and Second Conference on Biometeorology, American Meteorological Society, Philadelphia*: 144–46.

OHTAKE, T. 1967. Alaskan ice fog. *Proceedings of the International Conference on the Physics of Snow and Ice (Sapporo, Hokkaido University), part* 1: 105–18.

OHTAKE, T. 1968. Freezing of water droplets and ice fog phenomena. *Proceedings of the International Conference on Cloud Physics (Toronto)*: 285–89.

OHTAKE, T. 1970. Studies on ice fog. *University of Alaska Report UAG R-211, reprinted as Office of Air Programs Pub APTD-0626*. Research Triangle Park, Environmental Protection Agency.

OHTAKE, T. AND EATON, F. D. 1982. Removal processes of aerosols in ice fog. *Preprint volume of extended abstracts, Conference on Cloud Physics, November 15–18, Chicago, Ill*. Boston, American Meteorological Society: 57– 60.

OLIVER, V. J. AND OLIVER, M. B. 1949. Ice fog in the interior of Alaska. *Bulletin of the American Meteorological Society*, 30: 23–26.

PACK, D. H. 1964. Meteorology of air pollution. *Science* 146(3648): 1119–28.

RAATZ, W. AND SHAW, G. 1984. Long-range tropospheric transport of pollution aerosols into the Alaskan Arctic. *Journal of Climate and Applied Meteorology* 23: 1052–64.

REICHART, P. B. and REIDY, S. K. 1980. Atmospheric polycyclic aromatic hydrocarbons: an aspect of air pollution in Fairbanks, Alaska. *Arctic* 33: 316–25.

ROBINSON, E. AND BELL, G. B. 1956. Low-level temperature structure under Alaskan ice fog conditions. *Bulletin of the American Meteorological Society* 37: 506–13.

ROBINSON, E. AND OTHERS. 1957. Ice fog a a problem of air pollution in the Arctic. *Arctic*, 10(2): 89–104.

SHAW, G. 1985. Alaskan and Arctic-scale air pollution. *Northwestern Environmental Journal*, 1(2): 85–95.

THUMAN, W. C. AND ROBINSON, E. 1954. Studies of Alaskan ice-fog particles. *Journal of Meteorology*, 11: 151–56.

WELLER, G. 1969. Ice fog studies in Alaska; a survey of past, present, and proposed research. *Geophysical Institute Report, UAG R-207*. College, University of Alaska.

WEXLER, H. 1936. Cooling in the lower atmospere and the structure of polar continental air. *Monthly Weather Review*, 64: 122–36.

WINCHESTER, J. W. AND OTHERS. 1967. Lead and halogens in pollution aerosols and snow from Fairbanks, Alaska. *Atmospheric Environment*, 1: 105–19.

WHO'S POLLUTING THE ARCTIC? WHY IS IT SO IMPORTANT TO KNOW? AN AMERICAN PERSPECTIVE

KENNETH A. RAHN AND DOUGLAS H. LOWENTHAL

ABSTRACT. Deducing the exact sources of Arctic haze remains one of its most critical problems, despite research of the past few years. Understanding the sources is essential in determining the nature of transport to the Arctic, validating meteorological methods for deriving these pathways, and guiding future control strategies. The University of Rhode Island's new regional elemental tracer system has shown quantitatively that North America is not an important source of Arctic pollution aerosols, but that during winter 1979–80, at least half to three-quarters of the pollution aerosol observed at Barrow, Alaska, originated in the Soviet Union. Recent application of the tracer system to samples from the Arctic Gas and Aerosol Sampling Program (AGASP) of March-April 1983 has confirmed these general results and improved our understanding of a major pollution episode seen on both sides of the Arctic.

Contents

Introduction

Since the mid-1970s it has become apparent that, during the winter half-year, the Arctic troposphere contains large amounts of pollution aerosol. Mean concentrations of a variety of constituents in the lower Arctic troposphere are 10–100 times higher in winter than in summer, and can approach mid-latitude values. During episodes of high concentration, the situation is even more dramatic. At Barrow, Alaska, for example, peak concentrations of As reach 6 ng m^{-3}, which is higher than peak values seen in many mid-latitude areas. Peak sulfate is just under 5 μg m^{-3}, well above winter average concentrations of New England, for example. It is not an exaggeration to say that the Arctic is often more polluted than mid-latitudes are.

Much can be learned from Arctic haze, both with respect to its effects on the Arctic environment, and as a nearly unique example of transport and aging of polluted air masses on scales larger than are generally available in mid-latitudes. Over the long term, the latter may be as important as the former.

A key aspect of Arctic haze is its source regions. Knowing its sources is critical to determining the duration of transport to the Arctic and the pathways taken, validating meteorological methods for deriving these pathways, and guiding the direction of possible control strategies if it eventually becomes necessary to reduce Arctic pollution levels.

Experience has shown, however, that (1) even the best current meteorological analysis is still only partially reliable, and can often misplace the sources of Arctic haze, even during episodes when flow patterns are most clear-cut; (2) recent advances in synoptic and trajectory analysis for the Arctic have created an over-optimistic sense of how well they can determine sources of Arctic haze; (3) chemical techniques offer a critical constraint on meteorological analysis; (4) by combining chemistry with meteorology, it is now possible to determine the source areas of Arctic haze with relative confidence.

This paper summarizes the University of Rhode Island's new regional elemental tracer system for pollution aerosols, reviews current understanding of sources and transport of Arctic haze as derived from these chemical tracers, and offers the Arctic Gas and Aerosol Sampling Program (AGASP) of March–April 1983 as a case study of how chemical and meteorological techniques are currently being combined into a whole which is greater than either alone could be.

A regional elemental tracer system for pollution aerosols

Our current regional elemental tracer system (Rahn and Lowenthal 1984) is based on the seven elements As, Se, Sb, Zn, In, noncrustal Mn, and noncrustal V, which are measured by neutron activation. (Noncrustal Mn and noncrustal V refer to the mass present in excess of that expected from the earth's crust, as calculated from the following formula:

$$\text{Noncrustal } X_{aer} = \text{total } X_{aer} - Al_{aer}(X/Al)_{ref}$$

where X and Al refer to the abundances of the signature elements (Mn, V) and the crustal reference element (Al) in the aerosol and crustal reference material. At most sites, one-half the Mn and 90% of the V are noncrustal.) In addition to the tracer elements, sulfate is measured in all samples. Each of the seven tracer elements was chosen from the 40 or so that can be routinely determined by neutron activation because it was measured precisely, was strongly pollution-derived, and was largely fine-particle (submicron) in the aerosol.

The seven elements are used as six ratios to Sb to form signatures for the Arctic, but individually in subsequent operations. Ratios of elements are nearly independent of dispersion and removal, and much more stable during transport that are individual concentrations. Ratios appear to change by less than 25% during transport of 5,000–10,000 km from Eurasia to Barrow, Alaska (Lewis 1985).

A regional signature is the collection of six X/Sb ratios, each with its standard deviation. Wherever possible, signatures are determined from repeated sampling within the region of interest. When it is impossible to sample within the region of interest, samples as near to it as feasible are used. In cases where local or intervening influences are minimal, the distant signature can usually be determined satisfactorily.

Seven current regional signatures from North America, Europe, and Asia which can be applied to Arctic aerosol are shown in Table 1. Those from the central east coast of North America (CEC), the American Midwest (MW), the United Kingdom (UKS), Europe (EUR), and the central Soviet Union (CUSSR) have been described by Lowenthal and Rahn (1985). The provisional signature of the western Soviet Union (WUSSR) was derived from 14 samples taken at Ähtäri, Finland during an episode in March 1982 when the air blew strongly from the east and southeast. Its validity was confirmed by similar episodes in September and October 1982. The signature of northeastern China came from five samples taken in the Beijing-Tianjin area. It has not yet been possible to confirm this signature.

Table 1. Current regional elemental signatures for use in the Arctic.

	CEC(N=12)	MW(N=7)	UKS(N=11)	EUR(N=5)
As	0.74±0.34	2.6±0.4	1.74±0.48	3.7±0.6
Sb	1.00±0.30	1.00±0.30	1.00±0.29	1.00±0.36
Se	1.03±0.45	4.4±0.9	1.16±0.34	2.2±1.0
Noncr. V	26±10	1.54±0.17	6.7±2.0	11.2±5.0
Zn	28±11	46±4	44±11	106±44
Noncr. Mn	3.4±1.2	11.8±1.7	7.8±2.2	5.4±2.8
In(×10³)	4.0±2.9	7.9±0.4	14.9±8.5	–

	WUSSR(N=14)	CUSSR(N=5)	CHINA(N=5)
As	3.0±0.7	10.0±0.6	2.9±1.1
Sb	1.0±0.3	1.00±0.28	1.0±0.3
Se	0.86±0.19	1.03±0.48	1.92±1.13
Noncr. V	24±13	6.5±2.0	0.48±0.09
Zn	50±25	45±16	77±29
Noncr. Mn	10.5±4.9	4.4±1.3	16.6±9.2
In(×10³)	9.6±4.7	–	27±14

CEC = Central East Coast MW = Midwest
UKS = United Kingdom (from Sweden) EUR = Europe
WUSSR = Western Soviet Union
CUSSR = Central Soviet Union
CHINA = Northeastern China

Most aerosols sampled in the Arctic have mixed origins. Their components are resolved by chemical element balance (CEB) apportionment, which is essentially a weighted multiple linear regression with the signature elements in the sample as dependent variables and the same elements in the regional signatures as independent variables:

$$C_i = \sum_{j=1}^{p} X_{ij} S_j$$

where C_i is the concentration of the ith element in an ambient sample, X_{ij} is the concentration of the ith species in the jth regional source, S_j is the derived strength of the jth regional source, or the 'regional coefficient' of the source, and p is the number of regional sources. Each element is weighted inversely by its 'effective variance', which includes the element's variance in both the sample and the signatures (Watson and others 1984). Average concentrations of signature elements are usually accounted for to 25% or better. Details of the apportionment procedure are given elsewhere (Rahn and others 1985).

Sulfate, which can account for one-half or more of the pollution aerosol in the Arctic, must be apportioned differently because it is largely secondary in the aerosol (formed by oxidation of SO_2 in the atmosphere rather than being emitted directly as aerosol). It is apportioned indirectly among a suite of regional signatures by regressing its concentrations in a series of samples against the regional coefficients S_j over the series:

$$(SO_4^-)_i = \sum_{j=1}^{p} S_{ij} E_j$$

for p sources, where $(SO_4^-)_i$ is the concentration of sulfate in the ith sample, S_{ij} is the regional coefficient of the jth course in the ith sample, and E_j is a regression coefficient which represents the derived mean sulfate/Sb ratio for the jth source, which we term the 'effective sulfate'. The effective sulfate represents the mean sulfate (per unit Sb) associated with the jth signature over the series, ie the initial sulfate near the source plus that formed from SO_2 during transport to the receptor. The contribution of the jth source region to the sulfate of a sample is the product of E_j and S_{ij}, which is the effective sulfate of the signature multiplied by the coefficient of the signature.

Sources and transport of Arctic haze as derived from elemental tracers

The most detailed quantitative information on sources of Arctic aerosol was provided by Lowenthal and Rahn (1985), who analyzed and apportioned 100 daily samples of aerosol from Barrow during winter 1979–80. Table 2 is a simplified version of their Table 6, which shows that roughly three-quarters of the mass of the signature elements came from the Soviet Union (CUSSR) and one-quarter from Europe (UKS + EUR). North American contributions (CEC + MW) seemed to be negligible. These conclusions agreed with the weight of previous evidence that Eurasia is much more important than North America as a source of Arctic haze.

Table 2. Average regional apportionments of elements in the Barrow aerosol of winter 1979–80, five sources.

	Apportionment: % of mean predicted			Mean predicted ng m⁻³	Mean observed ng m⁻³	Mean obs./pred.
	East. N. Amer. (CEC + MW)	Europe (UKS + EUR)	CUSSR			
As	1.8 ± 0.7	4.9 ± 1.7	93 ± 3	0.85 ± 0.02	0.86	0.99 ± 0.09
Sb	6.4 ± 2.7	20 ± 6	74 ± 6	0.108 ± 0.008	0.124	1.19 ± 0.27
Se	19 ± 8	20 ± 7	61 ± 8	0.134 ± 0.015	0.102	0.84 ± 0.19
Noncr. V	6.8 ± 6.4	21 ± 7	72 ± 8	0.72 ± 0.07	0.55	0.79 ± 0.19
Zn	5.8 ± 2.3	21 ± 8	73 ± 8	5.0 ± 0.4	4.8	0.95 ± 0.18
Noncr. Mn	12 ± 2	28 ± 8	60 ± 6	0.58 ± 0.05	0.70	1.23 ± 1.14
Nonmar.SO₄	36 ± 11	25 ± 7	39 ± 5	1200 ± 140	1180	1.04 ± 0.47

The apportionment of sulfate in Table 2 differs from the other elements, with a full 36% given to North American sources. Lowenthal and Rahn considered this result a consequence of collinearity between North American and European signatures (see below) whereby small North American coefficients substituted for European coefficients and were assigned sulfate), and offered a number of reasons for rejecting it. As elements are added to the tracer system and the resolution among signatures improves, the contribution of North America to Arctic haze can be addressed more confidently. When North American sources were removed from the apportionment of sulfate, Lowenthal and Rahn (1985) found that Europe and the Soviet Union contributed comparably, ie that Europe was proportionately a greater source of sulfate than of the tracer elements.

How finely can sources be resolved in the Arctic? With the current seven elements, probably not more than indicated in Table 2, ie with European sources and North American sources lumped together. Without going into great detail, it can be seen from Table 1 that a number of the seven current signatures resemble one another moderately: MW, UKS, EUR, and to a lesser extent the newer WUSSR and CHINA. CEC with its high V and CUSSR with its high As and In are clearly distinct from the others. This collinearity restricted Lowenthal and Rahn (1985) to comparing CUSSR with European sources in general (collinearity with North American sources was rendered moot because their small contributions removed them from further consideration). Of the current signatures, a number of lines of reasoning agree that CUSSR, WUSSR, EUR, and CHINA or CEC form reasonably independent groups.

The mechanism of transport from Eurasia to the Arctic can be seen straightforwardly from meteorological maps. Transport from the central Soviet Union is controlled by

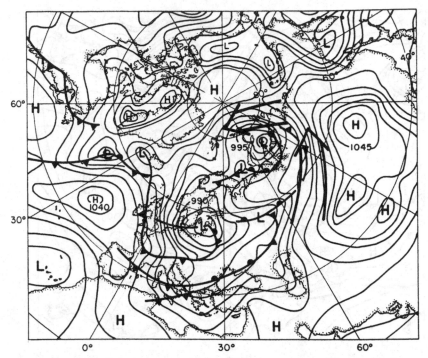

Figure 1. Surface synoptic map for OZ of 31 December 1977, during a typical period of transport of aerosol from the central Soviet Union to the Norwegian Arctic. Pressures in mb. Arrows indicate general airflow near the surface. (After Lowenthal and Rahn 1985.)

large low-pressure areas which come from the North Atlantic and follow the warm open water north of Scandinavia to the vicinity of Novaya Zemlya, where they meet pack ice, stagnate, fill, and eventually disappear. As their southern extent reaches into the Soviet Union, a strong pressure gradient is set up in combination with the Asiatic high, and air is moved rapidly northward near the Urals. Depending on the synoptic situation over the Arctic, this air can then follow the low westward to the Norwegian Arctic (Figure 1) or continue northward over the pole to Barrow (Figure 2). The retrograde course to the Norwegian Arctic has become known as the return-flow pathway. Of course, many variations on the above patterns are seen.

European aerosol reaches the Arctic differently. It tends to flow clock-wise into the Norwegian Arctic under the influence of a high-pressure area over Europe, which normally carries it over the United Kingdom and the Norwegian Sea, and occasionally even as far to the west as Iceland. This has become known as the direct-flow pathway. Several years ago, soon after the return-flow pathway was discovered, it was also thought to carry considerable European aerosol as well as Soviet Union aerosol, but this idea has not held up. Elemental tracers have shown that practically all events of European aerosol in the Arctic come via the direct pathway. Curiously, the direct pathways seem to hug the Scandinavian coastline, and often continue along the Arctic coast of the Soviet Union. Consequently, European aerosol commonly misses Spitsbergen, which is more influenced by aerosol from the central Soviet Union via the return-flow pathway.

Because the new WUSSR signature appeared so promising, we reapportioned the 100

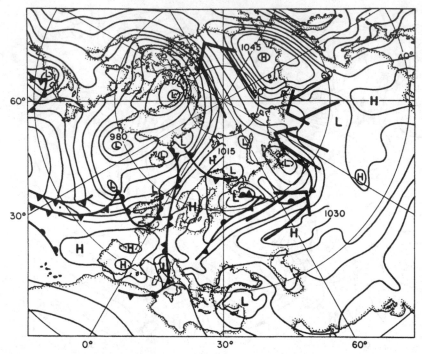

Figure 2. Surface synoptic map for OZ of 17 February 1980, during a period of unusually strong transport of aerosol from the central Soviet Union to the Alaskan Arctic. Pressures in mb. Arrows indicate general airflow near the surface. (After Lowenthal and Rahn 1985.)

Table 3. Average regional apportionments of elements in the Barrow aerosol of winter 1979–80, three sources.

| | Apportionment: % of mean predicted | | | Mean predicted ng m^{-3} | Mean observed ng m^{-3} | Mean obs./pred. |
	EUR	WUSSR	CUSSR			
As	4.6±1.1	2.8±0.6	93±2	0.86±0.02	0.85	0.95±0.15
Sb	10.8±2.7	8.2±1.9	81±4	0.099±0.005	0.123	1.37±0.64
Se	21±5	6.2±1.4	73±5	0.113±0.009	0.102	1.17±0.96
Noncr.V	14.3±3.6	23±6	62±3	0.83±0.06	0.54	0.65±0.17
Zn	22±5	7.8±1.8	71±4	5.2±0.4	4.8	0.87±0.19
Noncr.Mn	11.6±3.0	17.2±4.1	71±4	0.49±0.03	0.70	1.69±2.12
Nonmar.SO$_4$	27±4	16.0±2.3	57±2	1330±60	1180	1.11±0.78

Barrow samples of winter 1979–80 using it, CUSSR, and EUR. Table 3 summarizes the results. In general CUSSR contributions remained similar to those in Table 2 or increased slightly. WUSSR, however, was given roughly one-half the previous European and North American apportionments. As a result, the total contributions from the Soviet Union were 79–96%. With this combination of signatures, sulfate behaved similarly to the other constituents (84% from the Soviet Union). Thus, the new WUSSR signature has indicated that during winter 1979–80, the Soviet Union seemed to be responsible for 80–95% of the mass of pollution elements in the Barrow aerosol.

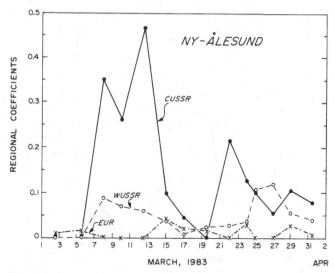

Figure 3. Regional coefficients for pollution aerosol from Europe (EUR), the western Soviet Union (WUSSR), and the central Soviet Union (CUSSR) at Ny-Ålesund during March 1983.

Chemistry and meteorology working together—the pollution episode of March 1983

The March 1983 pollution episode, which formed a major part of AGASP, provided a fine example of how the Arctic is ideally suited for chemistry and meteorology to work together to explain events which might otherwise be intractable. This episode was observed during the second week of March at the surface in Ny-Ålesund, Spitsbergen, and a few days later aloft and at the surface in Barrow, Alaska. It has been the subject of several articles. The paragraphs below first present our new chemical data and the resulting picture of the episode, then compare them with previous chemical and meteorological impressions.

We analyzed 13 samples from Ny-Ålesund covering 4 March-1 April 1983, and 22 samples from Barrow covering 2–27 March 1983. Each set was apportioned into contributions from EUR, WUSSR, and CUSSR, with the resulting regional coefficients shown in Figures 3 and 4, respectively. At Ny-Ålesund, the episode came strongly from the central Soviet Union and lasted from 7–14 March. During this time, European aerosol was practically absent, but aerosol from the western Soviet Union rose quickly to a gentle maximum and declined steadily thereafter.

Meteorological maps provide ready explanations for each of these observations. The big pulse of aerosol from the central Soviet Union was clearly linked with a low-pressure area which entered Europe and Asia from the Norwegian Sea, moved to the southeast, and stalled with the transport zone on its leading edge centered on the Urals. This is the classical return-flow pathway which has been seen many times before. The early pulse of aerosol from the western Soviet Union was caused as the leading edge of the low moved rapidly toward the Urals and temporarily propelled air from the western Soviet Union first north, then west as it became caught in the return flow. Because of the geometry of this particular low, aerosol from Europe moved eastward and southeastward, well away

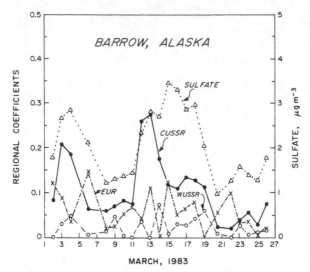

Figure 4. Sulfate concentrations and regional coefficients for pollution aerosol from Europe (EUR), the western Soviet Union (WUSSR), and the central Soviet Union (CUSSR) at Barrow, Alaska during March 1983.

from the northward transport zone. If this result is typical, as seems likely, low-pressure areas may routinely exclude European aerosol from the Arctic.

The smaller episode of 21–28 March contained a number of features similar to the earlier episode. It began with a peak of aerosol from the central Soviet Union on 21–23 March, which was associated with return flow around a low centered on Novaya Zemlya. As this low weakened, a much more internse low moved southeastward from Iceland, broke into three smaller systems, and stalled over Scandinavia. Its leading edge formed a transport zone which covered much of the western Soviet Union, with the result that a good-sized pulse of aerosol from there was seen on 25–28 March. When this system moved eastward, aerosol from the central Soviet Union replaced that from the western Soviet Union; ie, the episode followed the normal course of events.

At Barrow (Figure 4), the big episode was seen on 12–19 March 1983. This episode had two clearly defined phases: three days strongly from the central Soviet Union followed by five days of more mixed character in which the coefficient of the central Soviet Union decreased by a factor of two while those of Europe and the western Soviet Union remained roughly constant. Our analysis of aerosol samples from the haze layers aloft near Barrow early on the 12th and 14th, taken with the AGASP aircraft, show that they were also dominated by aerosol from the central Soviet Union (their As/Sb ratios were 8–17).

The aerosol of the first phase was transported from the central Soviet Union by the same low-pressure area which affected Ny-Ålesund. Once in the Arctic, however, the two paths diverged, with the Ny-Ålesund aerosol continuing around the low but the Barrow aerosol coming under the influence of a large high-pressure area centered over the Chukchi Peninsula. During the first days of the episode, this high moved toward Barrow and cut off direct transport from the central Soviet Union, thus decreasing its coefficients. Aerosol of the second phase of the episode can therefore be viewed as residue from the big pulse mixed with smaller, more normal contributions from other sources which affect the Arctic, such as Europe and the western Soviet Union. North American and other

Asian sources might also have contributed during this period; our tracer system is not yet sophisticated enough to resolve such complex mixtures as these, especially at the low concentrations seen here. On AGASP Flight 3, early in 16 March, the aerosol aloft did indeed appear to be European rather than from the central Soviet Union (As/Sb ratios of 3). The episode ended around the 20th or 21st, when the high had advanced far enough east to cut off Barrow completely from transported aerosol.

The duration of the episode at Barrow appears not to be fully resolved. Our chemical data show that the episode clearly lasted through the 19th. Nephelometer data from the surface, however, have been interpreted to restrict the episode to 12–16 March (Bodhaine and others 1984). Turbidity data (Dutton and others 1984) show a major event from 13–14th March, followed by an extremely weak tail which may be stretched through the 18th (our interpretation).

This problem may be resolved by segregating the data aloft from those at the surface and by carefully re-interpreting the scattering data of Bodhaine and others. All the data aloft, including our elemental data from the third and fourth AGASP flights and the turbidities of Dutton and others, which refer more to conditions aloft that at the surface, show aerosol decreasing rapidly after the second flight (14 March). The surface date, hwoever, including the sulfate concentrations of Hansen and Rosen (1984), show high concentrations through at least the fourth flight (18 March). Apparently, the aerosol aloft came early and left early, as might be expected for the more mobile air masses aloft decoupled from the surface inversion layer. Recall that the most concentrated aerosol aloft was found on AGASP 1 (12 March), three days before surface sulfate reached its maximum. The high-pressure conditions of the second phase of the episode would be ideal for bringing clean upper-tropospheric air down from aloft but not for cleaning the surface layer. In fact, the profiles of scattering and particle counts on Flights 3 and 4 (Schnell and Raatz 1984) show progressively cleaner air aloft but strongly polluted air remaining at the surface.

Considered in detail, the surface nephelometric data of Bodhaine and others (1984) agree well with this picture. They actually show three phases to the aerosol: a pre-episodic period with low scattering and low Angstrom index (coarse particles typical of less-polluted conditions), a first phase of the episode (12–14 March) with high scattering and increased Angstrom exponent (finer particles, or increased pollution character), and a second phase of the episode (14–18 March) with highly variable scattering but systematically higher Angstrom index (finest particles, or greatest pollution character). It is important to note that (1) the high Angstrom index persisted well after the pulses of aerosol had ceased, and (2) scattering during the 17th and 18th (when sulfate was still high) was considerably higher than before the episode. The particle counts reported by Bodhaine and others in their Figure 3 show feature (2) more intensely. In other words, all the data of Bodhaine and others show that the episode had not ended by the 18th. Note also that the time when the big pulses of aerosol disappeared, the 16th, was just the time when the high-pressure area moved eastward over Barrow. Apparently aerosol concentrations in transport zones vary on time scales shorter than can be properly represented by 24-hour samples; real-time data are needed.

Previous discussions of this episode have placed the sources somewhat differently. Based on the trajectories of Harris (1984), Bodhaine and others (1984) considered the first phase to be from western Asia and Europe, via the Kara Sea, and the second phase to be more from central Asia via the Taymyr Peninsula. (Their 'western Asia' seems to be similar to our 'central Soviet Union', however.) We find an opposite tendency, but note that both phases are mixtures of European and Asian sources. Raatz (1985), in a careful examination of synoptic data, including potential temperature, concluded that the first

Table 4. Average regional apportionments of elements in the Barrow aerosol of March 1983, three sources.

	Apportionment: % of mean predicted			Mean predicted ng m-3	Mean observed ng m-3	Mean obs./pred.
	EUR	WUSSR	CUSSR			
As	15.9±2.5	5.0±1.4	79±3	1.37±0.06	1.39	1.00±0.02
Sb	31±5	12.0±3.5	57±5	0.189±0.015	0.25	1.27±1.25
Se	50±10	7.5±2.2	43±5	0.26±0.03	0.172	0.70±0.19
Noncr. V	35±7	29±9	37±3	1.90±0.22	1.58	0.81±0.18
Zn	51±9	9.2±2.9	40±4	12.3±1.3	15.7	1.26±0.30
Noncr. Mn	31±6	23±7	46±4	1.03±0.11	1.17	1.19±0.22
Nonmar.SO$_4$	37±7	29±8	34±6	2200±300	2100	1.00±0.31

phase came from the Volga-Ural area of the central Soviet Union and the second phase came from central and western Europe. This result, which approaches the chemistry much more closely, has already been tempered by it, however; originally, Raatz did not consider central Soviet Union sources. Also, the tracer data show that the European contributions to primary pollution aerosol of the second phase were no larger than those of the first phase, ie that transport from Europe was no more efficient, at least at the surface. By contrast, European sulfate was higher during the second phase. (This shows how primary and secondary pollutants may give very different impressions about sources of aerosol.)

The average elemental apportionments for the Barrow aerosol of March 1983 among the European and Soviet Union signatures are shown in Table 4. The results are quite similar to those of winter 1979–80, but with the European contribution being 10–30% greater.

At the present moment, therefore, chemical and meteorological techniques are being fused in the Arctic to a degree hardly seen elsewhere. Patterns of interaction are being developed which will serve as models for the future. All this is most appropriate, for the difficulty of unravelling sources and transport of Arctic haze is so great that no single technique can realistically expect to do it all.

Acknowledgements

This work was funded by the Office of Naval Research under Contracts N00014–76–C–0435 and N00014–84–C–0035. We thank NOAA/AGASP for extensive cooperation and travel funds, the staff at the NOAA/GMCC Observatory in Barrow for collecting samples, Dr. B. Ottar of the Norwegian Institute for Air Research for providing samples from Ny-Ålesund, R. Kartastenpää and K. Markkanen of the Finnish Meteorological Institute for providing samples from Ähtäri, and Dr. Zhao Dianwu of the Institute of Environmental Chemistry, Academica Sinica, for providing samples from the Beijing-Tianjin area of northeastern China. All samples were analyzed at the Rhode Island Nuclear Science Center in Narragansett.

References

BODHAINE, B. A. AND OTHERS. 1984. Surface aerosol measurements at Barrow during AGASP. *Geophysical Research Letters*, 11: 377–80.
DUTTON AND OTHERS. 1984. Features of aerosol optical depth observed at Barrow, March 10–20, 1983. *Geophysical Research Letters*, 11: 385–88.
HANSEN, A. D. A. AND ROSEN, H. 1984. Vertical distribution of particulate carbon, sulfur, and bromine

in the Arctic haze and comparison with ground-level measurements at Barrow, Alaska. *Geophysical Research Letters*, 11: 381–84.

HARRIS, J. M. 1984. Trajectories during AGASP. *Geophysical Research Letters*, 11: 453–56.

LEWIS, N. F. 1985. *Particle-size distributions of the Arctic aerosol.* University of Rhode Island, MS thesis.

LOWENTHAL, D. H. AND RAHN, K. A. 1985.. Regional sources of pollution aerosol at Barrow, Alaska during winter 1979–80 as deduced from elemental tracers. *Atmospheric Environment*, 19: 2011–24.

RAATZ, W. E. 1985. Meteorological conditions over Eurasia and the Arctic contributing to the March 1983 Arctic haze episode. *Atmospheric Environment*, 19: 2121–26.

RAHN, K. A. AND LOWENTHAL, D. H. 1984. Elemental tracers of distant regional pollution aerosols. *Science*, 223: 132–39.

RAHN, K. A. AND OTHERS. 1985. *Elemental tracers applied to transport of aerosol from Midwest to Northeast.* Final report, Cooperative Agreement CR- 810903. Atmospheric Sciences Research Laboratory. Office of Research and Development, US Environmental Protection Agency, Research Triangle Park, NC 27711.

SCHNELL, R. C. AND RAATZ, W. E. 1984. Vertical and horizontal characteristics of Arctic haze during AGASP: Alaskan Arctic. *Geophysical Research Letters*, 11: 369–73.

WATSON, J. G. AND OTHERS. 1984. The effective variance weighting for least squares calculations applied to the mass balance receptor model. *Atmospheric Environment*, 18: 1347–55.

PART 2.

LOCAL, REGIONAL, GLOBAL, ECOLOGICAL AND CLIMATIC IMPLICATIONS

Back: R. Delmas; R. Schnell; Ø. Hov; F. P. J. Valero
Front: T. P. Ackerman; H. Rosen; B. Bodhaine

PART 2. INTRODUCTION

Man the meddlesome ape is present all over the earth, and has altered its surface in ways that can be seen from space. Now for the first time, by burning fossil fuels and pouring their by-products into the atmosphere, he is meddling on a scale that could change climates, and start a train of consequences that will end beyond his control. Having established the reality of arctic haze, its composition and origins, the symposium considered this man-made phenomenon in more detail, paying particular attention to how it might affect polar climates, the ice caps, and possibly the rest of the world.

H. Rosen and A. D. A. Hansen's paper, representing arctic haze as probably the best-documented example of a global-scale perturbation of the balance of solar radiation, examined the distribution in Arctic aerosols of the particles most strongly implicated in the perturbation–graphitic carbon particles derived from incomplete combustion. Their preliminary studies indicate that particle concentrations, though unevenly distributed, are dense enough over the highly reflective surface of the Arctic to affect solar radiation balance significantly. Their estimates of input fluxes of graphitic carbon, total carbon and sulphur particles indicate clearly the origin of these pollutants from mid-latitude sources.

Valero and Ackerman presented in detail two aspects of arctic haze – its absorption of solar radiation and its vertical structure – deducing changes in the atmosphere–surface system directly due to haze, and determining haze-induced changes in atmospheric heating rates at different altitudes. They conclude that changes detected could modify local climate significantly, but that more critical measurements are required to confirm the magnitude of the effect. Schnell reported on the first International Arctic Gas and Aerosol Sampling Program (AGASP-I) of March 1983 (to which many other contributors also referred) and a successor program planned for 1986: his paper summarizes the main findings and refers students to the main sources of information. These are clearly the kinds of international study, involving co-ordinated measurements at ground stations and in aircraft, that will throw most light on the huge international problems arising from arctic haze.

Glenn Shaw, the main architect of this symposium, drew attention to the significance of clouds in arctic energy budgeting, the importance of cloud condensation nuclei on which water droplets accumulate, and the possible role of haze particles in providing nuclei. Pollution from mid-latitudes could in principle alter cloud reflectivity in winter, but is unlikely to in summer, when tropospheric air flow patterns tend to be zonal, and incoming air masses are more likely to have been cleansed of their particles by rain. His paper, in tune with many others, stresses the need for further critical measurements of what could be a climatically significant effect. Ackerman, Stenback and Valero attempt an energy budget for the Arctic basin (defined as the area north of 70 °N) from published figures. They find that on present estimates, energy lost by radiation appears to exceed that gained from transportation by some 25 to 30%. Reviewing possible sources of error, they conclude that in spring (March, April and May), when their budget is best balanced, arctic haze contributes a perturbation of about 10%. Despite the uncertainties this is a considerable fraction, justifying further observations that would provide data for more accurate budgeting.

In his discussion of the aerosol record from the GMCC observatory at Barrow, Alaska, Bodhaine summarized valuable data of a kind that are rarely available to polar scientists in any field – the results of a well-considered, long-term monitoring program. The first nine-year period (1976–84) clearly justifies the existence of the station, which is fortunately maintaining its excellent record of observations. Many are taken continuous or hourly, and all have proved illuminating in the study of arctic haze, tying in effectively with many short-term observations over the years, and building up to provide indications of seasonal and annual trends. R. Delmas, of the Laboratory of Glaciology, Grenoble, France, dealt with the chemical analysis of snow layers in Antarctica and Greenland, derived from ice-core analysis. Comparisons between the hemispheres are particularly valuable; while both northern and southern ice sheets show records of natural pollution events (for example volcanic eruptions) from their inception, man-made pollutants are as yet almost entirely confined to the north. In the final paper of this section Ø. Hov, of the Norwegian Institute for Air Research, presented the arctic atmosphere as a valuable laboratory for future monitoring of global pollution levels, discussing the measurements that need to be made if we are to use this asset to the full.

LIGHT-ABSORBING COMBUSTION-GENERATED PARTICLES OVER REFLECTING POLAR ICE

H. ROSEN AND A. D. A. HANSEN

ABSTRACT. Arctic haze probably represents the best-documented example of a substantial perturbation of solar radiation balance over a large region of the globe, due to combustion-generated particles. Over reflecting polar ice, the solar radiation balance is sensitive to the absorbing component of the aerosol, which has been identified on a molecular level by Raman spectroscopy as graphitic carbon. We have studied the vertical, horizontal and temporal distributions of graphitic carbon particle (GCP) concentrations and their associated absorption coefficients in the Arctic. During winter and spring substantial GCP concentrations occur throughout the arctic troposphere. In vertical profile they may be strongly layered or almost uniformly distributed; within layers the concentrations can be as large as those found in typical urban areas in the United States. Preliminary studies indicate that associated absorption optical depths are large enough to cause substantial change in solar radiation balance over a highly reflecting surface. GCP concentrations show both large vertical inhomogeneities and also large horizontal variations with characteristic scales of 50–100 km. These distributions have been used in conjunction with a box model calculation to make preliminary estimates of springtime fluxes of carbon and sulfur particles entering the arctic troposphere. These fluxes are large, and suggest major sources of sulfur and soot south of 60°N latitude.

Contents

Aerosols and climate

Aerosol particles can change the radiation balance of the earth, leading to cooling or heating effect dependent on aerosol optical properties, aerosol concentrations, and the albedo of the underlying surface (Atwater 1970; Rasool and Schneider 1971; Mitchell 1971, Chylek and Coakley 1974). A purely scattering aerosol reflects back to space energy that would otherwise be absorbed by the earth-atmosphere system, leading to a cooling effect. Absorbing components added to the aerosol lead to a heating of the earth-atmosphere system if the reflectivity of the underlying surface is sufficiently high. Aerosols can also effect a redistribution of solar and terrestrial radiation within the earth-atmosphere system. Even a neutral aerosol (ie one that does not change the overall energy balance of the earth-atmosphere system) can effectively warm the atmosphere at the expense of ground-level absorption. Such a redistribution leads to changes in atmospheric stability and circulation (Ackerman 1977; Tanre and others 1984). Aerosols have only recently been incorporated into general circulation models (Joseph 1977; Randall and others 1985; Coakley and Cess 1983; Tanre and others 1984) and little is known of their atmospheric dynamic effects. They act as condensation or ice nuclei for clouds (Shaw, these proceedings). Changes in aerosol concentration or composition could change cloud droplet concentrations or size distributions, strongly affecting the optical properties of clouds (Twomey 1977; Charlock and Sellers 1980). Small concentrations of absorbing aerosols in clouds can affect cloud albedo and cloud dynamics (Twomey 1977; Ackerman and Baker 1977). Absorbing particles can also significantly affect snow and ice albedo (Warren and Wiscombe 1980; Grenfell and others 1981; Warren 1982; Chylek and others 1984).

Most studies of of aerosols and global atmospheric radiative transfer assume a dominantly scattering aerosol with only a small absorbing component, which seems well to represent naturally occurring aerosols (ie soil, maritime, stratospheric). These studies in general indicate that an assumed naturally-occurring haze of imaginary refractive index n_i close to zero and optical depth 0.1–0.2 depresses the earth's surface temperature by 2–4°C (Rasool and Schneider 1971; Yamamoto and Tanaka 1972; Wang and Domoto 1974; Reck 1976; Pollack and Cuzzi 1979; Charlock and Sellers 1980; Coakley and others 1983). This cooling effect is similar in magnitude and latitudinal distribution, but opposite in sign, to the effects of doubling the CO_2 concentration (Coakley and others 1983).

Combustion-generated aerosols and climate

Combustion-generated aerosols differ from naturally-occurring aerosols in containing a substantial aborbing component due to graphitic carbon particles (GCPs), produced from incomplete combustion. This component, identified on a molecular level by Raman spectroscopy (Rosen and others 1978), strongly affects optical properties of urban aerosols, leading to imaginary refractive indices of 0.02 to 0.1 (Gerber 1982). This range of values is due partly to different chemical compositions and partly to differences in measurement techniques. Such aerosols transported on a global scale would provide significant atmospheric absorption, even in remote regions. In recent years, unexpectedly large aerosol absorption has been observed at ground-level stations throughout the western Arctic (for example NOAA-GMCC station at Barrow, Alaska, Mould Bay, Igloolik and Alert in the Canadian Arctic, Bear Island and Spitsbergen in the Norwegian Arctic: Rosen and others 1981; Heintzenberg 1982; Patterson and others 1982; Rosen and Novakov 1983). As in the urban samples, the absorbing particles have been identified as combustion-derived graphitic carbon.

These results indicate long-range transport of a least 2,000 km from the closest significant source region. Trace element and meteorological analyses (Rahn and McCaffrey 1980,; Barrie and others 1981; Ottar 1981) suggest even longer-range transport from mid-latitude sources 5,000–10,000 km away. Recent fluxes of soot and sulfur particles entering the arctic troposphere have been compared with fossil-fuel burning at various latitudes; the haze produced cannot be due to arctic sources: source regions below 60°N or even 50°N latitude are implicated.

In March–April 1983 the NOAA airplane sampling program (Schnell and others 1984) provided vertical and horizontal distributions of GCPs and other components of arctic haze. Substantial concentrations of GCPs were found throughout the arctic troposphere (ground level to tropopause, Alaska, Canada, Norway, North Pole: Rosen and Hansen 1984; Hansen and Rosen 1984). Vertical profiles generally showed upper aerosol layers with GCP concentrations significantly higher than at ground level, and sufficient to absorb 5–10% of the incident solar flux (Shaw and Stamnes 1980; Porch and MacCracken 1982; Cess, 1983; Rosen and Hansen 1984). Net flux measurements (Valero and others 1984) on these flights indicated excess daytime tropospheric heating rates of 0.7 to 1.5% per day, attributed to aerosol absorption and consistent with the graphitic carbon profiles (Ackerman and Valero 1984; Rosen and Hansen 1984). These atmospheric heating effects are substantial and need to be incorporated into climate models for their potential impact to be estimated.

Combustion-generated particles at ground level in the western Arctic

Recent ground level studies in Alaska (Rahn and McCaffrey 1980; Rosen and others 1981) show the presence of substantial concentrations of particles containing carbon and sulfur, that seem to be characteristic of the whole arctic region (Barrie and others 1981; Ottar 1981). These particles both scatter and absorb visible radiation (Bodhaine and others 1981; Rosen and others 1981) and appear to be responsible for the phenomenon of arctic haze reported by Mitchell (1957) and by Shaw and Wendler (1972). Trace element analysis has suggested anthropogenic origins from mid-latitude sources (Rahn and McCaffrey 1979, 1980). Large GCP concentrations in the western Arctic have been identified by by Raman spectroscopy (Rosen and others 1981; Rosen and Novakov 1983). Concentrations in winter and spring at the GMCC-NOAA observatory near Barrow,

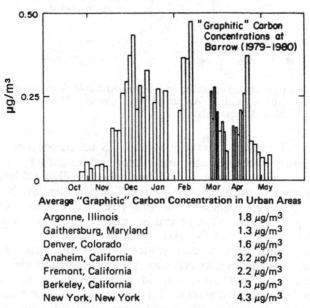

Fig 1. Comparison of ground-level graphitic carbon concentrations at Barrow, Alaska, with annual average concentrations in various urban areas.

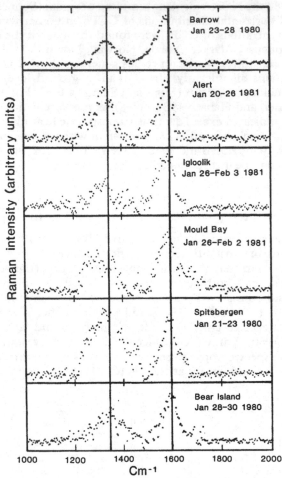

Fig 2. Raman spectra of particles collected in Alaskan Arctic (Barrow) compared with samples collected in the Canadian Arctic (Alert, Igloolik, Mould Bay) and the Norwegian Arctic (Svalbard, Bear Island).

Alaska, are only about a factor of 4 less than typical US urban concentrations (Figure 1). Natural fires are rare in these seasons, and recent carbon-dating (Currie, personal communication) and trace element analyses indicate only a small wood-burning component; industrial coal burning is the most likely source. The particles have large absorption cross sections (about 10 m²/g), suggesting imaginary refractive indices of 0.02 to 0.08 with significant heating implicationa (Shaw and Stamnes 1979; Porch and MacCracken 1982; Cess 1983; Valero and others 1986; Ackerman and others 1986).

Graphitic structures, with trivalent carbon atoms occupying lattice sites in a two-dimensional hexagonal honeycomb network, have intense Raman modes but very weak infrared vibrational spectra. The Raman modes identify graphitic structures even in the presence of a complex mixture of substances. Solvent extraction, heat treatment, optical absorption, and morphology studies (Wolff and Klimisch 1982) can provide indirect evidence for a graphitic component; but Raman spectroscopy appears to be the only current method yielding unambiguous identifications on a molecular level. The technique has been applied to concentrations of GCPs in combustion effluents, urban air (Rosen and others 1978; Rosen and others 1980) and the Alaskan Arctic (Rosen and others 1981). Spectra from these samples show the presence of two Raman modes located at 1350 cm^{-1} and 1600 cm^{-1}, identified as due to phonons propagating within graphitic planes. In Figure 2 these measurements are extended to samples collected from three sites in the Canadian Arctic (Mould Bay, Igloolik, Alert) and two sites in the Norwegian Arctic (Spitsbergen, Bear Island). These samples were obtained with the cooperation of B. Ottar of the Norwegian Institute for Air Research and L. Barrie of the Canadian Atmospheric Environmental Service; all were collected at similar times of year in 1980 and 1981. The spectra show significant concentrations of GCPs at all sites. Assuming fixed optical constants (Raman cross sections, absorption cross sections), one can use the integrated intensity of the 1,600 cm^{-1} Raman mode as a measure of relative GCP concentrations. Concentrations at all the sites are similar, the largest and smallest within a factor of about 3 with relative ordering for Spitsbergen, Bear Island, Barrow, Mould Bay, Alert, and Igloolik 2.1/1.7/1/0.8/0.8/0.7. In summary, these results show that the large GCP concentrations Barrow are not a local phenomenon, but characteristic of ground-level stations across the western Arctic.

Vertical distributions of combustion-generated particles: radiative transfer implications

A major uncertainty in modelling effects of arctic haze on solar radiation balance has been our limited knowledge of the vertical distribution of the haze, in particular the vertical distribution of GCPs and their associated absorption coefficients. The AGASP flights of March and April 1983, organized by NOAA, explored both vertical and horizontal distributions of the haze. One of the instruments used was the aethalometer (Hansen and others 1982), capable of determining GCP concentrations on a real-time basis using the calibration of Gundel and others (1984).

For vertical profiles the atmosphere was divided into altitudinal bins and average aethalometer response was determined in each over a time period. These were the first vertical profiles obtained in the Arctic, probably in the atmosphere in general (Rosen and Hansen 1984; Hansen ad Rosen 1984). Figures 3 and 4 show vertical profiles of GCP concentrations vs. altitude are shown for two flights in the Norwegian Arctic. Similar vertical profiles are observed in the Alaskan Arctic, Canadian Arctic, and at the North Pole. Figure 3 shows a vertical profile near Bear Island at 74°N, 24°E. also shown for comparison are the average ground-level concentrations found at the Barrow GMCC-NOAA observatory for April 1982, and mean annual ground-level concentrations in various US urban areas.

The vertical profile has considerable structure, with at least three layers at approximately 1 km, 2.5 km, and 4.5 km, and what appears to be a very narrow layer at 1.7 km. Haze layers can have sharp boundaries with dramatic changes in concentration over distances less than 100 m. Concentrations within these layers are large; the peak at 1 km is as high as typical ground-level concentrations in the US (Berkeley, Denver, Gaithersburg) and only a factor of about 2.5 lower than those of New York City. Concentrations in this band are a factor of about 10 higher than the 0.15–μg/m^3 ground-level concentrations recorded at Barrow in April 1982. The profile also indicates a relatively clean region at low altitudes, consistent with Alaskan Arctic results where simulataneous ground-level measurements were made (Hansen and Rosen 1984).

Figure 4 shows a vertical profile over the Norwegian Sea in 70°N, 0°E. This profile

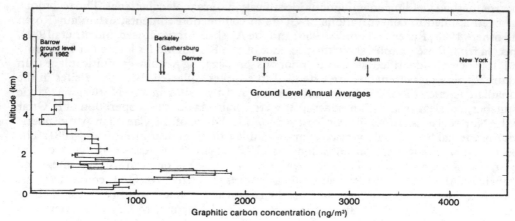

Fig 3. Vertical profile of graphitic carbon concentrations (ng/m³) on 31 March 1983 at about 74°N, 25°E. Shown for comparison are the annual average ground-level graphitic carbon concentrations at US urban locations, and the mean April 1982 ground-level values at the NOAA-GMCC observatory, Barrow.

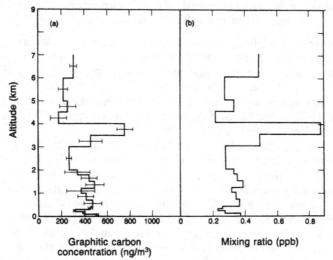

Fig 4. (a) Vertical profile of graphitic carbon concentrations (ng/m³), 5 April 1983 at about 70°N, 0°E. (b) As above, with graphitic carbon concentrations expressed as a mixing ratio.

is quite distinct from the one shown in Figure 3: it has much less structure with only one sharp band at 3.75 km superimposed over an almost uniform haze layer, decreasing only slightly with altitude even close to the top of the troposphere. If this profile is plotted vs the mixing ratio, as in Figure 4b, it is essentially flat to the top of the troposphere with an intense band located at about 3.75 km. This profile may indicate significant transport at high altitudes above the planetary boundary layer; alternatively it may show changes in vertical distribution that have occurred after the air mass had reached the Arctic.

Recent modelling of possible effects of arctic haze on arctic radiation budgets has indicated substantial changes in heat balance if the optical depth due to absorption is

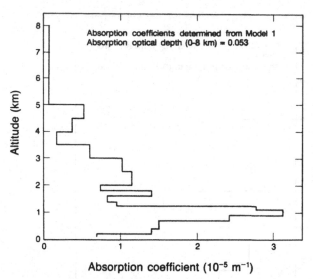

Fig 5. Vertical distribution of absorption coefficients obtained from the graphitic carbon profile shown in Figure 1 using aerosol Model 1. The corresponding absorption coefficients for Models 2 and 3 are obtained by multiplying the distribution by factors of 0.66 and 0.46 respectively.

sufficiently large (Shaw and Stamnes 1980; Porch and MacCracken 1982; Cess 1983). From the results shown in Figures 3 and 4 one can calculate absorption coefficients and optical depth for various models of the arctic aerosol. It is important in modelling to distinguish whether GCPs are mixed internally or externally with other non-absorbing components. Ackerman and Toon (1981) showed that such differences can lead to significant changes in aerosol absorption.

We will consider three possible models. In Model 1, GCPs are mixed internally with all the major submicron aerosol components (ie sulfates, organics). In Model 2 they are mixed internally only with submicron organic aerosol components, and in Model 3 they are externally mixed with the other aerosol components. Of these three possibilities, the first two are the most likely combinations. For them we treat the particles as homogeneous spheres and mix in the various components by volume-mixing of dielectric constants. In these Mie calculations, refractive indices of GCPs were chosen to be 1.94 to 0.66 (Ackerman and Toon 1981), with a density of 1.5; the refractive index of the non-absorbing components was taken to be 1.5. Relative concentrations of graphitic carbon and suphates were determined from chemical analyses of filters collected on the flights for appropriate time periods, and relative concentrations of organic component were obtained from ground-level measurements in April at Barrow. The particle size distribution for the submicron aerosol is asumed to be log-normal with $r_{gr} = 0.2$ μm and μ_g = 2, which appears to represent the urban aerosol well. For the external mixture we use the photoacoustic measurements of Roessler and Faxvog (1979) and Szkarlat and Japar (1981), who obtain an average value of 8.3 m^2/g for a wavelength of 0.5μ.

For the vertical profile in Figure 3, the three models yield specific absorption coefficients respectively of 18.1 m^2, 12.0 m^2, and 8.3 m^2 per gram of graphitic carbon; corresponding results for Figure 4 are 14.6 m^2, 12.0 m^2, and 8.3 m^2 per gram of graphitic carbon. Vertical profiles of the absorption coefficient using Model 1 are shown for flights 8 and 10 in Figures 5 and 6. Absorption coefficients for the other models can be

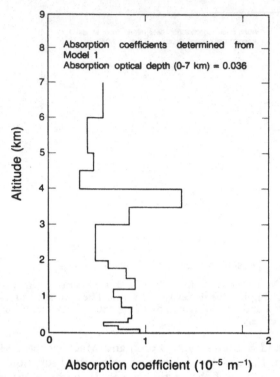

Fig 6. Vertical distribution of absorption coefficients obtained from the graphitic carbon profile shown in Figure 2 using aerosol Model 1. The corresponding absorption coefficients for Models 2 and 3 can be obtained by multiplying the distribution by factors of 0.82 and 0.57 respectively.

determined from these figures from ratios of specific absorption coefficients, which can be integrated over altitude to determine absorption optical depths for the various flights and aerosol models. For flight 8, the three aerosol models yield optical depths at 0.5μ of 0.053, 0.035, and 0.024; for flight 10, the corresponding optical depths are 0.036, 0.029, and 0.020. It is interesting to compare these optical depths with those obtained by Valero and others (1982) using a net flux radiometer. Such a comparison can be made for the vertical profile from flight 8 shown in figure 5, where Valero and others (1984) have completed their analyses. They obtain a best-fit optical depth due to absorption of 0.06; this is very close to the largest value we report, and could suggest that the aerosol is competely internally mixed, as represented in Model 1.

The optical depths reported here are large enough to produce significant changes in radiation balance over a highly reflecting surface; this is illustrated by the calculations of Porch and MacCracken (1982) and Cess (1983) who modelled the arctic aerosol for an absorption optical depth of 0.021 (close to our minimum estimate), cloud-free conditions, zenith angle 72° and surface albedo 0.8. For these parameters they obtained a change in the noontime solar radiation balance at the top of the troposphere of about 20 W/m^2. Averaged over the day for 15 March at 70°N, the surface-atmosphere energy absorption would increase by about 7 W/m^2. These changes are substantial, corresponding to an increase in energy absorbed by the earth-atmosphere system of approximately 5% of incident solar flux at the top of the troposphere. Correspondingly larger changes would be expected for the internally mixed aerosol models.

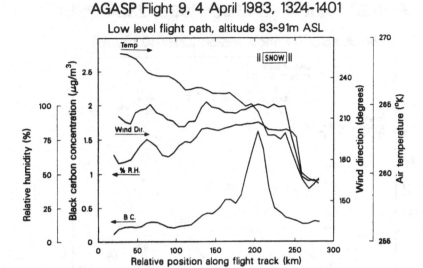

Fig 7. Flight data, AGASP low-level flight 4 April 1983; for details see text. The bar at upper right indicates snow flurries.

Horizontal inhomogeneities of arctic haze

During AGASP 1983, in addition to the vertical layering already noted, we found for the first time evidence of horizontal variations of aerosol black carbon concentration (BCC). Some correlated with meteorological phenomena; others with a characteristic scale of 50 to 100 km occurred in the absence of meteorological activity. Flights included both slow and fast ascents and descents, as well as level flying at various heights down to 30 m asl. Meshing the aethalometer data with the aircraft systems tape gave a data base containing position, altitude and mean concentration of black carbon at one-minute intervals (estimated error ± 0.09 μg/m^3), together with wind speed and direction, air temperature relative humidity and other variables (Schnell 1984).

Figure 7 shows data from AGASP flight 9, over a period of about 35 minutes in which the aircraft flew steadily north at 83–91 m asl. The track covered roughly 250 km from 75.5°N, 26.2°E to 78.2°N, 27.2°E, east of Spitsbergen. The first 50 km were flown over ocean with broken ice, thereafter over more or less continuous ice with leads. Air temperatures were steady at around 270°K before the ice was reached, gradually decreasing to 265°K at 200–km along the track. Winds were constant at 200° to 210°, dew point remained constant at 261°K (ie relative humidity increased from 60% to 80%), the sky was overcast and the air hazy. The aircraft seemed to be flying in a single moist air mass of slowly falling temperature. BCC increased from about 0.2 μg m^{-3} to a peak of 1.6 μg m^{-3} at the 200–km mark. At this point snow flurries began and continued until, by the 250–km mark, the aircraft had flown through a front into clear skies. Wind direction shifted by 55°, air temperature fell 5°, and dew point fell by 10°. The new colder, drier air mass had a reduced BCC of about 0.25 μg m^{-3}.

Figure 8 shows 47 minutes of data from AGASP flight 3, encompassing almost a complete loop 130 km long at heights from 30 m to 1800 m. Ascending in 71.5°N, 157.4°W, the aircraft flew a low-level southerly track from 200 km to 50 km north of Barrow and descended in 72.7°N, 156.8°W . During the low flight wind speed and direction were steady at 4.3 $\pm$ 0.4 m sec^{-1} and 51° $\pm$ 8° respectively. Realtive humidity

AGASP Flight 3, 15 March 1983, 2304-2351

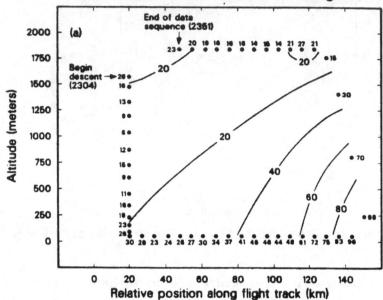

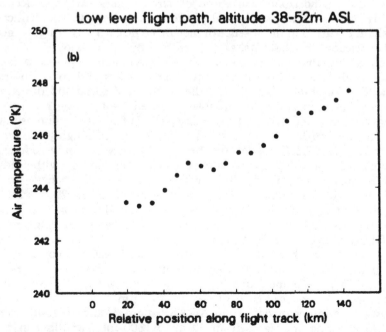

Fig 8. Flight data , AGASP loop flight 15 March 1983; for details see text. (a) Aerosol black carbon concentration vs altitude and relative horizontal position; contours are estimated from perimeter measurements. (b) Air temperature for the low-altitude flight track portion from 2317 to 2335 UT.

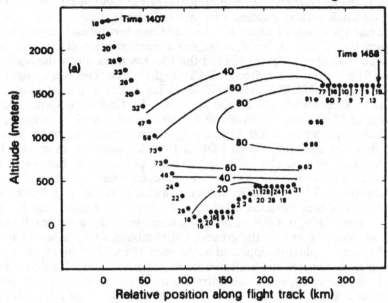

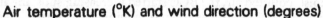

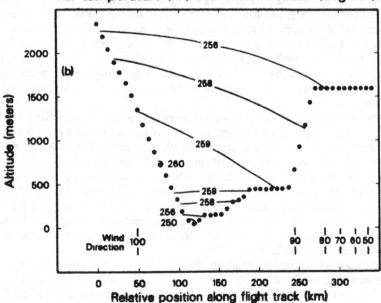

Fig 9. Flight data, AGASP flight 21 March 1983; ascent and descent patterns; for details see text. Contours are estimated from perimeter measurements. (a) Aerosol black carbon concentration vs altitude and horizontal position. (b) Air temperature and wind direction vs altitude and position.

was 75% ± 5%; there was no evidence of precipitation, and no air mass boundary was crossed. However, mean BCC increased steadily from 0.25 μg m^{-3} to almost 1 μg m^{-3} along the flight path (Figure 8a), while air temperature increased by 4°K Figure 8b). At the end of the track, the aircraft climbed and back-tracked at 1850 m for about 100 km. Contours of aerosol black carbon concentration are indicated; these are not measured contours, but because there were neither sources nor sinks for the carbon, the implied continuity is reasonable. The data therefore suggest a concentration of black carbon at low altitudes, with a peak at some point beyond the 150–km mark along the track.

Figure 9 shows 51 minutes of data from AGASP flight 5; the track ran straight from 71.8°N, 141.4°W to 74.0°N, 137.3°W over the Beaufort Sea during as part of a flight from Anchorage to Thule, Greenland, 100 to 400 km from the Canadian coast. The data represent a descent to 30 m followed by about 17 minutes of low-altitude flight, after which the aircraft climbed to 1,550m asl. Figure 9b (air temperature and wind direction) shows a strong temperature inversion with a fall of 10°K between 700 m and sea level; wind direction shifted very little along the first 250 km of the track, but then turned rapidly by 50 °soon after the climb-out. At this time, though air temperature had changed little, dew point rose from 237°K to 252°K, indicating a change of air mass. Black carbon data (Figure 9a) show a concentration of about 0.2 μg m^{-3} at height near the start, increasing to 0.73 μg m^{-3} at about 800 m on the descent and falling off rapidly to low values within the inversion layer near the ground. On climbing, BCC increased again to almost 1 μg m^{-3} but then rapidly disappeared at the start of the level flight, at the same time as meteorological measurements indicated an air mass boundary. The constructed contours shown in the figure suggest a concentration maximum close to this point.

Figure 7 indicates a situation similar to one described by Raatz and Schnell (1984) for another AGASP flight in the Norwegian Arctic. In the northward flight at constant low altitude in moist, polluted air, the BCC started increasing substantially at around the 150–km track mark but fell rapidly as soon as precipitation started. Shortly afterwards the aircraft passed through a front into an air mass of low aerosol BCC. Removal of aerosols by precipitation is expected; however, there was no current meteorological activity that would explain the five-fold change in BCC between the start of the track and the peak at the 200–km mark. Figure 8 shows a similar situation, in which aerosol BCC increased fourfold within the same air mass. Here air temperature gradually rose (see Figure 8b), but there was no clear evidence of recent meteorological activity or the approach of a front.

The maximum instantaneous BCC observed during the entire AGASP programme was approximately 1.5 μg m^{-3}, which we shall take as a reasonable value for the maximum concentration likely to be found at the center of an inhomogeneity. If we assume that the flight track of Figure 8 penetrated the edge of such a patch, we can estimate the size scale of the patch from the spacing of the contours. These projections suggest a horizontal scale of 50 to 100 km as the characteristic length over which aerosol BCC changes by a factor of two. We must strongly emphasize that this distance represents a characteristic length on the basis of partial perimeter measurements, rather than an actual complete mapping. The data shown in Figure 7 from the start of the track to the onset of precipitation also fit into this scale.

Figure 9 presents a more complex situation. Figure 9b indicates a surface temperature inversion layer extending to about 4,500 m and an air mass boundary at about 280 km along the track. The constructed contours in Figure 9a show strong aerosol BCC in the air above the surface layer but very low levels below the inversion, an observation in agreement with the expected weak coupling of these layers. The effect of the front at 280–300 km track is similar to that shown in Figure 7: in the new air mass, BCCs are low. The constructed contours indicate a very rapid change of aerosol BCC at the front and also show that the polluted air mass is resting on top of a clean surface temperature inversion layer. This contrasts with the situation often found in urban atmospheres, in which the emissions are trapped in the surface layer beneath the inversion. This observation in the Arctic may imply long-range transport of combustion emissions (Rahn and McCaffrey 1980; Iversen 1984).

The correlation between aerosol concentration changes and meteorological changes is expected; however, we see that these examples also show substantial BCC changes in steady meteorological conditions. In these instances, the aerosol black carbon appeared to be in patches with a characteristic horizontal scale of the order of 50–100 km. There may be several possible explanations for these observations. One is that the air masses were at one time uniformly polluted but that patchy precipitation or other scavenging had removed aerosols from portions of the flight track. Another is that the injection of carbonaceous aerosol into the air mass had been confined in an area at the source, and that these are plume edge profiles. However, this would require only a small amount of horizontal dispersion over the long transport paths (Harris 1984) and imply a diffusion coeffficient of the order of 10^9 cm^2 sec^{-1}. At present, we cannot unambiguously identify either these or other possible mechanisms as explanation for the data.

Springtime fluxes of soot and sulfur: possible source regions

The vertical profiles previously described can be used in conjunction with a box model calculation to estimate input fluxes of soot and sulfur entering the arctic troposphere. These are large and can only be accounted for by major sulfur and soot sources. Comparison with the amount of fuel burned in various latitude bands indicates that arctic haze cannot be due to pollution from arctic sources and strongly suggests that the dominant source regions are below 60°N latitude. Comparisons of arctic sulfur fluxes with sulfur emissions on a regional and global basis indicate that significant fractions enter the Arctic. These preliminary results suggest global-scale transport of combustion-generated aerosol particles.

Box model calculation.
For this we view the arctic troposphere as a cylinder with radius extending from the North Pole to 70°N latitude and a height of 7 km, the approximate level of the tropopause. For a conserved species like graphitic carbon that is neither produced nor destroyed in the atmosphere, rate of change of the mass of particles within this cylinder is equal to the flux of particles entering minus the flux leaving:

$$\frac{dM(t)}{dt} = F_M - \frac{M(t)}{\tau} \tag{1}$$

where M(t) is the mass of the component within the box at time t; dM(t)/dt is the rate of change of this mass within the box; F_M is the input flux through all surfaces of the box; and M(t)/dt is the flux leaving the box, which we assume can be represented by by an exponential decay with effective residence time τ. This residence time includes both losses by both deposition and transport of air masses out of the arctic basin. For equilibrium, dM/dt = 0, and the input flux equals the decay rate:

$$F_M = \frac{M(t)}{\tau}. \tag{2}$$

For nonequilibrium situations the input flux can be greater or less than the decay rate. If the input flux at time t is less than $M(t)/\tau$, it is easy to show that at some previous time the input flux had to be at least as large as this decay rate. In other words, even for a nonequilibrium situation, Equation 2 can be used to get a lower limit on the maximum input flux.

If one can estimate the mass of graphitic particles in the arctic troposphere, then using Equation 2, the input flux as a function of residence time can be determined. The mass of graphic particles G(t) is given by:

$$G(t) = \int_{\text{Arctic troposphere}} G(\text{lat., long., ht., time}) \, dV \qquad (3)$$

where G(lat, long, ht, time) is the concentration of graphitic particles at the specified latitude, longitude, height and time. The profiles of graphitic particles obtained during AGASP can be used to estimate a mean value of G(t) from March to April 1983. In this analysis we neglect the dependence of the graphitic cncentrations on latitude and longitude and use an average vertical distribution obtained from the nine tropospheric AGASP flights to characterize the whole arctic region. These flights included four in the Alaskan Arctic, three in the Norwegian Arctic, one from Anchorage to Thule via the Canadian Arctic and one from Thule to Bodø, Norway, via the North Pole. We assume that eastern and western Arctic are equally dirty, which may underestimate soot concentrations since the primary sources appear to be in the eastern sector (Rahn and McCaffrey 1980). With these assumptions:

$$G = \langle G(t) \rangle = \int_{\text{lat}=70°}^{\text{lat}=90°} 2\pi R^2 \cos \text{lat. d lat.} \int_{h=0}^{h=7 \text{ km}} \langle G(h) \rangle \, dh \qquad (4)$$

where R is the radius of the earth and $\langle G(h) \rangle$ is the average vertical distribution of graphitic particles. Substituting into this equation, we determine the average mass of graphitic particles in the arctic troposphere during AGASP to be $G = 2.45 \times 10^{10}$ g, or 2.45×10^4 metric tons. Soot is composed of both graphitic particles and organics. As shown by Novakov (1981), one can make a good estimate of soot concentrations from the graphitic component by assuming it represents approximately 25% of the soot mass. Reliable total carbon determinations could not be obtained from the airborne filters on these flights. However, ground-level determinations at the Barrow NOAA-GMCC station indicate that during March-April approximately 20% of the total carbon is graphitic (Rosen and others 1981). This would suggest that most of the particulate carbon at this time of year in the Arctic is soot, a result that needs validation by sampling in other parts of the Arctic. For present purposes we assume that total carbon in the Arctic can be estimated from the Barrow ground-level measurements at Barrow by multiplying the graphitic component by 5, yielding a total of 1.2×10^5 tons.

It is possible also to estimate total sulfur in the Arctic from the 20 airborne filters collected in AGASP, for which the sulfur-to-graphitic carbon ratio is 3.26 ± 0.75. Assuming that vertical profiles of sulfur and graphitic carbon are similar, then total particulate sulfur is $3.26 \times$ the graphitic carbon mass, or 8×10^4 tons; this is a lower limit for total sulfur which does not include the gas-phase contribution.

In conjunction with Equation 2 these figures can be used to estimate input fluxes of graphitic carbon, total carbon and sulfur into the arctic troposphere. They appear as a function of residence time in Figures 10a and 10b; shaded regions represent our estimates of reasonable values for residence time of particles, ie 1 week to 1 month. This effective residence time includes the effects of deposition as well as transport of air parcels out of

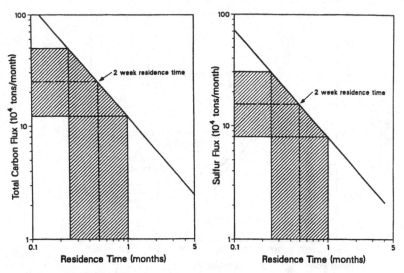

Fig 10. Flux of (a) soot and (b) sulfur entering the Arctic troposphere, estimated as a function of residence time. Shaded area; range of residence times of 1 to 4 weeks.

the arctic basin. The estimates may have to be revised as air parcel trajectories and deposition rates becomes better understood. The range of values is consistent with the estimates of 2–3 weeks by Rahn and McCaffrey (1980) and the very rapid transport (about 1 week) of aerosols from Eurasia to Barrow estimated by Raatz and Shaw (1983) and Harris (1984).

For comparing the fluxes shown in Figures 10a and 10b with fluxes from source regions, we chose a residence time of 2 weeks, which will give fluxes within a factor of 2 of the extreme values of residence time. This uncertainty should be kept in mind when comparisons are being made. Recently Marland and others (1984) estimated amounts of fossil fuel burned in 5° latitude bands. Table 1 compares the fluxes of soot (F_C) and sulfur (F_S) entering the Arctic to the carbon burned in various latitude ranges. Clearly, these cannot be due to sources north of 70° since the amount of carbon burned in this region is less than the fluxes; arctic sources cannot account for arctic haze. Between 65° and 70°N the fluxes are less than the carbon burned, but form too large a fraction of it; the fuel would have to contain about 20% sulfur, and one third of it would have to be burned incompletely. For comparison, the emission factor for soot as a fraction of the mass of

Table 1. Graphitic carbon, total carbon, and sulfur fluxes compared with the amount of fuel burned at specified latitudes.

Latitude	Carbon burned (tons month⁻¹)	F_G as % of carbon burned	F_C as % of carbon burned	F_S as % of carbon burned
75–90°N	1.5×10^3	3,200	16,300	10,600
70–75°N	5×10^4	98	490	320
65–70°N	7.6×10^5	6.4	32	21
60–65°N	3.3×10^6	1.5	7.4	4.8
55–60°N	2.5×10^7	0.2	1	0.6
50–55°N	7.1×10^7	0.07	0.35	0.2
45–50°N	5.3×10^7	0.09	0.45	0.3
0–45°N	2.3×10^8	0.02	0.1	0.07
Global	4.4×10^8	0.01	0.05	0.04

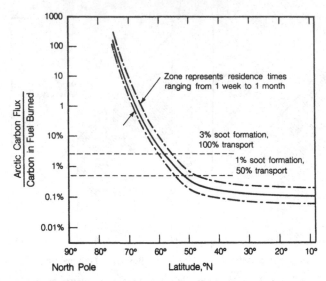

Fig 11. Arctic carbon flux estimated as a fraction of total carbonaceous fuel burned north of specified latitude, vs latitude. Solid curve: residence time of 2 weeks; dashed curves: residence times from 1 week to 1 month.

carbonaceous fuel burned ranges from zero for complete combustion to about 4% for very dirty burning of soft coal.

Significant source contributions in the 60° to 65°N latitude band are more convincing; however, even there it would require extremely dirty combustion of very high sulfur fuels, and almost all the emissions would have to be transported northward with minimal deposition along the way. Given uncertainties of the flux calculations, one cannot eliminate this latitude band as a significant source, but it does not appear to be a prime source region. Major contributors to the arctic haze in the 55–60°N latitude band become more reasonable, but even here it would require very effective transport of source emissions to the Arctic. As one goes below 55°N latitude, the fluxes become a very reasonable fraction of the fuel burned, which is consistent with the mid-latitude sources proposed on the basis of trace element analyses by Rahn and McCaffrey (1980).

Figure 11 presents these data by plotting the ratio of arctic carbon flux to total carbon burned as a function of increasing source region size, for various assumed residence times. The intersection of the curve and the horizontal line at a ratio of 0.03 indicates a latitude where arctic soot flux could be accounted for by 100% northward transport of soot, and a soot production rate that is 3% of all carbonaceous fuel burned north of that latitude. Such assumptions would imply source regions below 55° to 60°N with the spread on the curve representing the possible range of residence times. More reasonable conditions of 1% soot formation during combustion, and 50% transport efficiency to the Arctic, are shown by the lower horizontal line. These conditions may be over-optimistic in transport efficiency, yet clearly indicate that the source region for arctic haze must include areas south of 55°N latitude and probably below 50°N latitude.

In Figure 12, with a method similar to the interpretation of Figure 11, we have compared the arctic sulfur flux to the total carbonaceous fuel burned as a function of increasing latitude. The two lines represent emissions from fuel with 1.5% sulfur content with either 100% conversion to particulate sulfur and 100% northward transport, or a

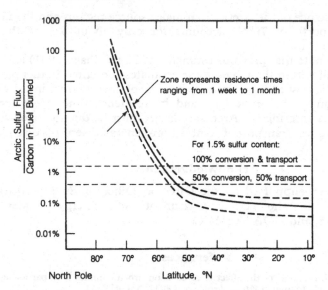

Fig 12. Arctic sulfur flux as a fraction of total cabonaceous fuel burned north of specified latitude vs latitude. Solid curve: residence time of 2 weeks; dashed curves: residence times from 1 week to 1 month.

Table 2. Arctic sulfur fluxes compared with sulfur emissions in specified regions.

Region	Sulfur emissions (tons month^{-1})	Arctic sulfur flux as % of emissions
Europe	2.5×10^6	6.4
Capitalist Europe	1.1×10^6	14.5
Socialist Europe	1.4×10^6	11.4
USSR	1.1×10^6	14.5
USA	1.3×10^6	12.3
Canada	2.5×10^5	64.0
Global	6.6×10^6	2.4

Sulfur emission data from Möller (1984).

more reasonable 50% conversion and 50% transport. For the more reasonable assumptions, the source region needs to extend southward to at least 50–55°N to account for the observed sulfur flux.

The flux of sulfur, F_S, entering the Arctic can be compared with estimated sulfur emissions in various source regions (Moller 1984). Based on the sulfur content of the fuels, these obviously represent a maximum possible aerosol sulfur input into the atmosphere because SO_2 may be scavenged before being converted to particulate sulfur. Table 2 shows this comparison with the arctic sulfur flux shown as a percentage of sulfur emissions from potential source regions. It is clear from the table that F_S is a substantial fraction of the sulfur emissions from each region individually (eg capitalist Europe, 14.5%; socialist Europe, 11.4%; USSR, 14.5%; Canada, 64%). These fractions are large, given the fact that the Arctic is only one of many possible receptor areas, and the likelihood of significant deposition of SO_2 and particulate sulfur along pathways to the Arctic. The results could indicate that arctic haze is due to the combined input of several major source regions. On a global scale, these calculations estimate that 2.5% of the northern hemisphere sulfur emissions enter the Arctic. This is a rather substantial fraction, given the distance of the Arctic from major source regions and the fact that the Arctic (ie the area north of 70°N) accounts for only about 6% of the northern hemisphere's surface area.

It is interesting to note that previous estimates of F_S by Shaw (1981) and Rahn and MacCaffrey (1979) fall within the range of the estimates presented in this paper. These conclusions can only be viewed as preliminary because of uncertainties in the vertical and horizontal distribution of the arctic haze and its residence time. However, they do indicate that the fluxes entering the Arctic are large, must be due to major combustion source regions, and suggest transport of combustion-generated aerosols on a global scale.

Acknowledgements

This work has been supported by the Director, Office of Energy Research, CO_2 Research Division of the US Department of Energy under contract DE-AC03–76SF00098, and NOAA GMCC.

References

ACKERMAN, T. P. 1977. A model of the effect of aerosols on urban climates with particular applications to the Los Angeles basin. *Journal of Atmospheric Science*, 34: 531–47.

ACKERMAN, T. P. AND BAKER, M. B. 1977. Shortwave radiative effects of unactivated aerosol particles in clouds. *Journal of Applied Meteorology*, 16: 63–9.

ACKERMAN, T. P. AND TOON, O. B. 1981. Absorption of visible radiation in an atmosphere containing mixtures of absorbing and nonabsorbing particles. *Applied Optics*, 21: 3661–68.

ACKERMAN, T. P. AND VALERO, F. P. J. 1984. The vertical structure of Arctic haze as determined from airborne net-flux radiometer. *Geophysical Research Letters*, 11: 469–72.

ATWATER, M. A. 1970. Planetary albedo changes due to aerosols. *Science*, 170: 64–6.

BARRIE, L. A. AND OTHERS. 1981. The influence of mid-latitudinal pollution sources on haze in the Canadian Arctic. *Atmospheric Environment*, 15: 1407– 19.

BODHAINE, B. A. AND OTHERS. 1981. Aerosol light scattering and condensation nuclei measurements at Barrow, Alaska. *Atmospheric Environment*, 15: 1375– 90.

CESS, R. D. 1983. Arctic aerosols: model estimates of interactive influences upon the surface-troposphere radiation budget. *Atmospheric Environment*, 17: 2555–64.

CHARLOCK, T. P. AND SELLERS, W. D. 1980. Aerosol effects on climate: calculations with time-dependent and steady-state radiative convective models. *Journal of Atmospheric Science*, 37: 1327–41.

CHYLEK, P. AND COAKLEY, J. A. 1974. Aerosols and climate. *Science*, 183: 75– 7.

CHYLEK, P. AND OTHERS. 1984. Albedo of snow. *Journal of Geophysical Research*.

COAKLEY, J. A. AND CESS, R. D. 198X. Response of the NCAR community model to the radiative forcing by the naturally occurring tropospheric aerosol. *Proceedings of the 5th AMS Conference on Atmospheric Radiation*, October-November 1983, Baltimore, Maryland.

COAKLEY, J. A. AND OTHERS. 1983. The effect of tropospheric aerosols in the earth's budget: a parameterization of climate models. *Journal of Atmospheric Science*, 116: 137.

GERBER, H.A. 1982. Absorption of light by atmospheric aerosol particles: review of instrumentation and measurements. In GERBER, H. E. AND HINDMAN, E. E. (editors) *Light absorption by aerosol particles*. Hampton, Virginia, Spectrum Press: 21–53.

GRENFELL, T. C. AND OTHERS. 1981. Spectral albedos of an alpine snowpack. *Cold Regions Science and Technology*, 4: 121–27.

GUNDEL, L. A. AND OTHERS. 1984. The relationship between optical attenuation and black carbon concentration for ambient and source particles. *Science of the Total Environment*, 36: 197–202.

HANSEN, A. D. A. AND ROSEN, H. 1984. Vertical distribution of particulate carbon, sulfur, and bromine in the Arctic haze and comparison with ground- level measurements at Barrow, Alaska. *Geophysical Research Letters*, 11: 381–84.

HANSEN, A. D. A. AND OTHERS. 1982. Real-time measurement of the absorption coefficient of aerosol particles. *Applied Optics*, 21: 3060–62.

HARRIS, J. M. 1984. Trajectories during AGASP. *Geophysical Research Letters*, 11, 453–56.

HEINTZENBERG, J. 1982. Size-segregated measurements of particulate elemental carbon and aerosol light absorption at remote Arctic locations. *Atmospheric Environment*, 16: 2461–68.

IVERSEN, T. 1984. On the atmospheric transport of pollution to the Arctic. *Geophysical Research Letters*, 11: 457–60.

JOSEPH, J. H. 1977. The effect of desert aerosol on a model on the general circulation. In BOLLE, H. J. (editor). *Proceedings of the International Radiation Symposium, Garmisch Partenkirchen*. Science Press: 489–92.

MARLAND, G. AND OTHERS. 1984. CO_2 from fossil fuel burning: global distributions of emissions. *Tellus*,

MITCHELL, J. M. Jr. 1957. Visual range in the polar regions with particular reference to the Alaskan Arctic. *Journal of Atmospheric and Terrestrial Physics*, special supplement: 195–211.

MITCHELL, J. M. Jr. 1971. The effect of atmospheric aerosols on climate with special reference to temperature near the Earth's surface. *Journal of Applied Meteorology*, 10: 703–14.

MÖLLER, D. 1984. Estimation of teh global man-made sulphur emission. *Atmospheric Environment*, 18: 19–27.

NOVAKOV, T. 1981. Microchemical characterization of aerosols. In MALISSA, H. AND OTHERS (editors). *Nature, aim and methods of microchemistry*. Vienna, Springer-Verlag: 141–65.

OTTAR, B. 1981. The transfer of airborne pollutants to the Arctic region. Atmospheric Environment, 15: 1439–45.

PATTERSON, E. M. AND OTHERS. 1982. Radiative properties of Arctic aerosol. *Atmospheric Environment*, 12: 2967–77.

POLLACK, J. B. AND CUZZI, J. N. 1979. Scattering by nonspherical particles of size comparable to a wavelength: a new semi-empirical theory and its application to tropospheric aerosols. *Journal of Atmospheric Science*, 37: 868–81.

PORCH, W. M. AND MacCRACKEN, M. C. 1982. Parametric study of the effects of Arctic soot on solar radiation. *Atmospheric Environment*, 16: 1365–71.

RAATZ, W. E. AND SHAW, G. E. 198?. Long range tropospheric transport of pollution into the Alaskan Arctic. *Journal of Climate and Applied Meteorology*,

RAATZ, W. E. AND SCHNELL, R. C. 1984. Aerosol distributions and an Arctic aerosol front During AGASP: Norwegian Arctic. *Geophysical Research Letters*, 11: 373–76.

RAHN, K. A. AND McCAFFREY, R. J. 1979. Long range transport of pollution aersol to the Arctic: a problem without borders. *WMO Symposium on Long Range Transport of Pollutants*. Geneva, WMO. (WMO publication 538.)

RAHN, K. A. AND McCAFFREY, R. J. 1980. On the origin and transport of the winter Arctic aerosol. *Annals of the New York Academy of Science*, 338: 486–503.

RANDALL, D. T. AND OTHERS. 1985. The sensitivity of a general circulation model to Saharan dust heating. In DEEPAK, A. AND GERBER, H. E. (editors). *Aerosols and their climatic effects*. Hampton, Virginia, Deepak Publishing.

RASOOL, I. S. AND SCHNEIDER, S. H. 1971. Atmospheric carbon dioxide and aerosols: effects of large increases on global climate. *Science*, 173: 138– 41.

RECK, R. A. 1976. Thermal and radiative effects of atmospheric aerosol in the northern hemisphere calculated suing a radiative convective model. *Atmospheric Environment*, 10: 611–17.

ROESSLER, D. M. AND FAXVOG, F. R. 1979. Optoacoustic measurement of optical absorption in acetylene smoke. *Journal of the Optical Society of America*, 69: 1699–1704.

ROSEN, H. AND HANSEN, A. D. A. 1984. Role of combusion-generated carbon particles in the absorption of solar radiation in the Arctic haze. *Geophsyical Research Letters*, 11: 461–64.

ROSEN, H. AND NOVAKOV, T. 1983. Combustionpgenerated carbon particles in the Arctic atmosphere. *Nature*, 306: 768–70.

ROSEN, H. AND OTHERS. 1978. Identification of the optically absorbing component of urban aerosol particles. *Applied Optics*, 17: 3859–61.

ROSEN, H. AND OTHERS. 1980. Soot in urban atmospheres: determination by an optical absorption technique. *Science*, 208: 741–43.

ROSEN, H. AND OTHERS. 1981. Soot in the Arctic. *Atmospheric Environment*, 15: 1371–74.

SCHNELL, R. C. (editor). 1984. Special issue on Arctic Haze. *Geophysical Research Letters*, 11: 359–472.

SHAW, G. E. 1981. Eddy diffusion transport of Arctic pollution from the mid- latitudes: a preliminary model. *Atmospheric Environment*, 15: 1483–89.

SHAW, G. E. AND STAMNES, K. 1980. Arctic haze: perturbation of the polar radiation budget. *Annals of the New York Academy of Science*, 338: 533–39.

SHAW, G. E. AND WENDLER, G. 1972. Atmospheric turbidity measruements at McCall Glacier in northeast Alaska. *Proceedings, Conference on Atmospheric Radiation*. Boston, American Meteorological Society: 1981–87.

SZHARLÁT, A. C. AND JAPAR, S. M. 1981. Light absorption by airborne aerosols: comparison of integrating plate and spectrophone techniques. *Applied Optics*, 20: 1151–55.

TANRE, D. AND OTHERS. 1984. First results of the introduction of an advanced aerosol radiation interaction in the ECMWF low resolution global model. In DEEPAK, A. AND GERBER, H. E. (editors) *Arosols and their climatic effects*. Hampton, Virginia, A. Deepak Publishing:

TWOMEY, S. 1977. The influence of pollution on the shortwave albedo of clouds. *Journal of Atmospheric Science*, 34: 1149–52.

VALERO, F. P. J. AND OTHERS. 1982. Radiative flux measurements in the troposphere. *Applied Optics*, 21: 831–38.

VALERO, F. P. J. AND OTHERS. 1984. The absorption of solar radiation by the Arctic atmosphere during the haze season and its effects on the radiation balance. *Geophsycial research Letters*, 11: 465–68.

WANG, W. C. AND DOMOTO, G. A. 1974. The radiative effect of aersols in the Earth's atmosphere. *Journal of Applied Meteorology*, 13: 521–34.

WARREN, S. G. 1982. Optical properties of snow. *Review of Geophysics and Space Physics*, 20: 67–89.

WARREN, S. G. AND WISCOMBE, W. J. 1980. A model for the spectral albedo of snow. II: Snow containing atmospheric aerosols. *Journal of Atmospheric Science*, 34: 2734–45.
WOLFF, G. T. AND KLIMISCH, R. L. (editors). 1982. *Particulate carbon: atmospheric life cycle*. New York, Plenum Press.
YAMAMOTO, G. AND TANAKA, M. 1972. Increase of global albedo due to air pollution. *Journal of Atmospheric Science*, 29: 1405–12.

ARCTIC HAZE AND THE RADIATION BALANCE

FRANCISCO P. J. VALERO AND THOMAS P. ACKERMAN

ABSTRACT. Airborne measurements of the absorption rate of solar radiation by arctic haze indicate atmospheric heating rates of 0.15 to 0.25 Kday^{-1} (24 hr weighted averages) at latitudes between 72.6° and 74°N in early spring (15 March—4 April 1983). Haze interacts with solar radiation to alter the radiative balance of the atmosphere- surface system, generally resulting in more solar energy being absorbed by the atmosphere and less by ground, ice or water. Planetary albedo is also affected, increasing for haze over water and decreasing for haze over ice. Haze interaction with infrared (planetary) radiation has not been measured, though the infrared component represents a most important contribution to overall energy balance. Calculations provide some information, but experimental evidence is needed both to validate the calculations and because of the enhanced concentration of greenhouse gases measured in the haze layers. Cumulative deposition of black carbon over the surface produces a change in optical properties of the ice, which may result in increased surface temperatures and accelerating ice melt. To evaluate its consequences, experimental evidence of the magnitude of this effect is needed. An extended monitoring program is suggested. Climatic effects of changes in energy budget depend on spatial and temporal properties of the haze layers, which need to be examined more thoroughly.

Contents

Introduction

One would expect polar regions to be clean, pristine environments. This is largely true of Antarctica; but the northern polar atmosphere, though clean and clear in summer, becomes heavily polluted with anthropogenic particles (aerosols) and gases in winter. Mitchell (1957) noted that the arctic aerosol was composed of non-ice particles of sizes less than 2 μm, and that a significant loss of visibility occurs during periods of heavy pollution (haze events). About 15 years later Shaw (1975), working near Barrow, Alaska, confirmed that the atmosphere became very turbid during haze events, though the origin of the haze was then uncertain. Recent arctic aerosol studies (Rahn and McCaffrey 1980; Shaw and Stamnes 1980; Rahn and others 1977; Barrie and others 1981; Heintzenberg and others 1982; *Geophysical Research Letters* special issue, May 1984) reported the presence of large concentrations of aerosols in the arctic atmosphere and discussed its possible impact on the earth's radiative balance and climate. Characteristically the haze increases remarkably in density from November to April, with a maximum between late winter and early spring. This behavior is governed by air-flow patterns, and lack of precipitation to wash the atmosphere in winter.

Rahn and McCaffrey (1980) showed that the pollution which permeates the Arctic in winter originates in industrialized regions of the Soviet Union, Europe and the United Kingdom, with minor contributions from North America. Haze events are accompanied by higher concentrations of combustion gases, probably from power and industrial plants (Khalil and Rasmussen 1984).

Interactions between solar and infrared radiation and the polluted atmosphere-surface system are climatically important. Arctic aerosols interact with solar radiation mostly in late winter and early spring, when aerosol concentration is at or near maximum and the sun is high enough for significant solar radiation to reach the Arctic. Radiation is scattered and absorbed in the atmosphere. Shaw (1982) and Bodhaine and others (1984) determined high optical depths, and Valero and others (1983, 1984) measured in situ absorption of radiation by the haze. Rosen and others (1981) reported that graphitic carbon (soot) is the component responsible for most of the solar energy absorption (Rosen and Novakov 1981; Rosen and Hansen 1984). Ice on the earth's surface reflects about 80% of incident solar energy, water only about 5–15%; much of the radiation impinging on ice in consequence passes twice through the haze layers, resulting in additional absorption. As more energy is retained by the soot, atmospheric warming is enhanced.

Increased atmospheric scattering and absorption result in less solar energy reaching the ground, and the changed energy absorption profile in the atmosphere (a function of vertical structure in the haze) further modifies the energetics of the atmosphere-surface system. Clearly this redistribution of radiation between atmosphere and surface must have consequences for arctic atmospheric dynamics and climate. Deposition of carbon over the ice-snow surface may be a further complication, resulting in more absorption by the surface, particularly in spring and summer when the haze season is over and sunlight reaches the ground with maximum intensity. This may lead to a self-accelerating rate of ice melt. As surface layers melt, soot deposited in previous seasons further darkens the surface and absorbs more radiation; cumulatively over several years this could shift the ice cap boundaries northward, reducing the extent of the insulating ice cover and opening up the ocean as an additional heating source. Little work has been done on the surface effects of the arctic haze, but the potential for significant changes is clearly present.

Finally, no measurements or calculations of infrared radiation have been made. The infrared contribution to the radiation balance is generally considered minor. However, higher concentration of pollution-related gases (nitrogen oxides, hydrocarbons, etc) with strong spectral bands in the 10–15 m region (at arctic atmospheric temperatures, close to the blackbody emission peak), may further influence the arctic energy budget. Furthermore, atmospheric infrared radiative effects of the haze take place through the winter months and the surface effects persist all year long. The time-scale of the infrared processes is such that their overall influence may not be trivial when compared to solar radiation.

In this paper we focus on two aspects of arctic haze—absorption of solar radiation, and its vertical structure as determined from net flux measurements. We deduce changes in distribution of solar energy in the atmosphere-surface system due to haze, and determine haze-induced changes in atmospheric heating rates as a function of altitude. New radiation measurements are also suggested to confirm and improve our present understanding of arctic energetics, and we emphasize the need for a program to monitor possible changes in ice cap albedo due to aerosol deposition.

Experimental concepts and measurements

Effects of arctic haze on the earth's energy budget can be studied by collecting samples of haze and measuring their optical properties in the laboratory. Alternatively, absorption can be measured in situ by comparing absorption values during haze events and in the haze-free atmosphere. In situ absorption by the aerosol was measured by monitoring radiative flux densities at different altitudes, and determining the energy absorbed in the atmospheric layers between flight levels. This was done by measuring upwelling (u) and downwelling (d) radiative flux densities at various altitudes (h) (Figure 1). The difference between d(h) and u(h) gave net flux density at altitude h; the difference between the net fluxes at two altitudes (A in Figure 1) was the radiative flux density absorbed by the layer between measuring altitudes.

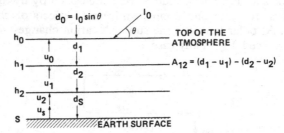

Fig 1. Method used to measure in situ the power absorbed by atmospheric layers.

Atmospheric gases including water vapor absorb light only at particular wavelengths (spectral features); haze on the other hand acts as a grey body, absorbing energy at practically all wavelengths. If the radiative flux densities in the atmosphere for a broad spectral bandpass (including all wavelengths) were measured, the result would be the radiative flux density absorbed by all components of the atmosphere, including gases and aerosols. During the first Arctic Gas and Aerosol Sampling Program (AGASP I), we wished to measure the absorption of solar energy by the haze itself within the atmosphere, independent of the contributions made by atmospheric gases to the total absorption. This can be done by the selection of wavelengths which atmospheric gases do not absorb; any absorption measured in these narrow wavelength intervals can be attributed exclusively to absorption by the haze. With data analysis and modelling, the absorption of radiation by the aerosol at all wavelengths can be determined from the measurements in narrow spectral intervals. Once one knows how much flux density is absorbed in the atmosphere, it is a simple matter to determine the resulting heat profiles in the atmosphere.

A way to verify the results from this approach is to measure simultaneously the radiation absorbed at all solar wavelengths. If the portion attributable to gases is subtracted from the total energy absorbed by the atmosphere, what is left is the absorption by the haze. The spectral parameters that describe the absorption by gases and water vapor are known from laboratory spectroscopy. With this information, the portion of the flux density absorbed by the haze can be calculated. A comparison of the two results (broadband and narrow spectral bandpasses approaches) offers a way to check the consistency of the results.

For the radiation measurements, we designed and built a special radiometric system that was installed in the NOAA WP-3D Orion research aircraft used during AGASP I. The radiometric system consists of three narrow, one intermediate, and one broad spectral bandpass channels. All channels have a hemispherical field of view. These hemispherically integrating radiometers were mounted in a rotor installed in the right wing tip of the aircraft. The rotating arrangement exposed each radiometer alternately to upwelling and downwelling radiative fluxes. Differencing the upward- and downward-looking measurements, taken with the same radiometer, minimizes systematic errors and provides a direct measurement of the net flux density crossing a plane parallel to the flightpath of the aircraft. The difference between net fluxes at different altitudes gives the absorption by the atmospheric column included between net flux measurement altitudes.

The wavelengths selected for the three narrow (0.01 μm) spectral bandpass channels were centered at 0.485, 0.680, and 0.860 μm; the one intermediate (0.08 μm) bandpass channel was centered at 0.940 μm (to monitor the concentration of water vapor), and the broad spectal bandpass channel covered the region from 0.26 to 2.6 μm.

Radiative net fluxes at different altitudes were measured by flying profiles similar to that illustrated in Figure 2, around noon to minimize effects of changes in the sun's elevation with time. At polar latitudes the sun's elevation changes very slowly, so these corrections did not exceed 2% in any case.

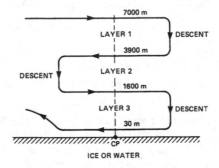

Fig 2. Flight profiles to measure net radiative fluxes.

Results

Narrow spectral bandpass radiometers

Figure 3 shows the narrowband flux density absorption measured during AGASP Flights 3, 4 and 8; for details of altitudes and layer thicknesses see Table 1. Table 2 shows the measured narrow spectral bandpass values that correspond to the radiative flux density absorbed by each atmospheric layer during each of the flights. Here we report only measurements obtained under cloud-free conditions. For flights 3 and 4, over the western Arctic on 15 and 17 March 1983, measurements were centered in 72.6°N, 156.9°W. Albedo (reflectivity) of the ice surface below was 0.75. For flight 8, over the eastern Arctic on 31 March 1983, measurements were centered in 74.0°N, 25.0°E. Some broken clouds were encountered during flight 8, but those portions of the flight are omitted from this analysis. The albedo of the open ocean was 0.13, a high value assumed to result from specular reflection at the large solar zenith angles.

To compute heating rates due to absorption of solar radiation we extended the narrowband spectral measurements to all solar wavelengths using the following procedure:

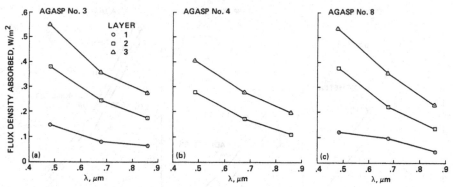

Fig 3. Radiative power absorbed in 0.01 μm spectral bandpasses as a function of wavelength. Layers are defined in Table 1.

Table 1. Flux measurement altitudes. h = altitude at which flux measurements were made: Δh = thickness of atmosphere layer: h_{mean} = mean altitude.

h km	Δh km	h_{mean} km	Layer number
7.0			
	3.1	5.45	1
3.9			
	2.3	2.75	2
1.6			
	1.57	0.81	3
0.03			

Table 2. Absorption of radiative flux densities. Altitude and thickness of layers are given in Table 1; spectral bandpass is 0.01 μm.

AGASP flight	Layer number	Flux density absorbed (wm^{-2}) at noon		
		Channel A (0.485 μm)	Channel B (0.680 μm)	Channel C (0.860 μm)
3	1	0.15±0.02	0.08±0.01	0.07±0.01
	2	0.38±0.03	0.25±0.02	0.18±0.03
	3	0.55±0.05	0.36±0.04	0.28±0.03
4	1	—	—	—
	2	0.27±0.03	0.17±0.01	0.11±0.01
	3	0.41±0.03	0.28±0.03	0.20±0.01
8	1	0.14±0.01	0.10±0.01	0.05±0.01
	2	0.38±0.03	0.23±0.01	0.14±0.01
	3	0.54±0.03	0.36±0.03	0.23±0.02

an approximate aerosol model, consistent with ground-based data on Arctic haze (Heintzenberg 1980; Heinzenberg and others 1981; Rosen and others 1981; Bodhaine and others 1981) was constructed. The aerosol, assumed to be an external mixture of sulfate and carbon, had a single-scattering albedo of 0.8 at 0.5 μm, was incorporated in a multilayer version of the delta-four-stream model described by Cuzzi and others (1982).

For simplicity, in the case of the narrow bandpass channels, we modelled the haze as a single layer, uniformly mixed with height. From particle counter data the haze layer was defined to be 3 km thick for flight 4, and 5 km thick for flights 3 and 8. In addition, a non-absorbing stratospheric aerosol layer of optical depth 0.1 (Dutton and others 1984)

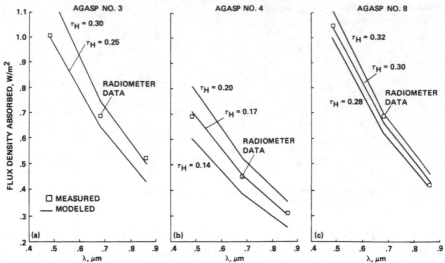

Fig 4. Measured and calculated total power absorbed by the atmosphere in 0.01 μm intervals. T_H is the extinction optical depth at 500 nm. Symbols are measured values, solid lines represent model results.

was placed at 10 km. Calculations were made (Figure 4) for various values of the aerosol optical depth for the three wavelengths, corresponding to the narrowband radiometer channels for each of the three flights. Surface albedo in each case was determined from the radiometer measurements at the lowest flight level.

There is good agreement in each case between measured and calculated absorption. For flight 3 ground optical depth values measured at the Barrow monitoring facility agreed well with the values determined here, though exact comparison is difficult because of stratospheric aerosol and spatial variations in the haze. It should be noted that calculation of absorbed energy is sensitive to the amount of black carbon included in the aerosol model. The carbon is relatively insensitive to the non-absorbing aerosol component and to the assumed aerosol microphysics. Consequently, the agreement exhibited in Figure 4 indicates only a correct choice of the amount of black carbon.

Using aerosol properties determined by fitting the measurements, the model was run for 26 spectral intervals covering the range of 0.25 to 4.3 m. Profiles of gaseous absorbers were taken from the sub-Arctic winter profile of McClatchey and others (1972). Instantaneous heating rates computed for the three flights are given in Table 3. For comparison we have also computed heating rate in the absence of haze aerosols, and the ratio of the two rates. To understand the role of the haze better, the results of a calculation

Table 3. Daily averaged heating rates as determined from narrowband measurements. [a]Haze optical depth at 500 nm; [b]If over ice. See text.

AGASP flight	Best Fit (τ_H)[a]	Heating rates K/day		Heat rate ratio, haze/no-haze
		Haze	No haze	
3	0.26	0.17	0.08	2.1
4	.17	.19	.07	2.7
8	.31	.21	.08	2.6
8[b]	.31	.27	.10	2.7

for flight 8, over a high-albedo surface ($A_s = 0.75$), are also included, representing the effects of a thick haze layer over the ice-covered Arctic basin.

Broad spectral bandpass radiometer

Broad spectral bandpass measurements were used to verify the narrow spectral bandpass results, and determine radiative transfer changes introduced by the arctic haze relative to an 'average' standard arctic atmosphere. For example, if water vapor concentrations are in some way correlated with haze concentrations or haze absorption properties (as may or may not be the case), our concern is to investigate in situ the combined effects of all atmospheric components on absorption during haze events, and their relationship to standard conditions.

Table 4. Total radiative flux density absorption and heating rates as determined from broad band measurements. [a]Layer 1 = 7 to 3.9 km; 2 = 3.9 to 1.6 km; 3 = 1.6 and 0.03 km. [b]Energy absorption is instantaneous at noon.

AGASP flight no	Measured broadband surface albedo	Layer[a]	Measured[b] flux density absorption, wm^{-2}	Measured haze event heating rate $Kday^{-1}$	Calculated haze-free heating rates	
					Standard atmosphere $Kday^{-1}$	Actual water vapour atmosphere, $Kday^{-1}$
3	0.78	1	20±2	0.16±0.02	0.10	0.09
		2	28±2	0.19±0.01	.08	.08
		3	29±2	0.25±0.02	.06	.06
8	.13	1	14±2	0.11±0.01	.09	.06
		2	36±3	0.24±0.02	.11	.14
		3	29±3	0.25±0.16	.06	.06

Table 4 shows the broadband measurements for flights 3 and 8. Values of absorbed power per unit area in Column 4 were used to calculate heating rates for each atmospheric layer. This was done using the conventional expression, in pressure coordinates, for the heating rate for an atmospheric layer thickness ΔP (Liou 1980):

$$\frac{\partial T}{\partial P} = \frac{g}{C_p} \frac{\Delta F(\Delta P)}{\Delta P}$$

where T is temperature (°K), t is time, g is the gravitational acceleration, C_p is the specific heat at constant pressure, ΔF is the radiative energy absorbed, and ΔP is the pressure differential between the top and bottom of the atmospheric layer under consideration. Heating rates calculated from Equation (1) are shown in Column 5 of Table 4. Figure 5 depicts the radiative power absorbed by the atmosphere as a function of altitude, and corresponding heating rates.

To compare the haze and no-haze heating rates we calculated heating rates for a representative arctic winter atmosphere in the absence of haze; results for individual atmospheric layers appear in Table 4, column 6. Water vapor profiles obtained from standard aircraft instruments (Schnell and Raatz 1984; Raatz and Schnell 1984) and data from the 940–nm spectral channel, dedicated to monitoring water vapor absorption (Valero and others 1983), were used to compute the values in Table 4, column 7, which show the heating rates that correspond to a haze-free atmosphere with water vapor concentrations determined as previously described. Figure 6 presents the flux density absorbed by the haze and the corresponding haze heating rates for each atmospheric layer.

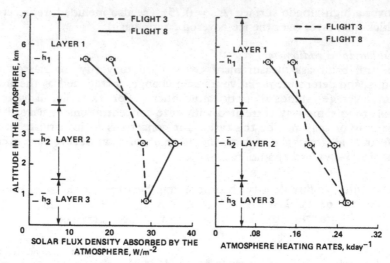

Fig 5. (a) Broadband (260–2600 nm) solar radiation absorbed by the polluted arctic atmosphere. (b) Averaged daily heating rates as a function of altitude.

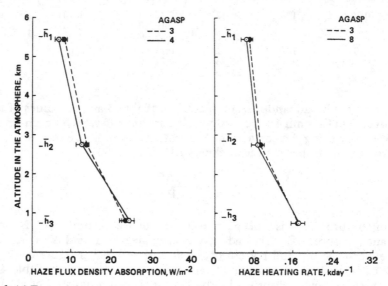

Fig 6. (a) Energy absorbed by the haze in the 260–2600 nm interval. (b) Corresponding heating rates.

Table 5 shows the broadband flux density absorbed in each layer and corresponding daily averaged heating rates. Table 6 presents a comparison of heating rates as determined from broad and narrow spectral bandpasses.

Vertical structure of arctic haze

By subtracting net fluxes at various altitudes and comparing the differences with values from model calculatios, we can deduce the aerosol optical depth for each of the three layers

Table 5. Haze flux density absoprtion and heating rates. [a,b]See Table 4.

AGASP flight no	Layer[a]	Haze heating rates, Kday-1	Haze[b] flux density absorption, wm-2
3	1	0.06±0.01	8.0±0.8
	2	0.11±0.05	13.8±1.0
	3	0.19±0.01	23.5±1.6
8	1	0.05±0.01	6.6±1.0
	2	0.10±0.01	12.7±1.0
	3	0.19±0.02	24.6±2.5

Table 6. Comparison of altitude averaged heating rates.

AGASP flight no	Average heating rate, Kday^{-1}	
	Broadband	Narrowband
3	0.20	0.17
8	.20	.21

defined by the flight altitudes. These values are also tested for consistency by comparing narrowband with broadband measurements. In addition, using the optical depth values and the carbon column values of Rosen and Hansen (1984), we can infer the in situ specific absorption of carbon.

The model described earlier was used to infer the aerosol optical depths for flights 3 and 8 by computing the absorbed power per layer in three narrowband channels. The calculations were repeated for several extinction optical depth profiles until good agreement with the measurements was obtained. Since the aerosol optical properties are fixed in the model, this is essentially equivalent to varying the absorption optical depth, τ_a. Multiple-scattering is included but is a second-order effect owing to the relatively modest aerosol optical depths being considered.

Figure 7 compares computed energy absorption in the three narrowband channels with measured values for flights 3 and 8. For flight 3, inferred total haze optical depth at 0.55 μm is 0.27. Of this, 0.03 is in layer 1 (3.9–7.0 km), 0.09 is in layer 2 (1.6–3.9 km), and 0.15 in layer 3 (0–1.6 km). For flight 8 the total optical depth is 0.31, with 0.03, 0.09, and 0.19 in layers 1, 2 and 3 respectively. The good agreement between data and model results indicates that choice of τ_a is correct, since the measurements constrain the absorption. The total optical depth is then constrained by the single-scattering albedo which we have fixed. Since this albedo may in fact lie roughly between 0.70 and 0.90, the total optical depth values have an uncertainty of 10–15%. As expected, the wavelength dependence of the absorption is consistent with a broadband absorbed such as black carbon. Thus, we can conclude from this comparison that absorption optical depth τ_a for the three aerosol layers was 0.007, 0.020, and 0.033 on flight 3, and 0.007, 0.020 and 0.041 for flight 8.

Computed and measured broadband (0.26 to 2.6 μm) energy absorption as a function of layer for both flights is presented in Table 7. The calculations were carried out using the same aerosol model and the optical depth values given above, which were determined from the narrowband measurements. Agreement between measurements and calculations is remarkably good, both for the total absorption and for absorption in the individual

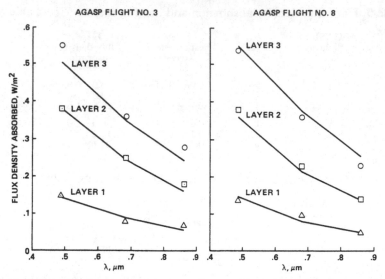

Fig 7. Power absorbed in bandpasses in each atmospheric layer. Symbols are measured values, solid lines model results.

Table 7. Broadband flux density absorption, W/m².

Layer number	Flight no 3		Flight no 4	
	Data	Model	Data	Model
1	20	21	14	15
2	28	28	36	37
3	29	30	29	32
Total	77	79	79	84

layers. This is further proof that aerosol absorption optical depths deduced from the narrowband data are correct, and that the wavelength dependence of the aerosol absorption is indeed broadband.

From these we conclude that during both flights 3 and 8 most of the absorbing aerosol was located between the surface and 1.6 km. For flight 8, 10% of the aerosol absorption optical depth is in layer 1, 29% in layer 2, and 61% in layer 3, the lowest layer. We have assumed that this absorbing aerosol is predominantly black carbon. Rosen and Hansen (1984) reported measurements of graphitic (or black) carbon concentration as a function of altitude for flight 8, covering the same altitude range (0 to 7 km) as our net-flux measurements and with higher vertical resolution. Integrating the vertical profile from Figure 3 of their paper we find that 14% of their total column concentration is in layer 1, 35% in layer 2 and 51% in layer 3. Given the considerable difference between the two experiments in terms of quantities being measured, temporal and spatial resolution and experimental techniques, as well as the uncertainties involved in both sets of measurements and in the model results, the agreement between the percentages is quite good.

Results given by Rosen and Hansen (1984) and those presented here can be used to infer a specific absorption for black carbon in the arctic atmosphere. The total column amount of carbon measured by Rosen and Hansen is 0.0028 gm^{-2}. From the model we infer a value of $\tau_a = 0.068$. Specific absorption is found by dividing τ_a by the total column

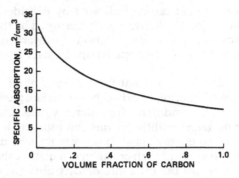

Fig. 8. Specific absorption per unit volume of carbon as a function of volume fraction of carbon.

amount, giving a value of 24 m² g⁻¹. This is somewhat surprising since the generally accepted value for black carbon is in the order of 8–10m² g⁻¹ (Roessler and Faxfog 1980; Szkarlat and Japar 1981). There are several possible explanations for this large specific absorption. One is that the model has grossly overestimated the value of τ_a because of an incorrect aerosol model. As we have already noted, this is unlikely. The measurements effectively constrain τ_a rather than the extinction optical depth. Thus a change in the aerosol model will have little effect on τ_a. Alternatively, the factor used by Rosen and Hansen (1984) to convert their optical attenuation data to black carbon concentration may be incorrect. In view of their laboratory studies (Gundel and others 1984), this also appears unlikely.

As a third possibility, Ackerman and Toon (1981) have shown that it is possible to enhance the absorption per unit volume (or mass) of an aerosol particle by surrounding it with a nonabsorbing shell. To illustrate this effect we made a series of calculations for particles composed of carbon cores and sufate shells, using the particle size distribution model assumed in this model for sulfate particles. Figure 8 shows the specific absorption per unit volume of carbon as a function of the column fraction of carbon in the particles. To find specific absorption per unit mass, one divides by the density of the carbon in units of g m⁻³ if the specific absorption of pure carbon spheres is to agree with the measured value of 8–10 m² g⁻¹. Such a density is reasonable if the carbon volume is an aggregate of small spheres rather than a single, solid sphere, as is often the case (Janzen 1980; Borghesi and others 1983). This density value converts the ordinate of Figure 7 to units of m² g⁻¹. We can then infer from the figure that our value of τ_a is consistent with the black carbon amount of Rosen and Hansen (1981) if the carbon is present as an internal mixture, as used in the model calculations above, and smaller that that deduced from the surface measurements of Rosen and others (1981) and Bodhaine and others (1981). The use of the internal mixture in the model leads to a reduction of aerosol extinction optical depth from 0.31 to 0.18, but causes essentially no change in the value of τ_a. The correct choice of aerosol model cannot be resolved at this time, but further analysis of the data collected on the AGASP flights may lead to new insights.

Discussion and conclusions

Changes in heating rate of the magnitude found could modify local climate significantly, and possibly influence climate at mid-latitudes. Currently we cannot assess completely the actual climatic effects of arctic radiation-haze interactions. However, their consequences on the energy budget can be indicated by considering the modifications introduced by haze, both in the total radiative energy absorbed by the atmosphere-surface system and in the resulting changes in distribution of absorbed energy between the atmosphere and the surface. This can be done using the data from two of the AGASP flights.

Flight 3 was over ice of albedo 0.78, flight 8 over ocean of albedo 0.13. Using these measured values, the measured energy absorbed by the atmosphere, and the calculated energy absorption for no-haze conditions (the same values used to determine the calculted heating rates for no-haze conditions), one can estimate how the energy budget and the distribution of absorbed energy between the atmosphere and the surface change with the haze and with the two kinds of surface. Such calculation shows that the presence of haze over ice (flight 3) *increased* the total flux density absorption in the atmospheric layer between 7 and 0.03 km from approximately 36 to 77 wm^{-2}, *decreased* the flux density absorbed by the ice surface itself from 54 to 43 wm^{-2}, *increased* the total absorption of radiative solar flux density by the atmosphere-surface system (7 to 0.03 km altitude) from approximately 90 to 122 wm^{-2}, and *reduced* the effective planetary albedo by about 9% (from 63 to 54%) at the central location.

Haze over water (flight 8) *increased* the total flux density absorption in the atmospheric layer between 7 and 0.03 km from about 40 to 79 wm^{-2}; *decreased* the flux density absorbed by the water surface from about 257 to 202 wm^{-2}; *decreased* the total absorption of solar radiation by the atmosphere surface system (7 to 0.03 km altitude) from 297 to 285 wm^{-2}; and *increased* the effective planetary albedo by about 2.5% (from 21.8 to 24.3%) at the central location.

Heating rates in these results are comparable to those computed in earlier purely theoretical studies (eg Porch and MacCraken 1982). However, our calculated values for absorbed energy are greater, primarily because of the larger optical depths found during AGASP. It is also important to note that the vertical extent of the haze layers during AGASP I was considerably greater than had been anticipated by those studies. The exact effect of these perturbations in the energy budget is difficult to assess using only a radiative model, because of the complicated interaction with ice physics and atmospheric dynamics. Furthermore, we neglected one of the most complicated aspects of the problem, the interaction of the haze with clouds; studies described here were all for cloud-free conditions. Calculations indicate that the heating may be reduced if clouds overlie the haze. However, if the haze is mixed with the clouds, there is a potential enhancement of the absorption within the clouds.

A new series of radiation measurements over the Arctic is necessary. Airborne sun-photometry, infrared radiometry, and a total:diffuse ratio radiometry should be included as well as particle size measurements, size-segregated samples for black carbon analysis, and filter micrographs to see what particles look like. The information from these filter micrographs will complement the flux density measurements, simplify the necessary corrections, and remove uncertainties from the model calculation. Also, a detailed and extended program to measure surface albedo and its seasonal and annual changes should be contemplated since temporal changes in albedo, owing to cumulative deposition of black carbon over the ice, may introduce significant radiative effects. The cumulative deposition of arctic haze components over the polar ice cap with its potential to alter the optical properties of the surface suggests the need for a detailed and extended program to monitor the deposition of pollutants over the surface.

References

ACKERMAN, T. P. AND TOON, O. B. 1981. Absorption of visible radiation in atmospheres containing mixtures of absorbing and nonabsorbing particles. *Applied Optics*, 20(20): 3661–68.

ACKERMAN, T. P. AND VALERO, F. P. J. 1984. The vertical structure of arctic haze as determined from airborne net-flux radiometer measurements. *Geophysical Research Letters*, 11(5): 469–72.

BARRIE, L. A. 1981. The influence of mid-latitudinal pollution sources on haze in the Canadian Arctic. *Atmospheric Environment*, 15(8): 1407–19.

BODHAINE, B. A. AND OTHERS. 1981. Aerosol light scattering and condensation nuclei measurements at Barrow, Alaska. *Atmospheric Environment*, 15(8): 1375–89.

BODHAINE, B. A. AND OTHERS. 1984. Surface aerosol measurements at Barrow during AGASP. *Geophysical Research Letters*, 11(5): 377–80.

BORGHESI, A. AND OTHERS. 1983. The absorption efficiency of submicron amorphous carbon particles between 2.5 and 40 μm. *Infrared Physics*, 23(2): 85–92.

CUZZI, J. N. AND OTHERS. 1982. The delta-four-stream approximation for radiative flux transfer. *Journal of Atmospheric Science*, 39(4): 917–25.

DUTTON, E. AND OTHERS. 1984. Features of total vertical aerosol optical depth at Barrow, Alaska, 10–20 March 1983. *Geophysical Research Letters*, 11(5): 385–98.

GUNDEL, L. A. AND OTHERS. 1984. The relationship between optical attenuation and black carbon concentration for ambient and source particles. *Science of the Total Environment*.

HEINTZENBERG, J. 1980. Particle size distribution and optical properties of the arctic haze. *Tellus*, 32(3): 251–60.

HEINTZENBERG, J. 1982. Size-segregated measurements of particulate elemental carbon and aerosol light absorption of remote arctic locations. *Atmospheric Environment*, 16(10): 2461–69.

HEINTZENBERG, J. AND OTHERS. 1981. The chemical composition of arctic haze at Ny-Alesund, Spitzbergen. *Tellus*, 33(2): 162–71.

JANZEN, J. 1980. Extinction of light by highly nonspherical strongly absorbing colloidal particles: spectro-photometric determination of volume distributions for carbon black. *Applied Optics*, 19(17): 2977–85.

KHALIL, M. A. K. AND RASMUSSEN, R. A. 1984. Analysis of trace gases in arctic haze. *Geophysical Research Letters*, 11(5): 437–40.

LIOU, K. N. 1980. *An introduction to atmospheric radiation*. New York, Academic Press.

McCLATCHEY, R. A. AND OTHERS. 1972. *Optical properties of the atmosphere*. AFCRL Report, 3rd ed, 72–0197.

MITCHEL, J. M. Jr. 1957. Visual range in the polar regions with particular reference to the Alaskan Arctic. *Journal of Atmospheric and Terristrial Physics*, Special Supplement pt 1: 195–211.

OTTAR, B. 1981. The transfer of airborne pollutants to the arctic region. *Atmospheric Environment*, 15(8): 1439–45.

PORCH, W. M. AND MacCRACKEN, M. C. 1982. Parametric study of the effects of arctic soot on solar radiation. *Atmospheric Environment*, 16(6): 1365–71.

RAATZ, W. E. AND SCHNELL, R. C. 1984. Aerosol distributions and an arctic aerosol front during AGASP, Norwegian Arctic. *Geophysical Research Letters*, 11(5): 373–76.

RAHN, K. AND OTHERS. 1977. The Asian source of arctic haze bands. *Nature*, 268(5622): 713–15.

RAHN, K. AND McCAFFREY, R. J. 1980. On the origin and transport of the winter arctic aerosol. *Annals of the New York Academy of Science*, 338: 486–503.

ROESSLER, D. M. AND FAXFOG, F. R. 1980. Photoacoustic determination of optical absorption to extinction ratio in aerosols. *Applied Optics*, 19(4): 578–81.

ROSEN, H. AND NOVAKOV, T. 1983. Combustion-generated carbon particles in the arctic atmosphere. *Nature*, 306(5945): 768–70.

ROSEN, H. AND HANSEN, A. D. A. 1984. Role of combustion-generated carbon particles in the absorption of solar radiation in the arctic haze. *Geophysical Research Letters*, 11(5): 461–64.

ROSEN, H. AND OTHERS. 1981. Soot in the Arctic. *Atmospheric Environment*, 15(8): 1371–74.

SCHNELL, R. C. 1984. Arctic haze and the Arctic Gas and Aerosols Sampling Program (AGASP). *Geophysical Research Letters*, 11(5): 361–63.

SCHNELL, R. C. AND RAATZ, W. E. 1984. Vertical and horizontal characterization of arctic haze during AGASP, Alaskan Arctic. *Geophysical Research Letters*, 11(5): 369–72.

SHAW, G. E. 1975. The vertical distribution of atmopheric aerosols at Barrow, Alaska. *Tellus*, 27(1): 39–50.

SHAW, G. E. 1982. Atmospheric turbidity in the polar regions. *Journal of Applied Meteorology*, 21(8): 1080–88.

SHAW, G. E. AND STAMNES, K. 1980. Arctic haze: perturbation of the polar radiation budget. *Annals of the New York Academy of Science*, 338: 535–39.

SZKARLAT, A. C. AND JAPAR, S. M. 1981. Light absorption by airborne aerosols: comparison of integrating plate and spectrophone techniques. *Applied Optics*, 20(7): 1151–55.

VALERO, F. P. J. AND OTHERS. 1983. Radiative effects of the arctic haze. *Geophsyical Research Letters*, 10(12): 1184–87.

VALERO, F. P. J. 1984. The absorption of solar radiation by the arctic atmosphere during the haze season and its effects on the radiation budget. *Geophysical Research Letters*, 11(5): 465–68.

THE INTERNATIONAL ARCTIC GAS AND AEROSOL SAMPLING PROGRAM

RUSSELL C. SCHNELL

ABSTRACT. Following the successful International Arctic Gas and Aerosol Sampling Program (AGASP-I) of 1983, arctic air pollution (arctic haze) will again become the subject of intensive ground-based and airborne studies (across the non-Soviet Arctic) in spring 1986. A multifaceted international program, using aircraft of the National Oceanic and Atmospheric Administration (NOAA), the University of Washington nd the Atmospheric Environment Service of Canada will study vertical and horizontal characteristics of the haze in relation to surface-based measurements conducted at baseline stations located at Barrow, Alaska, Alert, Canada, and Ny-Ålesund, Svalbard. Baseline measurements will be augmented by observations at Poker Flat, Alaska, and Mould Bay, Canada. Concentrations and composition of gases and aerosols composing the haze will be measured, together with changes that the gases and aerosols produce in solar radiation flux within the atmosphere. Overall results expected will include greater understanding of chemical and physical properties of arctic haze, interactions occurring, the manner in which air pollution is transported into and across the Arctic, and information on the time and space representativeness of continuous surface-based haze measurements. Modelling studies will then determine long-term climatic effects of the pollution, possible health and ecological effects, and the applicability of using arctic haze studies as models for nuclear winter scenarios.

Contents

Introduction

Arctic haze originates in a number of contiguous countries, traverses international boundaries, and is transported from one continent to another over the north polar regions on time scales as short as one or two weeks. Research and monitoring of this anthropogenic air pollution is enhanced when it is both international in scale and three-dimensional in scope.

In spring 1983 six research aircraft—one from Norway, two from the Federal Republic of Germany and three from the United States—studied arctic haze in a concentrated international effort coordinated with established fixed-base ground measurement programs. One portion of this effort, the Arctic Gas and Aerosol Sampling Program (AGASP), used a long-range National Oceanic and Atmospheric Administration

135

(NOAA) WP-3D Orion research aircraft to study haze during flights spanning the Arctic from Alaska to Norway. Scientists from five countries representing 17 different research institutions conducted 35 different experiments on these AGASP flights (see Hileman 1983 for a list of participants and programs).

In spring 1986 an expanded and fine-tuned version of the 1983 field program will again be conducted across the Arctic. This broad-based international program will be built upon research contributions from at least 60 scientists representing eight countries, will involve three atmospheric research aircraft, and will include measurements from a range of surface baseline and background monitoring operations such as those at Ny-Ålesund, Svalbard; Alert, Canada; and Barrow, Alaska.

Arctic Gas and Aerosol Sampling Program (AGASP-I)

The NOAA AGASP project was designed to extend into three dimensions the arctic haze measurements conducted at the surface stations of Barrow, Alert, and Ny-Ålesund. In so doing, a much broader picture of haze over the arctic ice cap would be obtained and the representativeness of the ground-based measurements ascertained.

The WP-3D (Figure 1) flew 144 flight hours on 12 separate research missions in

Fig 1. The National Oceanic and Atmospheric Administration WP-3D Orion research aircraft used in the Arctic Gas and Aerosol Sampling Program. This aircraft is capable of 10–12 hour research missions with a range of 3,000+ nautical miles.

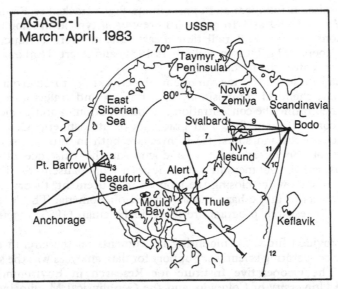

Fig 2. Approximate flight tracks of the NOAA WP-3D during the AGASP-I project. Flight 6 was into the stratosphere in a tropopause fold event, and Flight 12 was over the center of the Icelandic low into which 11 dropwindsondes were released in a regular pattern.

Fig 3. Research flight crew (less two members) of the NOAA WP-3D before a flight from Anchorage, Alaska, north over the polar ice cap. Besides the 18 aircrew, personnel consisted of 12 additional aircraft ground support staff and support scientists.

AGASP-I; for mission numbers and approximate flight tracks see Figure 2. Figure 3 shows the NOAA WP-3D and 16 of the 18 crew on a typical mission. In support of aircraft operations, aerosol and radiation measurements were conducted at Barrow (Bodhaine and others 1984; Dutton and others 1984) and Alert (Hoff and Trivett 1984) before, during, and after the flights.

AGASP-I had these central measurement objectives: (1) to determine the spectra, optical properties, chemical composition, distribution, and trajectories of arctic haze aerosols; (2) to determine the concentration and distribution of arctic trace gases; (3) to determine the concentration, flux, and gradients of atmospheric CO_2 in relation to important arctic sources and sinks; (4) to measure both in situ and surface-observed radiative effects of the haze; and (5) to document the existence of, and conduct stratospheric gas and aerosol measurements in, polar tropopause folds. Taken together, such measurements allow for closing the loop on measurement, theory, and modelling of the climate effects of arctic haze and associated phenomena. The 35 scientists who joined the WP-3D research program are listed in Hileman (1983) together with their programs.

AGASP-I provided funds for mounting instruments on the aircraft and paid field expenses for the cooperating scientists. Support for data analyses was the scientists' own responsibility. The Cooperative Institute for Research in Environmental Sciences (CIRES) of the University of Colorado, and the Geophysical Monitoring for Climatic Change (GMCC) program of the NOAA Air Resources Laboratory, NOAA, provided personnel and support services for organizing and directing the program, and for data archiving and publishing. The AGASP program was supported by NOAA (GMCC and Environmental Research Laboratories), the Office of Naval Research, and the National Aeronautics and Space Administration.

Results of the AGASP-I program

The main results of this research program include 28 papers published in the May 1984 issue of *Geophysical Research Letters, Proceedings of the Third Symposium on Arctic Air Chemistry* of May 1984 (see References), and a special Arctic Haze issue of *Atmospheric Environment* (December 1985). These papers show that the springtime arctic troposphere contains high concentrations of anthropogenic gases and aerosols mainly of Eurasian and Soviet origins. They further show that these enhanced concentrations are advected across the Arctic in distinct layers, on time scales of 10 days or less, over path lengths of up to 10,000 km, and that the haze layers have an appreciable impact on the flux of solar radiation reaching the ground in the high Arctic. A synthesis of these data sets on arctic haze has (1) confirmed a number of earlier observations, (2) provided answers to some outstanding questions, and (3) led to surprising new findings and thus to new questions; hence the second AGASP effort in 1986.

Results of the recent airborne studies confirm the long record of ground-based measurements of the composition of the arctic haze aerosols; they also confirm the presence of preferred transport pathways for these aerosols (Raatz and Shaw 1984).

Totally new information has been obtained on the horizontal extent and distribution of the haze over the polar ice; haze had previously been measured only from sites on the periphery of the arctic basin. Similarly, in the vertical, a many-layered, banded structure to the haze has been documented. High concentrations of anthropogenic gases have been measured in association with the haze aerosols, while between the layers the air has pristine background qualities. These differences are maintained over vertical heights of a few tens of meters for times up to at least 10 days. Such a banded structure is shown

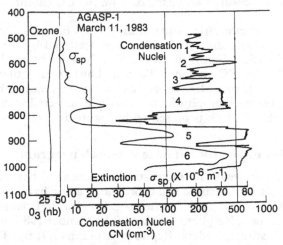

Fig 4. Profile of ozone (nb), scattering of light by haze aerosols (σ_{sp}, $\times 10^{-6}$ m^{-1}), and concentration of small combustion-generated smoke-like particles (condensation nuclei, cm^{-3}) from Flight 1 over Barrow, Alaska, 11 March 1983. Six distinct layers of haze pollution are in evidence, and haze concentrations are greater aloft than near the surface.

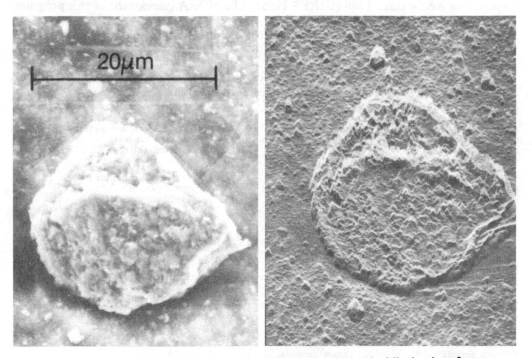

Figure 5. Arctic haze aerosol collected at 1,500 m above ice level while the aircraft was flying over the North Pole, 28 March 1983. Scanning electron microscope image is shown on the left with corresponding Y-mod (topographic image) on the right. The large particle is composed of elements lighter than sodium and has characteristics of black carbon or soot.

in Figure 4 for arctic haze measured above the Barrow GMCC baseline station, 11 March 1983.

Finally, some of the most interesting and significant new findings from these programs relate to the amount of solar radiation absorbed and scattered by the haze layers. For instance, haze-induced optical depths in excess of 0.25 were observed during the 1983 AGASP program at the GMCC station at Barrow (Dutton and others 1984), as were significant absorptions of solar radiation by the carbon-containing haze (Valero and others 1984). An electron micrograph of a soot particle collected in haze in the region of the North Pole, 28 March 1983, is presented in Figure 5.

1986 international haze research program

Within the past three years an international consensus has developed on the desirability of cnducting a well-focused and coordinated arctic haze research program in spring 1986. This consensus and plans for the program evolved from scientific meetings at the Arctic Haze Workshop, Ringberg Castle, 4–9 September 1983 (hosted by Dr Paul Crutzen, Max-Planck Institute, Federal Republic of Germany); the Third Symposium on Arctic Air Chemistry, Toronto, Canada, 7–9 May 1984 (hosted by Dr Len Barrie, Atmospheric Environment Service, Canada); the Planning Symposium on Arctic Air Pollution, Arctic Science Conference, Anchorage, Alaska, 2–5 October 1985 (hosted by Dr Glen Shaw, University of Alaska); and two arctic haze research program planning meetings held in Boulder, Colorado (March 1984 and March 1985). Full plans are presented in Arctic Haze-1986 (CIRES 1985). The NOAA component of this program is referred to as AGASP-II.

The main measurements and ground programs involved in the 1986 program are presented in Table 1.

Table 1. Aircraft, ground stations and responsible agencies/scientists involved in the 1986 arctic haze research program.

Facility	Measurement	Agency/Contact
Alert baseline station; other Canadian Arctic sites; AES Twin Otter research aircraft	Gases, aerosols and solar radiation	Atmospheric Environment Service, Canada Dr Neil Trivett
Ny-Ålesund baseline station; Norwegian island stations	Gases, aerosols and solar radiation	Norwegian Institute for Air Research Dr Brynjulf Ottar
Barrow baseline station	Gases, aerosols, and solar radiation	National Oceanic and Atmospheric Administration, USA Mr Bernard Mendonca
Poker Flat Optical and Atmospheric Chemistry M Facility	Gases, aerosols, and solar radiation	University of Alaska USA Dr Glenn Shaw
Dye-3	Gases and aerosols	Carnegie-Mellon University, USA Dr Cliff Davidson
Convair C-131A research aircraft	Meteorology, gases, aerosols, and cloud physics	University of Washington USA Dr Larry Radke
WP-3D Orion research aircraft	Meteorology, gases, aerosols, and solar radiation	National Oceanic and atmospheric Administration, USA Dr Russell Schnell

Central objectives

The central scientific measurement objectives of the 1986 program are to:

1. Characterize the vertical and horizontal distributions, chemical composition, and radiative properties of background air and arctic air pollutant haze throughout the troposphere in relation to continuous surface baseline station measurements. This will be accomplished by extensive use of aircraft, balloon-borne instruments, and lidar measurements tied into the operations at Barrow, Alert, and Ny-Ålesund.

2. Determine, on a fine-scale grid, the meteorological processes that produce and maintain the haze bands over 10,000 km travel distances. This will be accomplished by using ground-based and airborne lidars to map the haze structures, and research aircraft, dropwindsondes and radiosondes to measure the relevant in situ physical parameters.

3. Conduct a detailed set of ground-based precipitation chemistry observations to obtain a measure of the haze removal mechanisms and rates. This will be accomplished by analyzing precipitation collected at the baseline stations and at a number of other sites across the Arctic.

Within these three main objectives are a number of sub-objectives that will be addressed as meteorological and operational activities allow. For instance, extensive use of dropwindsondes in the course of flights will provide data for the study of features such as the Brooks Range-North Slope Jet discovered during AGASP-I (Shapiro 1985). Dedicated research flights into the stratosphere will be made to study the present status of the El Chicon debris cloud observed on AGASP-I (Shapiro and others 1984).

Additional meteorological and chemical tracer data will be collected and used to validate trajectory models, and to define the synoptic meteorology of the Arctic during the 1986 research period.

The planned flight tracks, bases of operation, and ground measuring sites are presented in Figure 6.

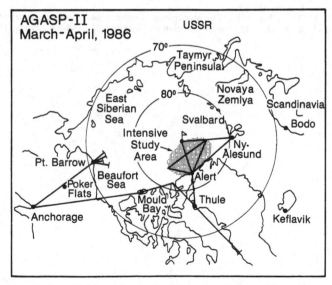

Fig 6. Operational area, flight tracks, and surface sampling and base operations for the 1986 arctic haze research project.

142	SCHNELL

Data analyses, publications, and archives

Each program, through its participating principal investigators, will be responsible for analysis of its data sets. The NOAA WP-3D program will provide aircraft meteorological, gas and aerosol data, as well as ground support data, to all cooperating scientists. Coordination of data gathering, analyses and interpretation will be aided and encouraged by the Haze Program Office.

An issue of *Geophysical Research Letters* will be prepared for May 1987 release, in the same manner as for the 1983 Arctic Haze Program. This will probably be followed by a Fourth Arctic Air Chemistry Symposium, possibly in Norway or Boulder, Colorado, in spring 1988. The symposium proceedings would logically appear in a special issue of *Atmospheric Environment* as in the past.

The CIRES-NOAA Haze Program Office will act as a long-term repository for the arctic haze research data sets submitted, and will provide support data and processing of selected NOAA-collected data tapes as required by cooperating scientists.

References

BODHAINE, B. A. AND OTHERS. 1984. Surface aerosol measurements at Barrow during AGASP. *Geophysical Research Letters*, 11: 377–80.

CIRES. Haze Program Office. 1985. Arctic haze – 1986: research program plan. University of Colorado, Boulder.

DUTTON, E. G. AND OTHERS. 1984. Features of aerosol optical depth observed at Barrow, 10–20 March 1983. *Geophysical Research Letters*, 11: 385–88.

HILEMAN, B. 1983. Arctic haze. *Environmental Science and Technology*, 17: 233–36.

HOFF, R. M. AND TRIVETT, N. B. A. 1984. Ground-based measurements of Arctic haze made at Alert, NWT, during the Arctic Gas and Aerosol Sampling Program (AGASP). *Geophysical Research Letters*, 11: 389–92.

RAATZ, W. E. AND SHAW, G. E. 1984. Long-range transport of air pollution aerosols into the Alaskan Arctic. *Journal of Climatology and Applied Meteorology*, 23: 1052–58.

SHAPIRO, M. A. 1985. Dropwindsonde observations of an Icelandic low and Greenland mountain lee wave. *Monthly Weather Review*, 113: 680–83.

SHAPIRO, M. A. AND OTHERS. 1984. El Chichon volcanic debris in an Arctic tropospheric fold. *Geophysical Research Letters*, 11: 421–24.

VALERO, F. P. J. AND OTHERS. 1984. The absorption of solar radiation by the arctic atmosphere during the haze season and its effects on the radiation balance. *Geophysical Research Letters*, 11: 465–68.

THE ARCTIC HAZE–ARCTIC CLOUD CONNECTION

GLENN E. SHAW

ABSTRACT. The reflectivity of arctic clouds is extraordinarily sensitive to the concentration and chemistry of cloud condensation nuclei (CCN) upon which cloud water droplets condense. To what extent are such nuclei affected by the presence of arctic air pollution? Studies in Alaska have demonstrated that the relevant CCN are soluble particles a few hundredths of a micron in diameter. It is deduced that pollution from middle latitudes can in principle alter cloud reflectivity in winter but not in summer. At present, insufficient data exist to allow estimation of the magnitude of climatic perturbation brought about by the arctic haze-cloud connection.

Contents

Introduction

This contribution discusses possible alterations in cloud albedo, and the climatic changes that could result from them, which might be caused if arctic pollution aerosol affects the cloud condensation nuclei (CCN) upon which cloud water droplets condense. The haze-cloud climatic linkage, first suggested by Keith Bigg, has so far received little attention. It could however be an important cause of climatic change, similar to the amplifier action of a Fleming valve circuit. The submicron CCN are analogous to the grid which modulates a much larger signal, in this case the size spectrum of cloud water droplets. Variations in cloud droplet size spectrum will show up as alterations in cloud reflectivity (Twomey and Seton 1980).

The aerosol-cloud connection is relatively simple to examine for arctic clouds in winter, primarily because for these kinds of stratiform clouds, there is a reasonably well developed theory—the Twomey activiation theory— relating cloud droplet size spectrum to CCN supersaturation spectrum. We use this theory, supported by measurements of aerosol size distribution in Alaskan air masses, and of the supersaturation spectrum of cloud condensation nuclei, to develop some ideas about possible connections between arctic air pollution and arctic clouds.

The microphysics of arctic clouds

Layer-clouds of the type found within the region of polar night are created by humid air parcels rising at updraft speeds of 1 to 10 cm s^{-1}, which correspond to cooling rates of approximately 0.1 to 1°C/hr. Just after the air reaches saturation, supersaturation increases linearly with time, but the rate slows down as vapor is progressively depleted by condensation on previously nucleated and growing water droplets. Supersaturation passes, therefore, through a maximum. The change of supersaturation S with time is:

$$\frac{dS}{dt} = \alpha v - \beta S \sum r_i \tag{1}$$

where α and β are constants, v is the updraft speed, and the sum is over the radii of cloud droplets which have previously been nucleated and are now growing.

Twomey (1959) has developed analytical solutions providing underestimates and overestimates of S(t) under the assumption that the supersaturation spectrum of cloud condensation nuclei can be approximated as a power law function:

$$N(S) = C S^k \tag{2}$$

where N is the number of drops nucleated (cm^{-3}) at supersaturation S and C and K are constants. For arctic-derived air masses in winter, values of K lie in the range 0.5 to 1.0 (Shaw 1985).

Twomey's upper estimate for the maximum supersaturation S_{max} is:

$$S_{max} = \left\{ \frac{(\alpha v)^{\frac{3}{2}}}{\beta \sqrt{G} CKB(\frac{3}{2},\frac{K}{2})} \right\}^{\frac{1}{K+2}} \tag{3}$$

where α and β are the constants in equation 1, C and K are the CCN supersaturation power law parameters in equation 2, v is updraft speed, and B is the complete Beta function $B(x,y) = \Gamma(x) \Gamma(y) / \Gamma(x + y)$, Γ being the Gamma function and G the cloud droplet growth rate dr/dt (Fukuta and Walter 1970).

Calculated values of S_{max} for updraft speeds of 1 cm/s and 10 cm/s are listed in Table 1, for several values of K. The maximum supersaturation expected during these conditions is seen to be less than about 0.5%

Figure 1 shows the CCN supersaturation spectra, calculated on the basis of the measured aerosol particle size distribution spectrum; it assumes (a) that the aerosol particles are soluble salts (the calculation was performed for ammonium sulphate but would not vary significantly for other salts), and (b) that the aerosols are composed of insoluble but wettable particles. By comparing measured spectra (stippled region in Figure 1) with computed CCN spectra, one deduces that the active cloud condensation nuclei in Alaskan air masses are soluble salts.

Now the application of supersaturation S_{max} will nucleate all soluble salts of dry radius larger than r_{nuc}, where $r_{nuc} = 1.54 \times 10^{-6} S^{-2/3}$ (S is percent). From values of r_{nuc} listed in Table 1 we see that soluble aerosol particles larger than 0.02 to 0.04 microns are expected to be nucleated during the formation of clouds in the arctic regions. If mid-latitude pollution sources alter particles $r > r_{nuc}$, in the Arctic, then one might expect to find an anthropogenically-induced alteration in cloud droplet size.

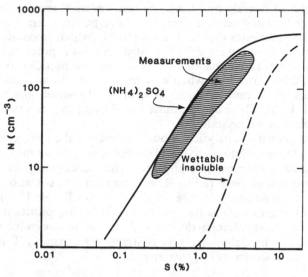

Fig 1. Supersaturation spectra of cloud condensation nucleus measured in Alaska during winter (stippled) and calculated from the measured aerosol size distribution, assuming the aerosol particles to be ammonium sulphate (solid line) and wettable but insoluble particles (dashed line).

Table 1. Maximum supsaturation during the formation of clouds in the Arctic. K is the power low exponent in CCN supersaturation spectrum (equation 2); r_{nuc} is the minimum-size aerosol particle (assumed to be ammonium sulphate) which will nucleate.

K	Updraft speed = 1 cm s^{-1} S_{max} (%)	r_{nuc} (10^{-6}cm)
0.5	0.07	8.9
0.7	0.09	7.5
1.0	0.11	6.7
1.5	0.17	5.0

K	Updraft speed = 10 cm s^{-1} S_{max} (%)	r_{nuc} (10^{-6}cm)
0.5	0.26	3.8
0.7	0.29	3.5
1.0	0.33	3.2
1.5	0.40	2.8

The smallest arctic aerosol available as CCN

Arctic air pollution builds up throughout the polar night, reaching a maximum in late winter and early spring. This seasonal behavior is documented for the Scandinavian Arctic (Ottar 1981) and North American Arctic (Barrie, this symposium): one can imagine in the Arctic a pond of stagnant, polluted air containing aerosol particles with a mean 'age' measured from the time that the air mass first entered the region of polar night. This section discusses the relationship between that age and the size of the smallest particles of arctic haze in the air mass, and explores its connection with cloud condensation nuclei.

Before a polluted air mass flows into the Arctic Basin, it is illuminated by solar radiation, and particles are produced actively by photolytic conversion of trace gases. From then on the aerosol size distribution spectrum alters only in response to removal mechanisms. Though several physical processes deform the spectrum, (which is assumed to begin as a power law distribution, $dn/d \log r \propto r^{-n}$), the most significant process, ie the one that proceeds fastest, is attachment and consolidation of the smallest particles to cloud droplets by Brownian motion; this is true even when we consider that the airmass spends only a relatively small fraction of its time in cloud. It applies, of course, only to the smallest particles of less than 0.1 μm. These are the most mobile, and the ones that are lost most rapidly to cloud droplets or ice crystals.

Shaw (1986) has developed an analytical model to describe the progressive loss of the smallest, most mobile, aerosol particles to cloud droplets, and the attendant deformation of the aerosol particle size spectrum. The physics of the model incorporates attachment of aerosols to cloud droplets of mean radius Rc and number concentration Nc, diffusing to the cloud droplet at the Maxwellian rate $dn_a/dt = -4 \pi$ Rc Nc D (r) n_a, where n_a is the number concentration of aerosol particles (cm^{-3}) having diffusion coefficient D (cm^2/sec). For the case of the smallest particles ($r < \lambda$, λ = mean free collision path-length = 0.067 μm), D = kT $\lambda/6\pi\eta^*r^2$ where k is Boltzmann's constant, T is absolute air temperature and η is the viscosity of the carrier gas.

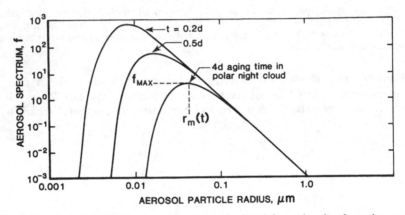

Fig 2. Aerosol size distribution spectrum calculated from the cloud erosion model for three values of tc.

Figure 2 shows a family of aerosol size distributions calculated from the cloud erosion model. Mode radius of the aerosol size distribution function increases with tc, the time the aerosol-laden air mass has spent in the presence of cloud since it entered the Arctic. The theory shows that a parabolic relationship exists between $r_m(t)$ (the mode radius) and tc:

$$r_m^2(tc) = \frac{4Rc\,Nc\,kT\lambda}{3\nu\eta}\,tc \qquad (4)$$

where, it will be recalled, Rc and Nc are the mean radius and number concentration of cloud droplets, and ν is the 'Junge' power law exponent.

It is interesting to consider separately (1) the particular size class of particles which are nucleated (ie r is greater than r_{nuc}) and thus incorporated into cloud droplets, and (2) the particular size class of particles which would coexist with an air mass *containing* cloud droplets, $r_{min} < r < r_{nuc}$, where, from Figure 1 it is apparent that $r_{min} \sim r_m(tc)$. Equation 4, for representative conditions in arctic stratiform clouds, $N_c \simeq 90$ cm^{-3},

Rc = 6.7 μm (Jayaweera and Ohtake 1973), may be expressed as $r_m^2 \sim 3.25$ tc (units 0.01 μm) for tc in units of days. Notice that r_m would grow to $r_{nuc} \sim 0.02$ to 0.04 μ in tc $\simeq 1$ to 2 days. One would expect, therefore, some kind of fundamentally different aerosol–cloud connection for airmasses entering the Arctic from mid-latitudes, depending upon whether they have coexisted for greater or for lesser times than 1 or 2 days in the presence of cloud water droplets.

If a pollution-charged air mass enters the Arctic and coexists with cloud droplets for the relatively long time of 1 to 2 days in winter darkness, then, in this hypothetical case, the CCN spectrum (consisting of the pollution by-products) would be strongly affected by cloud scavenging. The situation would be rather complex, approaching a limit where, upon the formation of cloud, *all* aerosol particles would be nucleated and incorporated in cloud. With precipitation or other removal mechanisms, memory would slowly be lost of the original pollution aerosol, and one could say that there would, in later times, be relatively minor effects of pollution originating from the middle (sunlit) latitudes on arctic clouds.

If on the other hand tc is less than about 1 day, then one would have the condition $r_m \ll r_{nuc}$, and the pollution aerosol would in principle be available to serve as CCN. In this case mid-latitude pollution could conceivably alter the microphysics of arctic clouds.

Variables relating to arctic clouds: experimental results

Aerosol size distributions were measured in Alaska at the Poker Flat Rocket Range using a diffusion battery, with the objective of deriving the parameter r_m. Experiments performed in late winter and spring 1985, and studies at Ester Dome Observatory in 1983 and 1984, showed that the mean aerosol radius, or equivalently r_m, decreased as the temperature of the air mass increased. The inverse relationship between r_m and T is thought to be related to aerosol age: warm air masses of Pacific marine origin have coexisted for relatively short times in the presence of cloud since entering the Arctic, whereas cold, Arctic-derived air mass systems have resided in the region of the polar darkness and subject to cloud particle erosion for much longer.

If an air mass, originally at temperature T_0, is imagined to drift poleward into the polar night region, cooling at an average rate of σ(°C/d) and coexisting in the presence of cloud for fractional time f_c, then from Equation 4 the following expression is derived between r_m^2 and air mass temperature T:

$$r_m^2(t) = \frac{3.25 \, f_c}{\sigma}[T_0 - T(t)]. \tag{5}$$

In Figure 3, r_m^2 is plotted against T for an assumed value of $\sigma = 0.50$ °C/d and for the condition $T_0 = 0$°C (top) and $T_0 = 10$°C (bottom). The straight lines in Figure 3 are Equation 5 plotted for several numerical values of f_c (f_c = fractional time during which the air mass has coexisted in cloud). The data suggest that $f_c \simeq 0.01$ to 0.1 (geometric mean 0.03). This means that a representative arctic air mass has spent a few percent of its time travelling in cloud since entering the Arctic.

The geometric mean value of t_c (time spent in cloud derived from equation 4 is 7.2 hours for the air mass systems found in central Alaska in winter. The corresponding transport time is about 10 days, comparable to direct transport times to Alaska from sources in central Eurasia.

In summary, experiments on air masses in the Alaska subarctic region indicate that:

(1) Active cloud condensation nuclei are predominantly *soluble* particles (Figure 1) of size r_{nuc} greater than 0.03 μm.

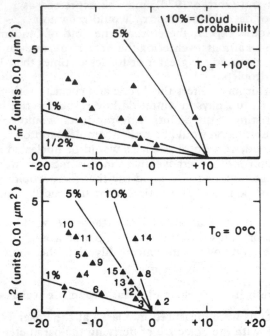

Fig 3. The parameter r_m^2 plotted against air temperature (°C) for fifteen 4-hour periods of aerosol measurement, Poker Flat Rocket Range, Alaska, during times when the aerosol concentration was steady (ie no contamination). Solid lines show the theoretical relationship between r_m^2 and T assuming air mass cooling rates in the polar night region of 0.5 °C/d. The initial temperature condition of the air mass as it enters the Arctic is assumed to be +10°C (top) and 0°C (bottom).

(2) Particles likely to have been injected into the Arctic Basin at mid-latitudes are present in size $r_m < r_{nuc}$ and can serve as condensation nuclei when clouds form.

(3) Fractional cloudiness in the Arctic Basin during winter is in the order of 0.03, a relatively low value that allows small pollution-derived particles to exist all over the arctic regions in winter. If fractional cloudiness were higher, the small-CCN size aerosols would be scavenged by clouds.

Discussion and speculations

The size distribution of droplets in arctic clouds is controlled by soluble aerosol particles (these are the particles which nucleate when the humid air supersaturates). In arctic clouds the supersaturation is expected to be less than about 0.5%, which implies that soluble particles larger than a few hundredths of a micron would be nucleated. At the same time it has been shown that Alaskan air masses contain aerosol particles a factor of several times smaller than those which would be expected to serves as active cloud condensation nuclei. Apparently both can occur because of the relatively small fractional cloudiness in the winter arctic troposphere. Otherwise, the small particles would have been removed by Brownian diffusive attachment to cloud droplets. These statements assume that the rate of production of new aerosol particles within the darkness of the polar night is nil.

Typical trans-Arctic transport times in winter (for example the time for pollution generated in central Eurasia to be carried to Alaska), are of the order of ten days. If the air mass during these ten days is in the presence of cloud 3% of the time (as was previously deduced) one would expect the sub-micron particles serving as CCN to contain a significant amount of mid-latitude pollution aerosol. In other words, cloudiness is insufficient and transport too rapid to remove particles several hundredths of a micron in radius. Therefore, it seems likely that pollution injected into the atmosphere in the middle latitudes has the possibility to affect arctic clouds during winter.

In summer, arctic cloudiness increases and one would expect a decoupling of the midlatitude aerosols and the CCN which form in the Arctic, for two reasons: (1) because the tropospheric air flow patterns are primarily zonal in summer (hence there is less direct transport of midlatitude air to the higher latitudes), and (2) because air masses from lower latitudes would undergo much more cloud scavenging.

The alteration in cloud albedo (which might occur in the Arctic from pollution aerosol for the mid latitudes in winter) is difficult to predict when we know so little about the microphysics of arctic clouds and the related CCN. One can surmise that increased pollution aerosol would lead to an increased C (in Equation 2) and possibly to an increase in the power law exponent K, although this is highly speculative.

An increase in C (with no change in the shape of the size distribution) by itself would be associated with an increased number concentration of cloud droplets, finer droplets and probably an increased cloud albedo. An increase in the constant K by itself, however, would increase S_{max}, decrease the concentration of cloud droplets and probably reduce cloud albedo. It is not profitable to speculate, therefore, as to the possible change in cloud albedo which might be induced by the incursion of mid-latitude pollution by-products into the Arctic Basin, since the two effects work in opposite directions. There is a pressing need for further investigations in this interesting area of possible human impact on climate.

Acknowledgement

This research was supported in part by US National Science Foundation, Division of Atmospheric chemistry (Grant ATM83–00155).

References

FUKUTA, N. AND WALTER, L. A. 1970. Kinetics of hydrometeor growth from a vapor-spherical model. *Journal of Atmospheric Science*, 27: 1160–72.
JAYAWEERA, K. AND OHTAKE, T. 1973. Concentration of ice crystals in arctic stratus clouds. *Journal de Recherches Atmospheriques*, 7: 199–207.
OTTAR, B. 1981. The transfer of air-borne pollutants to the arctic region. *Atmospheric Environment*, 15(8): 1439–45.
SHAW, G. E. 1985. Cloud condensation nuclei associated with arctic haze. *Atmospheric Environment*, in press.
SHAW, G. E. 1986. On the physics of aitken particles in polar air mass systems. *Journal of Aerosol Science*, in press.
SHAW, G. E. AND STAMNES, K. 1980. Arctic haze: perturbation of the polar radiation budget. *Annals of the New York Academy of Science*, 338: 533–39.
TWOMEY, S. 1959. The nuclei of natural cloud formation. *Geofisica pura e applicata*, 43: 227–49.
TWOMEY, S. AND SETON, K. J. 1980. Inferences of gross microphsyical properties of clouds from spectral reflectance measurements. *Journal of Amospheric Science*, 37: 1065–69.

THE IMPORTANCE OF ARCTIC HAZE FOR THE ENERGY BUDGET OF THE ARCTIC

THOMAS P. ACKERMAN, JANINE M. STENBACK
AND FRANCISCO P. J. VALERO

ABSTRACT. Estimated energy lost annually through radiative processes from the Arctic Basin exceeds energy gained through atmospheric and oceanic transport by about 30%. Due to uncertainties in the available data, the source of the difference cannot be resolved. Arctic haze contributes a perturbation to the radiation balance in the order of 10% of the net radiation averaged over the spring season. The climatic impact of such a perturbation cannot easily be ascertained. However, previous climate studies suggest that it cannot be dismissed as unimportant.

Contents

Introduction

Detailed studies of the nature and origin of the arctic haze have been carried out only since about 1980 (Barrie 1986). Theoretical studies of the climatic impact of the arctic haze reported by Porch and MacCracken (1982) and Cess (1983) were hampered by lack of data on the optical properties of the haze. However, they made it clear that the dominant climatic effect of the haze is an enhancement of solar absorption due to the carbonaceous component. The first direct measurements of solar absorption of the haze, carried out by Valero and Ackerman (1986) as part of AGASP-1 flight series (Schnell 1986), showed that haze increased the absorption of solar radiation in the atmosphere by a factor of 2 to 3. Over an ice surface this increased absorption translated into an overall reduction in the planetary albedo, and a consequent heating of the surface-atmosphere system.

Measurements during AGASP were essentially instantaneous observations of haze absorption, taken on three or four days over a two-week period. To assess the importance of haze on climatological scales, it is necessary to (1) estimate the energy fluxes within the Arctic basin on annual and seasonal timescales, and (2) extrapolate the local, instantaneous measurements to diurnal and seasonal timescales and to spatial scales covering the Arctic. The energy fluxes obtained in these two steps can then be compared, in order to estimate the climatic importance of arctic haze.

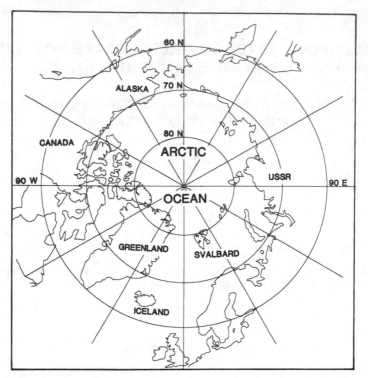

Fig 1. The Arctic Basin: location map. Latitude circles are at 10° intervals.

Arctic energy budget

To compute Arctic Basin energetics we define the Arctic as the area north of 70°N latitude. As shown in Figure 1, this parallel crosses the northern edges of Alaska, the Canadian islands, Scandinavia, and the Soviet Union and encloses an area primarily of ocean covered by pack ice. The only significant land mass included is the northern two-thirds of Greenland, most of which is permanently snow-covered. From both meteorological and oceanographic perspectives it would be more appropriate to define the Arctic in terms of geophysical boundaries such as the polar front in the atmosphere or ocean currents, but unfortunately these boundaries are neither precise nor constant in time; furthermore, atmospheric data are routinely analyzed on latitudinal grids, forcing our adoption of a latitude circle.

The Arctic energy budget can conveniently be separated into three terms: (1) net radiation (solar input minus infrared emission) at the top of the atmosphere, (2) poleward atmospheric energy flux, and (3) poleward oceanic heat flux. Assuming a current steady-state climate, these three terms must sum to zero on an annual basis. The assumption of steady-state implies that there is no increase or decrease in annual extent and thickness of sea ice, or in the average temperature of the Arctic Ocean. Obviously, the energy fluxes need not sum to zero on a monthly or seasonal basis because heat is stored or released on seasonal timescales by atmosphere, sea ice and ocean.

Table 1. Seasonal and annual values of net radiation (NR), absorbed solar radiation (SA), and emitted infrared radiation (IR) at the top of the atmosphere for the area north of 70°N latitude. Units are W m⁻². (From Stephens and others 1981.)

Season	NR	SA	IR
DJF	−156	1	157
MAM	−82	94	176
JJA	−20	189	209
SON	−154	19	173
Annual	−102	76	178

Net radiation

Values of net radiation at the top of the atmosphere were taken from an analysis of satellite data carried out by Stephens and others (1981). Seasonally and annually averaged values for the Arctic are given in Table 1. We have defined both absorbed solar and emitted infrared radiation to be positive, and the net radiation to be solar minus infrared; thus a negative net radiation implies an energy loss from the system. As expected, the Arctic is a net radiation sink in all seasons of the year. The season of greatest interest from the perspective of arctic haze is spring (MAM), during which the radiation budget is roughly the same as the annual average.

Atmospheric energy flux

The total energy transported into the Arctic by the atmosphere is the sum of three components: sensible heat, latent heat, and potential energy. Sensible heat is related simply to the actual temperature of the air which is being transported, latent heat to the moisture content of the air, and potential energy to potential air temperature (actual temperature corrected for adiabatic heating or cooling induced by changes in pressure as the air moves vertically). In polar latitudes the sensible heat flux is the dominant component of energy flux.

At mid- and polar latitudes, atmospheric energy transport is predominantly due to eddy motions (motions associated with large-scale waves in the atmosphere) rather than mean meridional motions (motions associated with a zonally-averaged, direct polewards flow of air). Mean meridional motions are the result of large-scale overturning of the atmosphere, which produces a net transport across latitude circles. By definition, eddy motions time-average to zero but produce quasi-horizontal transports due to non-zero correlations between the eddy velocity and the gradient of the transported quantity. Conventionally, eddy motions are classified mathematically as either 'standing' or 'transient'. The former are wave patterns forced by surface features such as mountains or land-sea temperature contrasts, while the latter are temporally or spatially varying features.

Oort (1983) has compiled circulation statistics of the atmosphere using a variety of routine data sources which have been carefully analyzed and averaged over various time periods. His values of seasonally and annually averaged energy transport across 70°N latitude by standing eddies, transient eddies, and mean meridional motions are given in Table 2; averages are for the 10–year period 1963–73. Energy transports have been divided by the area of the polar cap north of 70°N (15.4 x 10¹² m²) to give the atmospheric heat flux per unit area in W m⁻², which can then be compared directly with the values in Table 1.

Table 2 shows that transport is dominated by transient eddies. Standing eddies and

Table 2. Poleward transport of energy by the atmosphere due to transient eddies
(TE), standing eddies (SE), and mean meridional motions (MM) across 70°N
latitude. Energy transport has been divided by the surface area north of 70°N. Units
are W m⁻². (From Oort 1983).

Season	TE	SE	MM	Total
DJF	56	30	18	104
MAM	50	7	9	66
JJA	52	2	0	54
SON	64	8	9	81
Annual	54	9	9	72

mean motions contribute roughly equal amounts, although it should be noted that the
calculation of mean medional transport is considerably less certain. On an annual basis,
51 W m⁻² or about 70% of the total energy flux is sensible heat. Latent heat is responsible
for about 20% of the total, with potential energy accounting for the remainder.

Oceanic heat transport

The amount of oceanic heat transport into the Arctic Basin is not so well established.
Direct measurements reported by Aagaard and Griesman (1975) suggest that poleward
heat flow is of the order of 7 W m⁻² for the annual average and ranges from 5 W m⁻²
in winter to 10 W m⁻² in summer. These values agree reasonably with oceanic heat
transports deduced from a general circulation model (Slingo 1982). In this latter case
oceanic heat transport was computed on an annual basis as the residual between net
radiation and atmospheric heat transport predicted by the model, and was of the order
of 10 to 15 W m⁻². However, Oort and Vonder Haar (1976) find an equatorward transport
of heat north of 60°N latitude. Their analysis is based on ocean temperature profiles
which are used to compute oceanic heat storage. The time divergence of the heat storage
is then used to compute heat fluxes. This equatorward transport is counter-intuitive, and
the authors themselves recognize its implausibility. It is likely that their results are in
error due to the sparseness of oceanic data and the difficulty of computing relatively small
divergences from heat storage estimates.

Arctic energy balance

A comparison of annual values in Tables 1 and 2 shows that energy lost by radiation
exceeds energy transported by the atmosphere by about 30 W m⁻². While ocean transport
may account for some 10 to 15 W m⁻², this still leaves approximately 15 to 20 W m⁻²
unaccounted for. Unfortunately, all of the terms potentially have substantial errors
associated with them, making it difficult to decide how to balance the energy budget.
Some of the options are:

Increase the assumed solar absorption in the Arctic, thereby reducing the net radiative loss.
Solar absorption is deduced from satellite measurements by differencing incident and
reflected solar radiation. However, reflection measurements are made at particular
viewing angles and the total reflectance is computed using an assumed reflectivity
function at angles not observed by the satellite. Given high reflectivities in the Arctic,
uncertain reflectivity functions for snow and ice surfaces, and the oblique angles at which
polar latitudes are typically observed from satellites (because viewing geometries are
usually optimized for mid- and tropical latitudes rather than polar latitudes), measured
solar absorption is likely to include substantial errors.

Increase the polewards atmsopheric energy flux. Oort's analysis of circulation statistics, though very careful, relies on routine atmospheric soundings. Since only a handful of sounding stations exist in the Arctic, substantial errors may be introduced into the analysis of atmospheric temperatures and windfields, and thus into calculations of energy transport.

Increase the oceanic heat transport. Given the large uncertainties in the oceanic data, it is clearly possible that heat flux estimates have large errors. However, current estimates would have to be increased by relatively large factors of two to three to balance the energy budget.

Relax the assumption of annual equilibrium. As noted earlier, heat could be stored or released in the Arctic annualy as changes either in the amount of ice, or in water temperature. With regard to ice, the rate of energy release needed to balance the terms above would require freezing an excess of more than a metre of ice annually over the entire Arctic Basin. For water, it would raise the temperature of a 50 m column more than 2 K in a year. Such large trends seem implausible, especially when considered over a period of years, and would easily have been detected had they occurred.

A comparison of the data for the spring (MAM) season alone can also be made from Tables 1 and 2. In this case, assuming that the oceanic heat transport is on the order of 10 W m^{-2}, the imbalance is only about 5 W m^{-2}, which is well within the uncertainties in the individual terms. However, considering the imbalance in the annual energy budget, it is difficult to say whether this means that the Arctic is in 'equilibrium' during the spring season or that the agreement is simply fortuitous. Given the uncertainties of the data, the latter explanation is perhaps the more appropriate.

Absorption of solar radiation by arctic haze

Arctic haze is essentially a winter and spring phenomenon with peak concentration from March to mid-May (Bodhaine 1986). As noted above, its principal climatic impact is to alter atmospheric solar absorption. Under clear skies, over a snow or ice surface in the arctic spring, only 35 to 40% of incident solar radiation is absorbed by the atmosphere-surface system. Of that absorption, some 50–60% is atmospheric, due to water vapor and carbon dioxide; the remainder occurs at the snow or ice surface. If a haze layer is present, absorption of incident radiation increases to about 50%, approximately 65% of the absorption occurring in the atmosphere.

The exact effect of haze on solar radiation depends on the total optical depth and the single-scattering albedo of the haze. Total optical depth τ is defined as the total amount of haze (mass per unit area) in a vertical column multiplied by the specific extinction cross-section (area per unit mass) of the haze particles. A beam of photons traversing this column of haze is exponentially attenuated according to the relationship e$^{-\tau}$. The interaction of a photon with a haze particle can result in either absorption or scattering of the photon, the sum of which is defined to be the extinction. The probability of whether scattering or absorption will occur is usually expressed in terms of a single-scattering albedo ω, defined as the ratio of scattering to extinction. For arctic haze, the dominant absorbing material is black carbon or 'soot' produced by combustion processes (Rosen and Hansen 1986). Thus, the value of ω depends on the fraction of the haze which is black carbon. Obviously, the larger the fraction of black carbon, the smaller the value of ω and the greater the absorption for a given optical depth of haze. Similarly for a fixed ω, larger τ will result in more absorption.

Porch and MacCracken (1982) performed a parametric study of the impact of arctic haze on solar absorption, concluding that a haze layer with $\tau = 0.038$ and $\omega = 0.44$ over a highly reflecting ice surface would increase solar absorption in the atmosphere-surface system by 21 W m^{-2} while decreasing absorption at the surface by 4 W m^{-2}. Cess (1983) performed a similar analysis, considering monthly variations in such quantities as surface reflectivity and haze optical depth. His results show that, for a haze layer with τ about 0.8 and $\omega = 0.70$, the diurnally-averaged increase in absorbed radiation in spring is about 10 W m^{-2}, and the decrease in absorption at the surface is about 4 W m^{-2}.

Valero and Ackerman (1986) reported actual measurements of haze absorption from flight 3 (15 March 1983) of AGASP-1 over pack ice near Barrow, Alaska. At about local noon the haze layer ($\tau = 0.25$, $\omega = 0.78$) increased solar absorption by 30 W m^{-2} in the system and reduced absorption at the surface by 10 W m^{-2}. They obtained similar results on several other flights during AGASP-1. These results are consistent with the model studies cited above. Porch and MacCracken's (1982) haze model has an absorption optical depth τ_{abs} (defined as $\tau_{abs} = (1 - \omega)\tau$) of about 0.021, which is about half that measured by Valero and Ackerman. This would suggest that the measured absorption should be twice the computed absorption or about 40 W m^{-2}. However, the model aerosols also had a much smaller scattering optical depth. As a result the actual haze had a higher reflectivity than the model haze, which reduced the absorptivity of the actual layer relative to that of the model layer.

Comparison with Cess's results is more difficult since his calculations represent a diurnal average over some 100 solar zenith angles. However, as a rough guide for the spring period, his calculations suggest that the diurnally-averaged value is about half to one-third of the noontime value. Using the latter value, Valero and Ackerman's result would then give a diurnally-averaged increase in solar absorption of about 10 W m^{-2}. This is marginally greater than Cess's result of 7.5 W m^{-2} computed for 15 March at 75°N, assuming $\tau = 0.09$ and $\tau_{abs} = 0.027$. However, considering the differences in τ_{abs} and the diurnal integration, the model and the measured values are in reasonable agreement.

Arctic haze and the energy budget

The figures for solar absorption in Table 1 show that average solar energy absorbed in the Arctic spring is of the order of 90 W m^{-2}. The presence of haze increases this by about 10 W m^{-2} or approximately 10%, reducing net radiation loss by an equivalent amount. There are three immediate ways in which the system could respond to such a perturbation.

Heat the Arctic Basin. On a seasonal basis, the additional heating might simply warm the arctic atmosphere. An additional 10 W m^{-2} distributed uniformly over a column from surface to 5 km would heat the column at a rate of 0.2°K day^{-1}. Spread over the 60 to 90 days of spring, this amounts to some 12 to 18 K. Presumably this heating would increase the amount of ice melted during the following summer.

Reduce the atmospheric heat flux. The eddy motions which transport heat into the Arctic are driven in part by the temperature gradient between equator and pole. Reducing this gradient by direct heating could reduce the strength of the eddies and thus the amount of heat transported. Although a reduction of 10 W m^{-2} is fairly trivial on a global basis, it represents about 10 to 20% of the transport into the Arctic (Table 2). The effect of such a change on atmospheric circulation and arctic climate cannot be intuitively estimated.

Increase the infrared emission. According to Table 1, the average infrared emission in spring is 176 W m^{-2}, corresponding to an equivalent black-body temperature of 236 K. If it is assumed that the increased solar absorption is simply re-radiated, then the average infrared emission increases to 186 W m^{-2}, which corresponds to a temperature of 239 K or an increase of 3 K.

The preceding estimates were based on seasonal energetics. On an annual basis the haze contribution should be divided by a factor of 4, which gives a perturbation on the order of 2.5 W m^{-2}. This of course represents a change of only a few percent in the annual energy input to the Arctic. While so small a perturbation might be judged negligible, we suggest caution; climate studies have shown that small perturbations in planetary energetics can cause significant climatic responses. Doubling the concentration of CO_2 changes mean global outgoing infrared flux by 4 W m^{-2} and increases the mean global surface air temperature by about 2 K (Ramanathan and others 1979). Stratospheric aerosol layers typically enhance planetary reflectivity by only a few percent, but can produce coolings of a few degrees K (Chou and others 1984). Clearly, these cases represent a larger perturbation than arctic haze, due to their global extent and longer time-span, but haze could prove to be as important on a regional basis. Furthermore, these same climate studies have suggested that the Arctic is a climatically sensitive region. For global perturbations such as doubled CO_2, variations in insolation, or injection of stratospheric aerosols, temperature changes in the Arctic are greater than those of any other region. Thus even a small perturbation might produce a noticeable effect in the Arctic.

Several caveats should be added to the discussion above. Information available on arctic haze is very limited. Some ground stations, notably those at Barrow and Ny-Ålesund, have data records spanning several years. A few research flights such as AGASP-1 have operated, primarily in a survey mode. Consequently, little is known about the temporal and spatial extent of the haze over the basin as a whole. For instance, the haze disappears from Barrow in late May. Does it also disappear from the rest of the Arctic Basin at the same time? Furthermore, almost half of the Arctic Basin lies within the boundaries of the Soviet Union (Figure 1), and virtually no studies of arctic haze are known from this area. Conclusions that have assumed uniform distribution over the basin can hardly be justified without some data from the Soviet Arctic.

Conclusions

From this discussion it is clear that our understanding of arctic climate is by no means complete. Current estimates of energy lost and gained annually by the Arctic Basin (defined as the area north of 70°N latitude) are out of balance by 25 to 30%. The origin of the discrepancy is not clear, but it could arise from a number of plausible sources. In attempting to address the issue of the climatic impact of pollution in the Arctic, one of the top priorities is obviously to obtain better data on the current climate. Due to the remoteness of the Arctic and the difficulty of maintaining surface-based observing networks, the use of satellite platforms designed with polar observations in mind should be encouraged.

The magnitude of perturbations in solar absorption by the haze appears to be significant both seasonally and regionally; it is likely to be less important on an annual basis, and certainly on a global basis. Better information on the spatial and temporal extent and variability of the haze is needed in order to make a better assessment of the climatic impact. In addition, modelling of the rather complicated interaction between atmospheric heating and surface processes in the Arctic is needed in order to understand the probable consequences of heating due to the presence of haze.

Finally, it should be noted that whatever the impact of arctic haze on the climate, that impact is already occurring. The earliest reference to arctic haze in the scientific literature was in the mid 1950s. Ice core data (Barrie 1986) indicates yearly variations in acidity beginning at about the same time. However, almost all of the available climatic data on the Arctic have been obtained in the last 20 years, and must reflect the presence and influence of the haze. Thus we are forced to ask whether the arctic climate has changed or is changing, rather than asking whether it will change as a result of the haze. The former is harder to answer than the latter, because we lack knowledge about the unperturbed baseline and because we are forced to deal with past and future variations in pollutant concentrations, as well as global climatic trends. Since the needs of modern society will bring pressure for further development of arctic resources, and modern technology will enable this development to take place, it is vital that we begin to understand the Arctic as it now is, rather than be faced with trying to unravel the effects of increased pollution in another 20 or 30 years.

Acknowledgements

The authors thank Drs L. Pfister, D. Westphal, and C. McKay who read an earlier version of this manuscript and provided helpful comments and suggestions.

References

AAGAARD, K. AND GRIESMAN, P. 1975. Toward new mass and heat budgets for the Arctic Ocean. *Journal of Geophysical Research*, 80: 3821–27.

BARRIE, L. A. 1986. Arctic air pollution: an overview. In STONEHOUSE, B. (editor). *Arctic Air Pollution*. Cambridge, Cambridge University Press: 5–23.

BODHAINE, B. 1986. The Barrow Aerosol Record, 1976–1984. In STONEHOUSE, B. (editor). *Arctic Air Pollution*. Cambridge, Cambridge University Press: 159–173.

CESS, R. D. 1983. Arctic aerosols: model estimates of interactive influences upon the surface-atmosphere clear-sky radiation budget. *Atmospheric Environment*, 17: 2555–64.

CHOU, M.–D. AND OTHERS. 1984. Climate studies with a multilayer energy balance model. Part III: climatic impact of stratospheric volcanic aerosols. *Journal of the Atmospheric Sciences*, 41: 759–67.

OORT, A. H. 1983. *Global atmospheric circulation statistics*, 1958–1973. Washington, NOAA. (NOAA Professional Paper 14; available from US Government Printing Office, Washington DC).

OORT, A. H. AND VONDER HAAR, T. H. 1976. On the observed annual cycle in the ocean-atmosphere heat balance over the northern hemisphere. *Journal of Physical Oceanography*, 7: 781–800.

PORCH, W. M. AND MACCRACKEN, M. C. 1982. Parametric study of the effects of Arctic soot on solar radiation. *Atmospheric Environment*, 16: 1365–71.

RAMANATHAN, V. AND OTHERS. 1979. Increased atmospheric CO_2: zonal and seasonal estimates of the effect on the radiation energy balance and surface temperature. *Journal of Geophysical Research*, 84: 4949–58.

ROSEN, H. AND HANSEN, A. D. A. 1986. Light-absorbing combustion particles over reflecting polar ice. In STONEHOUSE, B. (editor). *Arctic Air Pollution*. Cambridge, Cambridge University Press: XXX-XX (this volume).

SCHNELL, R. C. 1986. The International Arctic Gas and Aerosol Sampling Program. In STONEHOUSE, B. (editor). *Arctic Air Pollution*. Cambridge, Cambridge University Press. 135–142.

SLINGO, J. 1982. A study of the earth's radiation budget using a general circulation model. *Quarterly Journal of the Royal Meteorological Society*, 108: 379–405.

STEPHENS, G. L. AND OTHERS. 1981. Earth radiation budgets. *Journal of Geophysical Research*, 86: 9739–60.

VALERO, F. P. J. AND ACKERMAN, T. P. 1986. Arctic haze and the radiation balance. In STONEHOUSE, B. (editor). *Arctic Air Pollution*. Cambridge, Cambridge University Press: 121–133.

THE BARROW AEROSOL RECORD, 1976–1984

BARRY A. BODHAINE

ABSTRACT. The Geophysical Monitoring for Climatic Change (GMCC) program under the National Oceanic and Atmospheric Administration (NOAA), operates an atmospheric monitoring station at Barrow, Alaska. Levels of carbon dioxide, ozone, aerosols, and other background pollutants are monitored so that their possible effects on climate may be understood. The aerosol measurement program involves continuous monitoring of condensation nuclei concentration (CNC) and aerosol scattering extinction coefficient (σ_{sp}), for which records are available from 1976 to the present. The σ_{sp} data show a strong annual cycle with maxima exceeding 10^{-5} m^{-1} in winter and spring (arctic haze), and minima of about 10^{-6} m^{-1} in summer and fall. CNC values have a half-yearly cycle, with peaks of several hundred cm^{-3} during the winter and spring σ_{sp} peaks, secondary maxima in August, and minima of about 100 cm^{-3} in summer and late fall. Neither shows significant diurnal cycles. Measurements during the Arctic Gas and Aerosol Sampling Program (AGASP) of March-April 1983 showed that ground-based aerosol measurements are strongly correlated with aerosol loading in the vertical column, and that a series of aerosol events detected at the ground were caused by rapid long-range transport to the vicinity of Barrow from Eurasia.

Contents

Introduction

The Barrow Observatory (71.32°N, 156.61°W), operated by the Geophysical Monitoring for Climatic change (GMCC) programme under the National Oceanic and Atmospheric Administration (NOAA), has become an important location for ground-based measurement of representative arctic atmosphere. Figures 1 and 2 show the observatory and surroundings at Point Barrow, the northernmost tip of Alaska: Figure 3 shows its location relative to Barrow village and the Naval Arctic Research Laboratory (NARL). Observation began in September 1971 at the US Geological Survey Geomagnetic Observatory approximately 300 m southwest of the present site. The present building was completed in October 1972 and the first station chief began observations in January 1973. Barrow has now accumulated a 9–year data set which clearly shows the annual event popularly referred to as arctic haze. The Naval Arctic Research Laboratory is no longer in operation, though some activity continues at the site.

Fig 1. The Barrow GMCC observatory and surrounding area.

Fig 2. Closeup of the Barrow GMCC observatory complex.

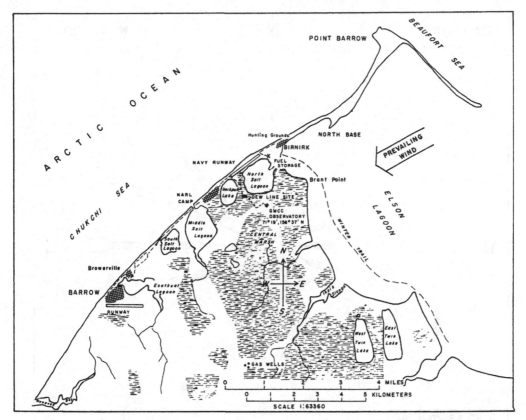

Fig 3. Point Barrow, Alska, showing the GMCC observatory in relation to other buildings and landmarks.

Figure 4 shows frequency, direction and strength of winds for the years 1977–1983 at the observatory. Winds blow from NNE-SE, the clean-air sector for the site, about 67% of the time. Bodhaine and others (1981) showed that observatory data can be screened to remove possible effects of pollution produced by local sources.

Mitchell (1957) was one of the first to publish a description of arctic haze; Duce and others (1966) and Radke and others (1976) were among the first to measure aerosol at Barrow, and Rahn and others (1977) were the first to give a detailed chemical analysis of an arctic haze episode. The first GMCC aerosol data, consisting of condensation nucleus concentration (CNC) measurements starting in 1971, were presented by Turner and Cotton (1975). Bodhaine and others (1981) described the site climatology and instrumentation, and presented the 1977–79 aersol data set, including a study of local pollution sources. Murphy and Bodhaine (1980) presented the 1976–79 data record in more detail, and Bodhaine (1983) reviewed aerosol research at Barrow and extended the long-term data record through 1981.

In March-April 1983 the Arctic Gas and Aerosol Sampling Program (AGASP) made four flights in the vicinity of Barrow (Schnell 1984). Intensive surface-based aerosol and optical measurements were also taken. Preliminary results were reported by Bodhaine and others (1984) and Dutton and others (1984), and *Geophysical Research Letters*, 11(5): 359–472, 1984 was devoted to the results of AGASP. This paper extends the GMCC

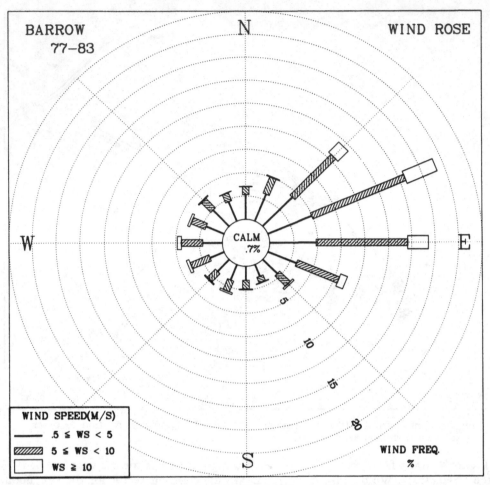

Fig 4. Frequency and direction of winds 1977–1983 at the Barrow GMCC observatory.

Barrow aerosol data set through 1984, relates some AGASP results to the regular cycles seen in the long-term record, and reassesses and corrects the 1976–79 data set.

Instrumentation

CNC is measured continuously with a General Electric automatic CN counter (Skala 1963) with modifications suggested by Norman Ahlquist of the University of Washington. This instrument produces an expansion in a cloud chamber, causing aerosol particles in a humidified air sample to grow into water droplets at a supersaturation of about 300%. The intensity of the resulting cloud is then measured optically. A Pollak photoelectric nucleus counter (Metnieks and Pollak 1959) is used as an on-site standard to obtain daily calibration points for the automatic CN counter. The Pollak counter is also an expansion instrument using optical means to detect the resulting water cloud. A review of CN counters was given by Miller and Bodhaine (1982). In practice, the automatic CN counter is routinely forced to agree with the Pollak CN counter, a calibration process described in detail by Bodhaine and Murphy (1980).

Aerosol scattering extinction coefficient is measured continuously using a four-wavelength nephelometer similar to that described by Ahlquist and Charlson (1969). Nephelometers used in the GMCC programme were specially designed to measure σ_{sp} at wavelengths of 450, 550, 700, and 850 nm. By using a real time subtraction to eliminate the molecular scattering of air, this instrument is capable of measuring σ_{sp} values as low as 10^{-7} m^{-1}, about 1% of the Rayleigh scattering by air molecules. The nephelometer is calibrated by filling it with carbon dioxide; for details of calibration and operation see Bodhaine (1979).

Three values of Angstrom exponent may be calculated from the four successive σ_{sp} values according to the formula $\alpha = -\Delta \log \sigma_{sp}/\Delta \log \lambda$, where α is the Angstrom exponent and λ is the wavelength. Interpretation of α for the multiwavelength nephelometer has been discussed by Thielke and others (1972). Assuming a power law relationship, $dN/dr \propto r^{-n}$, for an aerosol size distribution, then theoretically $\sigma_{sp} \propto \lambda^{-\alpha}$, and $\nu = \alpha + 3$, where r is the aerosol radius and ν is the slope of aerosol size distribution (Van de Hulst 1957). In general, a large α implies an aerosol size distribution biased toward small particles; small α implies a bias toward large particles. Although a power law aerosol size distribution is not a reasonable assumption over the entire range of aerosol sizes, and the above result cannot be strictly applied, the three-segment approximation for the Angstrom exponent shows significant variations and can be interpreted qualitatively in terms of variations in aerosol size distribution.

Air sampling intakes used at GMCC observatories were described by Komhyr (1983). At the Barrow observatory, a sampling stack with an inlet height of 10 m supplies air through isokinetic sampling nozzles to the instruments inside the building.

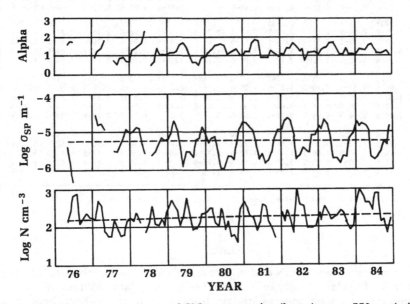

Fig 5. Monthly geometric means of CN concentration (lower), σ_{sp} at 550 nm (middle), and mean Angstrom exponent (upper), Barrow. Data are listed in Table 1; for details of trend lines see Table 2.

Table 1. Monthly geometric means of CN concentration (cm^{-3}) and σ_{sp} (m^{-1}) at 450, 550, 700, and 850 nm, Barrow.

	JAN	FEB	MAR	APR	MAY	JUN	JUL	AUG	SEP	OCT	NOV	DEC
1976 CN	-	-	-	-	147	272	710	815	121	179	246	188
σ_{sp}(450)	-	-	-	-	5.68-6	2.03-6	7.05-7	-	-	-	-	-
σ_{sp}(550)	-	-	-	-	4.16-6	1.40-6	4.71-7	-	-	-	-	-
σ_{sp}(700)	-	-	-	-	3.12-6	1.02-6	3.64-7	-	-	-	-	-
σ_{sp}(850)	-	-	-	-	2.09-6	6.67-7	2.44-7	-	-	-	-	-
1977 CN	177	160	518	365	77	60	61	156	86	61	65	147
σ_{sp}(450)	-	3.04-5	1.70-5	1.94-5	1.47-5	-	-	4.27-6	3.83-6	6.69-6	1.09-5	1.62-5
σ_{sp}(550)	-	2.84-5	1.49-5	1.65-5	1.15-5	-	-	3.12-6	3.02-6	4.94-6	7.61-6	1.21-5
σ_{sp}(700)	-	2.10-5	1.00-5	1.08-5	6.65-6	-	-	2.77-6	2.76-6	4.66-6	6.60-6	1.11-5
σ_{sp}(850)	-	1.68-5	7.45-6	7.90-6	4.46-6	-	-	2.69-6	2.80-6	4.62-6	6.13-6	1.05-5
1978 CN	181	158	254	138	-	76	147	366	122	113	167	101
σ_{sp}(450)	1.32-5	1.22-5	1.58-5	1.47-5	6.58-6	3.74-6	-	2.36-6	2.99-6	5.00-6	7.97-6	6.59-6
σ_{sp}(550)	9.83-6	1.07-5	1.33-5	1.22-5	5.24-6	2.67-6	-	2.34-6	2.92-6	4.25-6	7.35-6	6.15-6
σ_{sp}(700)	9.14-6	7.14-6	8.70-6	7.74-6	3.19-6	1.41-6	-	1.94-6	2.33-6	2.80-6	5.35-6	4.50-6
σ_{sp}(850)	8.85-6	5.25-6	6.23-6	5.39-6	2.19-6	8.41-7	-	1.71-6	1.94-6	2.02-6	4.12-6	3.44-6
1979 CN	337	521	388	272	111	182	585	199	87	94	125	229
σ_{sp}(450)	6.69-6	9.63-6	2.16-5	1.20-5	5.10-6	1.54-6	3.75-6	3.28-6	2.21-6	6.39-6	7.32-6	1.46-5
σ_{sp}(550)	6.15-6	8.81-6	2.01-5	1.09-5	4.26-6	1.21-6	3.15-6	2.99-6	2.13-6	6.29-6	7.18-6	1.41-5
σ_{sp}(700)	4.42-6	6.19-6	1.42-5	7.46-6	2.66-6	7.38-7	2.05-6	2.23-6	1.73-6	5.02-6	5.57-6	1.05-5
σ_{sp}(850)	3.35-6	4.60-6	1.04-5	5.42-6	1.84-6	5.15-7	1.49-6	1.79-6	1.46-6	4.11-6	4.48-6	8.07-6
1980 CN	245	314	323	128	101	311	86	149	58	65	40	310
σ_{sp}(450)	1.05-5	5.46-6	7.33-6	8.99-6	3.24-6	1.26-6	1.18-6	2.54-6	2.55-6	2.07-6	5.50-6	1.32-5
σ_{sp}(550)	1.04-5	5.12-6	6.86-6	7.60-6	2.81-6	1.05-6	9.96-7	2.34-6	2.22-6	1.83-6	4.83-6	1.26-5
σ_{sp}(700)	7.83-6	3.57-6	4.48-6	4.96-6	1.69-6	6.28-7	6.45-7	1.72-6	1.59-6	1.20-6	3.19-6	8.78-6
σ_{sp}(850)	5.98-6	2.57-6	3.06-6	3.36-6	1.14-6	4.30-7	4.56-7	1.35-6	1.21-6	8.99-7	2.29-6	6.53-6
1981 CN	459	341	193	172	83	139	399	341	133	93	54	-
σ_{sp}(450)	1.93-5	1.83-5	1.40-5	1.24-5	6.31-6	2.37-6	2.97-6	1.63-6	1.96-6	3.44-6	8.01-6	1.23-5
σ_{sp}(550)	1.86-5	1.68-5	1.30-5	1.04-5	5.08-6	1.90-6	2.76-6	1.47-6	1.83-6	3.01-6	7.22-6	1.16-5
σ_{sp}(700)	1.34-5	1.13-5	8.84-6	6.20-6	2.92-6	1.11-6	2.08-6	1.10-6	1.41-6	2.04-6	5.08-6	8.59-6
σ_{sp}(850)	1.01-5	8.20-6	6.51-6	4.18-6	1.91-6	7.62-7	1.71-6	9.28-7	1.13-6	1.50-6	3.85-6	6.60-6
1982 CN	274	143	268	138	69	972	394	368	97	98	106	110
σ_{sp}(450)	1.31-5	2.04-5	2.63-5	2.19-5	9.07-6	2.28-6	2.82-6	4.66-6	3.44-6	3.24-6	9.60-6	9.96-6
σ_{sp}(550)	1.20-5	1.90-5	2.33-5	1.93-5	7.51-6	1.84-6	2.42-6	3.95-6	3.19-6	2.86-6	8.49-6	8.78-6
σ_{sp}(700)	8.44-6	1.36-5	1.50-5	1.24-5	4.54-6	1.16-6	1.69-6	2.66-6	2.57-6	1.99-6	5.76-6	5.96-6
σ_{sp}(850)	6.35-6	1.02-5	1.05-5	8.79-6	3.07-6	8.77-7	1.30-6	2.00-6	2.16-6	1.50-6	4.31-6	4.47-6
1983 CN	225	151	332	247	159	186	213	272	241	143	94	107
σ_{sp}(450)	1.48-5	1.34-5	2.40-5	1.87-5	9.20-6	2.58-6	2.27-6	1.76-6	2.64-6	2.55-6	7.25-6	9.55-6
σ_{sp}(550)	1.37-5	1.19-5	2.12-5	1.62-5	7.48-6	1.99-6	1.88-6	1.53-6	2.34-6	2.15-6	6.69-6	8.80-6
σ_{sp}(700)	9.80-6	7.86-6	1.37-5	1.04-5	4.48-6	1.15-6	1.22-6	1.10-6	1.63-6	1.44-6	4.77-6	6.38-6
σ_{sp}(850)	7.44-6	5.66-6	9.55-6	7.22-6	3.08-6	8.41-7	9.21-7	8.82-7	1.29-6	1.04-6	3.64-6	4.90-6
1984 CN	398	952	697	435	456	273	776	761	139	156	70	173
σ_{sp}(450)	1.89-5	1.57-5	1.86-5	1.61-5	1.23-5	2.76-6	2.21-6	2.62-6	3.59-6	6.52-6	5.43-6	1.60-5
σ_{sp}(550)	1.76-5	1.40-5	1.65-5	1.44-5	1.03-5	2.14-6	1.84-6	2.30-6	3.14-6	5.73-6	4.71-6	1.50-5
σ_{sp}(700)	1.24-5	9.20-6	1.10-5	9.87-6	6.39-6	1.26-6	1.26-6	1.75-6	2.29-6	3.97-6	3.25-6	1.09-5
σ_{sp}(850)	9.22-6	6.51-6	8.01-6	7.30-6	4.44-6	9.45-7	9.25-7	1.27-6	1.75-6	2.99-6	2.39-6	8.36-6

A compact exponential format is used to save space; for example, 5.68-6 is equivalent to 5.68 x 10^{-6} m^{-1}.

Data

Monthly geometric means of CN and σ_{sp} are calculated from hourly data for the entire record (Figure 5); for clarity, only one channel of the σ_{sp} data (550nm) is shown. The straight dashed lines are least-squares straight lines fit to the logarithms (base 10) of the data, representing possible long-term trends on a logarithmic scale. Table 1 lists monthly means, Table 2 gives results of the linear regression, and Table 3 shows grand geometric

Table 2. Geometric least-squares trend analyses of CN and σ_{sp} (550) data presented in Figure 4.

	Slope	Intercept	SE	Trend yr^{-1}
σ_{sp}	3.31×10^{-6}	-5.25	0.400	0.01%
CN	1.46×10^{-3}	2.19	0.310	4.1%

The scale of the abscissa is such that January 1976 = 1 and December 1984 = 108.
SE = standard error about the regression line.

Table 3. Grand geometric means and standard deviations of the logarithms (base 10) of CN, σ_{sp} and α. Angstrom exponents were calculated from grand geometric means of σ_{sp}.

	Geometric Mean	Mean and Standard Deviation of Logarithms of Data Points
CN	186 cm^{-3}	
$\sigma_{sp}(450)$	6.39×10^{-6} m^{-1}	2.2713 ± 0.3107
$\sigma_{sp}(550)$	5.53×10^{-6} m^{-1}	-5.1948 ± 0.3736
$\sigma_{sp}(700)$	3.85×10^{-6} m^{-1}	-5.2571 ± 0.3851
$\sigma_{sp}(850)$	2.89×10^{-6} m^{-1}	-5.4150 ± 0.3873
α_{12}	0.71	-5.5398 ± 0.3920
α_{23}	-1.51	
α_{34}	-1.48	

means of the entire data set. Note that means, standard deviations and standard errors about the regression lines are given for the logarithms (base 10) of data points.

The long-term record of σ_{sp} (Figure 5, middle) shows clearly the annual occurrence of arctic haze, often peaking in March with a monthly mean value exceeding 2×10^{-5} m^{-1}; individual haze events often peak much higher. In summer the arctic region undergoes a cleansing process with σ_{sp} values often less than 10^{-6} m^{-1}. The Angstrom exoponent record (Figure 5, upper) also shows an annual cycle with a tendency towards smaller sized particles during late spring and early summer, probably because of photochemical particle production. In general, however, a large σ_{sp} episode will most likely be produced by an accumulation-mode aerosol, because an Aitken mode aerosol is a low-efficiency scatterer. A wind-blown soil-dust aerosol or a sea salt aerosol will give high σ_{sp} and small Angstrom exponents. The CN record (Figure 4, lower) shows a peak in early spring, coincident with the peak in σ_{sp}, and a second peak about August. These annual cycles are discussed below.

Steep trend line slopes for σ_{sp} were caused in early years by substantial year-to-year variations. As further annual records have accumulated the trend line has flattened, and is not statistically significant. Similarly the trend of about 4% yr^{-1} for CN (Table 2) is not statistically significant and due to high year-to-year variability. The apparent increasing trend is caused by unusually large CN concentrations in 1984.

Annual trends.

To study annual trends in the data, monthly geometric means of σ_{sp} and CNC for the entire data record appear in Figure 6. Long-term trends were not removed from the data because they are not statistically significant. The semi-annual CNC cycle is apparent, with peaks in March and August, minima in May and November. The geometric standard deviations (dashed lines in Figure 6, lower) suggest largest variability in July.

The remarkable σ_{sp} annual cycle (Figure 6, middle) is variable by a factor of more

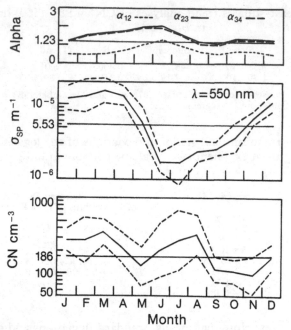

Fig 6. Geometric means by month of CN concentrations (lower), σ_{sp} at 550 nm (middle), and Angstrom exponent (upper), Barrow. Dashed lines: one standard deviation.

than ten. In general variability (dashed lines) is small compared to annual change, and substantially smaller than the variability in CNC. The largest σ_{sp} variability, like that of CNC, appears in July; the small August increase coincides with the large CNC peak, suggesting that the August aerosol maximum has particles primarily in the Aitken size range, probably produced locally by biogenic processes in the nearby ocean, or possibly the tundra. Local pollution sources have been eliminated from the data.

The annual cycle of Angstrom exponent (Figure 6, top) is interesting in increasing from January to July, implying an increase in the relative numbers of small particles during this period. This is probably caused by increased gas-to-particle conversion during transport from lower latitudes as the altitude of the sun increases. The Angstrom exponent peaks in June, coincident with the σ_{sp} minimum, implying an absence of large particles. The lowest values of Angstrom exponent, occurring during August and September, could be caused by a small influx of sea salt particles, but are not fully understood at present.

Diurnal effects.

To study possible diurnal effects, hourly geometric means of σ_{sp} and CNC were calculated monthly for 1983; March, June, September, and December means appear in Figure 7. To avoid confusion, standard deviations are not shown; however, geometric standard deviations are large, respectively about 0.26, 0.56, 0.54, and 0.35 for CNC, and 0.17, 0.51, 0.36 and 0.36 for σ_{sp}. Because of these large standard deviations, no significant diurnal cycle can be identified.

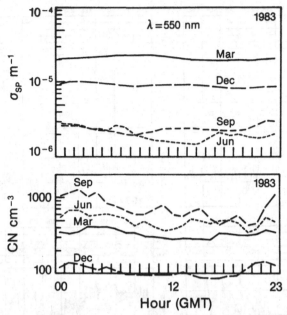

Fig 7. Monthly geometric means of CN concentrations (lower) and σ_{sp} (upper) by hour of day (GMT) for March, June, September and December 1983, Barrow.

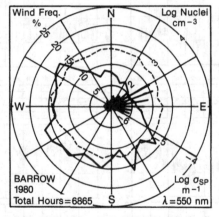

Fig 8. CN concentration (solid) and σ_{sp} at 550 nm (dashed) as a function of wind direction (36 points), 1980, Barrow. Wind direction frequency for the data set is shown by the bars in the center.

Wind direction effects.

Figure 8 shows CNC (solid line) and σ_{sp} at 550 nm (dotted line), plotted against wind direction; values are for 1981 only. As expected, CNC is significantly higher when winds are not from the clean-air sector as defined previously; effects of local pollution sources (Figure 3) show up clearly. Surprisingly, σ_{sp} is not obviously correlated with local pollution sources and is fairly independent of wind direction. Similar plots for individual

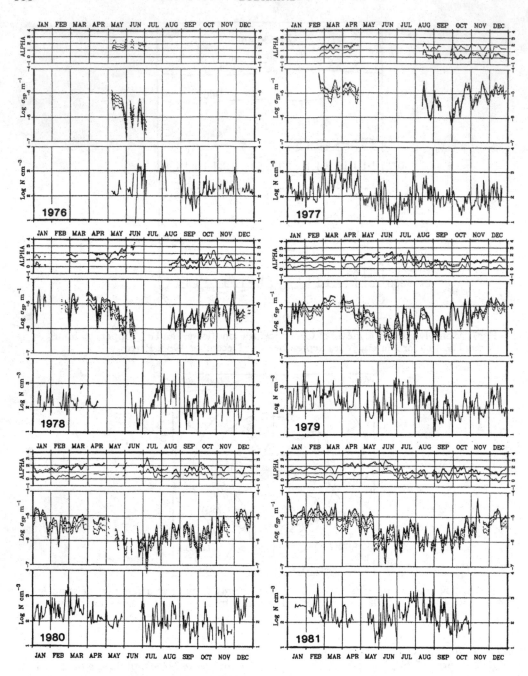

Fig 9. For caption see facing page.

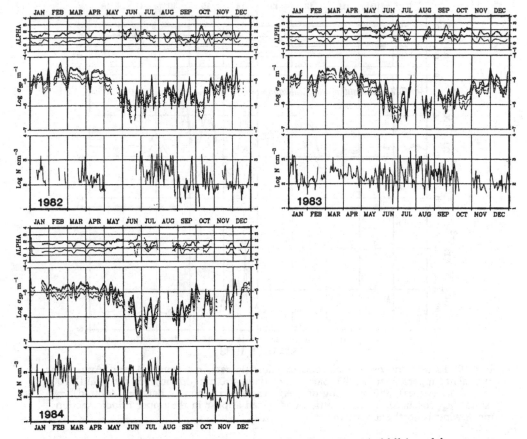

Fig 9. Daily geometric means of CN concentration (lower), σ_{sp} (middle), and Angstrom exponent (top), 1976–84, Barrow. σ_{sp} data are shown for 450 nm (dotted), 550 nm (solid), 700 nm (dashed), and 850 nm (long dashed). Angstrom exponent data are shown for α_{12} (dotted), α_{23} (solid), and α_{34} (dashed).

months again fail to show correlation between σ_{sp} and wind direction. Aerosol particles from local pollution sources do not contribute signficantly to σ_{sp}.

Complete data set, 1976–84.

This appears in Figure 9. Daily geometric means are plotted for CN concentration (lower), σ_{sp} (middle), and Angstrom exponent (upper). Though aerosol monitoring instruments were installed in May 1976, instrument problems made σ_{sp} measurements unreliable until 1977, and 1978 was the first full year of good data. All years show characteristic arctic haze in early spring, especially in the σ_{sp} data. In March and April the haze is quite persistent; however, usually sometime in mid-May, σ_{sp} values decrease sharply, reaching minimal values in June, a decline probably caused by increased cloudiness and scavenging at this time of year. Generally, the arctic haze season is marked by a series of events, or pulses, in which the aerosol appears at the surface at Barrow; records for March 1980 and 1981 show this clearly. In other years, for example March 1982 and 1983, the haze remains strong for as long as two weeks or more with only small variations.

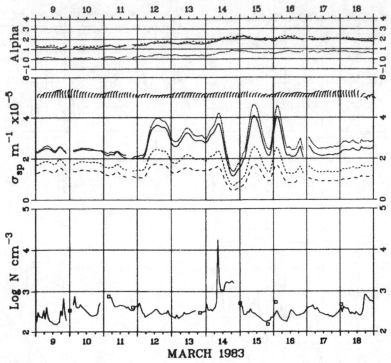

Fig 10. Hourly means of CN concentration (lower), σ_{sp} (middle), and Angstrom exponent (upper), March 1983, Barrow. σ_{sp} data: 450 nm (dotted), 550 nm (solid), 700 nm (short dashes), 850 nm (long dashes). Angstrom exponent data: α_{12} (dotted), α_{23} (solid), α_{34} (dashed). Wind direction and speed appear in the upper middle section: mean wind is north, about 5 m s^{-1}.

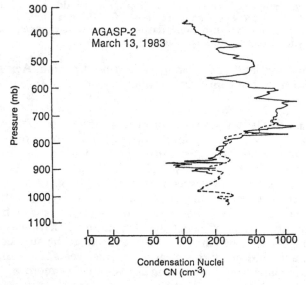

Fig 11. CN concentration as a function of altitude, 2204–2240 GMT 13 March 1983 (dashed), and 0232–0258 GMT 14 March 1983 (solid), near Barrow.

BRW TRAJECTORIES ARRIVING 83073 – 3/14/83 AT 0Z

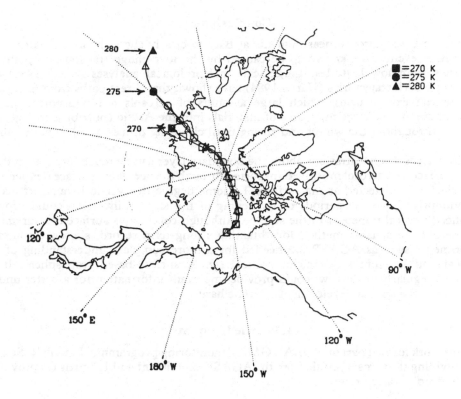

ARRIVING AT BRW

270 K: 1.8 KM   807 MB
275 K: 2.4 KM   742 MB
280 K: 3.2 KM   665 MB

Fig 12. Isentropic trajectories calculated backward from Barrow at 0000 GMT 14 March 1983 at 270 K, 275 K and 280 K potential temperature levels. Inset gives arrival heights of the trajectories at Barrow.

AGASP flights.

During the 1983 AGASP programme four aircraft flights in the Barrow vicinity on 11, 13, 15 and 17 March took vertical aerosol profiles (Schnell and Raatz 1984). The surface measurements taken at the same time were reported by Bodhaine and others (1984). Figure 10, presenting Barrow surface σ_{sp} data during the flights, shows a series of aerosol events during 12–16 March. Optical depth simultaneously recorded at the surface showed good correlation with these data, suggesting that surface measurements are representative of tropospheric phenomena above Barrow. Figure 11 presents an aircraft profile taken about 2300 GMT on 13 March, which clearly shows the haze layer between 600 and 800 mb; Figure 12 shows isentropic trajectories (Harris 1984) calculated backward from Barrow at three levels which bracket the aerosol layer shown in Figure

11. The trajectories, terminated when they fall below the 1000 mb level, clearly indicate a possible source region in Eurasia.

Conclusions

Surface-based aerosol measurements at Barrow clearly show the annual springtime arctic haze event, now known to be caused by the long-range transport of pollution materials from low latitudes. Large-scale meteorological analyses (Raatz 1984) and large-scale back trajectories (Harris 1984) have shown that direct paths from Eurasia to Barrow can exist, during which large amounts of aerosols are transported in the troposphere. These large injections of materials into the Arctic contribute to a general buildup throughout the winter and spring until cleansing processes take over in about May.

Since aerosol concentrations above the boundary layer can be much larger than those at the surface, the aerosol must be transported through the persistent surface temperature inversion to be detected at the surface. To some degree this vertical transport occurs continuously; however, periods of weakening of the surface inversion could provide enhanced vertical transport to the surface. Although continuous surface measurements are the only reasonable method for obtaining a long-term record, short-term aircraft experiments such as AGASP are needed for a more complete understanding of the vertical and horizontal extents of arctic haze. A surface-based tropospheric lidar, operated regularly at Barrow, could provide important information for a better understanding of the annual development of arctic haze.

Acknowledgements

This work formed part of NOAA's GMCC monitoring programme. I thank R. Schnell for providing the aerosol profile from the AGASP experiment and J. Harris for providing the isentropic trajectories.

References

AHLQUIST, N. C. AND CHARLSON, R. J. 1969. Measurement of the wavelength dependence of atmospheric extinction due to scatter. *Atmospheric Environment*, 3: 551–64.

BODHAINE, B. A. 1979. Measurement of the Rayleigh scattering properties of some gases with a nephelometer. *Applied Optics*, 18: 121–25.

BODHAINE, B. A. 1983. Aerosol measurements at four background sites. *Journal of Geophysical Research*, 88: 10753–68.

BODHAINE, B. A. AND MURPHY, M. E. 1980. Calibration of an automatic condensation nuclei counter at the South Pole. *Journal of Aerosol Science*, 11: 305–12.

BODHAINE, B. A. AND OTHERS. 1981. Aerosol light scattering and condensation nuclei measurements at Barrow, Alaska. *Atmospheric Environment*, 15: 1375– 89.

BODHAINE, B. A. AND OTHERS. 1984. Surface aerosol measurements at Barrow during AGASP. *Geophysical Research Letters*, 11: 377–80.

DUCE, R. A. AND OTHERS. 1966. Iodine, bromine, and chlorine in winter aerosols and snow from Barrow, Alaska. *Tellus*, 18: 238–48.

DUTTON, E. G. AND OTHERS. 1984. Features of aerosol optical depth observed at Barrow, March 10–20, 1983. *Geophysical Research Letters*, 11: 385–88.

HARRIS, J. M. 1984. Trajectories during AGASP. *Geophysical Research Letters*, 11: 453–456.

KOMHYR, W. D. 1983. An aerosol and gas smapling apparatus for remote observatory use. *Journal of Geophysical Research*, 88: 3913–18.

METNIEKS, A. L. AND POLLAK, L. W. 1959. Instruction for the use of photo- electric condensation nucleus counters. *Geophysical Bulletin No 16*, School of Cosmic Physics, Dublin Institute for Advanced Study. Dublin, DIAS.

MILLER, S. W. AND BODHAINE, B. A. 1982. Supersaturation and expansion ratios in condensation nuclei counters: an historical perspective. *Journal of Aerosol Science*, 13: 481–90.

MITCHELL, J. M. Jr. 1957. Visual range in the polar regions with particular reference to the Alaskan arctic. *Journal of Atmospheric and Terrestrial Physics*, Special Supplement: 195–211.

MURPHY, M. E. AND BODHAINE, B. A. 1980. *The Barrow, Alaska automatic condensation nuclei counter and four wavelength nephelometer: instrument details and four years of observations*. Silver Springs, MD, Air Resources Laboratory (Technical memorandum ERL ARL-90).

RAATZ, W. E. 1984. Tropospheric circulation patterns during the Arctic Gas and Aerosol Sampling Program (AGASP), March/April 1983. *Geophsyical Research Letters*, 11: 449–52.

RADKE, L. F. AND OTHERS. 1976. Observations of cloud condensation nuclei, sodium-containing particles, ice nuclei and the light-scattering coefficient near Barrow, Alaska. *Journal of Applied Meteorology*, 15: 982–95.

RAHN, K. A. AND OTHERS. 1977. The Asian source of Arctic haze bands. *Nature*, 268: 713–15.

SCHNELL, R. C. 1984. Arctic haze and the Arctic Gas and Aerosol Sampling Program (AGASP). *Geophysical Research Letters*, 11: 361–64.

SCHNELL, R. C. AND RAATZ, W. E. 1984. Vertical and horizontal characteristics of Arctic haze during AGASP: Alaskan Arctic. *Geophysical Research Letters*, 11: 369–72.

SKALA, G. F. 1963. A new instrument for the continuous measurement of condensation nuclei. *Analytical Chemistry*, 35: 702–06.

THIELKE, J. F. AND OTHERS. 1972. Multiwavelength nephelometer measurements in Los Angeles smog aerosols, II. Correlation with size distributions, volume concentrations. *Journal of the Colloid and Interface Sciences*, 39: 252–59.

TURNER, C. AND COTTON, G. 1975. *Summaries of Aitken concentrations and associated surface wind data at Barrow, Alaska, from September 1971 to February 1975*. Boulder, CO, Environmental Research Laboratories (NOAA Technical Report ERL 349–ARL 2).

VAN DE HULST, H. C. 1957. *Light scattering by small particles*. New York, Wiley.

PAST AND PRESENT CHEMISTRY OF NORTH AND SOUTH POLAR SNOW

ROBERT J. DELMAS

ABSTRACT. Chemical analysis of snow layers in Antarctica and Greenland has provided valuable information on present and past background aerosol composition, despite probable surface effects at the air-snow interface. Pre-industrial polar precipitation is chemically similar in central areas of the two ice sheets, with a weak primary aerosol component relatively free of the gas-derived acids (mainly H_2SO_4 and HNO_3) characteristic of later deposits. Large volcanic eruptions have contaminated polar snows, particularly by H_2SO_4 fall-out. Records of major volcanic events of the last 200 years are dissimilar in the two ice sheets: only the 1815 eruption of Tambora appears to be strongly and equally recorded. Anthropogenic pollution has not yet affected south polar regions: the case of heavy metals is very difficult to assess due to their extremely low concentrations in snow. Much more particulate matter reached polar regions during the last glacial age than at present; however, currently available glaciobiochemical data indicate that the effect was different both qualitatively and quantitatively for Greenland and Antarctica.

Contents

Introduction

Gaseous and particulate impurities in the atmosphere of polar regions currently attract growing interest. It has been pointed out that changes in the chemical properties of the polar atmospheres might have important climatic consequences, regionally or even globally. Moreover polar regions, particularly Antarctica, are remote from major industrial centres, and therefore the last locations where nearly pristine natural atmospheric conditions may still be investigated.

Atmospheric studies provide data over very limited and recent time periods. Glaciochemistry, which investigates chemical composition of polar precipitation, unveils atmospheric parameters of the past under a wide variety of global climatic conditions.

The capacity of polar ice sheets to lock up past environmental information over many millenia is possibly the most important justification for polar ice core studies.

A small but increasing number of reliable data sets has been recovered from fossil Greenland and Antarctic ice, with a time scale spanning entirely, though incompletely, the last glacial age and the Holocene (0 − 150,000 years BP). This paper assesses the extent to which information from the two hemi-spheres is comparable.

General features of polar snow chemistry

Substances present in polar snow include:

(1) Major soluble traces, able to liberate ions in meltwater, which determine the chemical composition of snow. Their concentrations are in the range 1 to 100 parts per billion (ppb) in most cases.

(2) Insoluble particles that participate in the mass budget, but not in the ionic budget, of snow impurities. They are a primary component in the aerosol phase.

Both may influence markedly the physical properties of snow and ice.

(3) Minor traces ('ultraces', whose concentrations may be arbitrarily fixed below 1 PBB) have no real influence on the physical or chemical properties of snow and ice, but may serve as 'tracers', particularly useful for identifying the various sources (anthropogenic, volcanic, etc) of the deposited matter.

(4) Gases contained in bubbles enclosed in ice do not strictly belong to the chemical compounds of ice since they are not incorporated in the lattice and are mostly released after melting. Moreover, they do not (in principle) interact with the impurities present in the solid. For instance, it has been shown that CO_2 is confined to bubbles and absent from the ice lattice. Permanent and trace gases are important testimonies of global atmospheric changes and of great interest in paleoenvironmental studies. Ice records of gases will not be discussed in this paper.

Snow layers record information about certain physical and chemical properties of the atmospheres in which they were deposited. It is generally accepted that no significant modification of the entrapped information occurs within the ice sheets. Discussion arises when the results of the chemical analysis of snow and ice samples are interpreted in terms of past atmospheric composition. The transfer of atmospheric gaseous or particulate impurities to snow is not a simple process. The scientific community involved in air chemistry studies is often very critical of glaciochemical studies. Interpretation of chemical ice records in terms of past atmospheric variations has often been empirical, without first establishing the physical phenomena at the air snow interface.

During wet deposition the aerosol or gas is scavenged by precipitation before reaching the ground. In polar atmospheres, scavenging within the clouds (rainout) is dominant in comparison with below-cloud processes (washout). However, how the impurities are physically included in snowflakes or ice crystals in polar conditions is poorly understood.

In dry deposition, the gas or particle is directly deposited by impact on the snow surface; the process is generally considered more efficient for gases than for aerosol particles. There is now some evidence that dry deposition is far from negligible in polar areas of very low snow accumulation rates (A less than 20 g cm^{-2}a^{-1}). At Dome C in Greenland (A = 3.4) it has been found that 60% to 70% of beta radioactivity and sulfate could be dry deposited. As areas of very low rates of accumulation are much more extensive in Antarctica than in Greenland, the dry deposition process should be especially important in Antarctica.

However the largest differences between aerosol and polar precipitation compositions

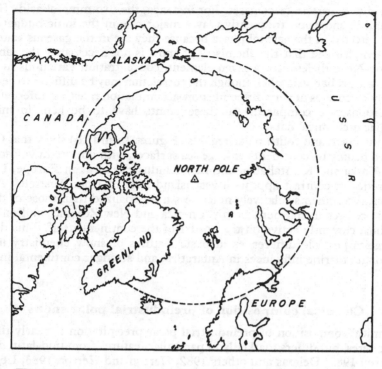

Fig 1. The north and south polar regions and adjacent continents.

are probably due to certain trace gases. For instance, the two mineral acids HNO_3 and HCl are found in snow and frequently play a major role in the ionic budget; however they are not detected in the aerosol phase because they are in the gaseous state. It must therefore be emphasized that the chemistry of snow is not necessarily the same as that of the aerosol. Nevertheless, for a given element, concentration variations in the snow and in the air are probably parallel, though the correlation may be difficult to demonstrate quantitatively when sites of very different snow accumulation rates, or different meteorological conditions, are compared. All these points have to be kept in mind when discussing glacio-chemical data.

Maps of the north and south polar regions (Figures 1a and 1b) show that Greenland is surrounded mainly by continents or large sea surfaces, with important volcanic centres in Iceland, Alaska and Kamtchatka, and major industrial areas in Europe, USSR and USA. Antarctica in contrast appears a vast island isolated from Australia, Africa and South America (continents relatively much less industrialized than those of the north), with volcanic centres in southern South America and New Zealand, and one volcano, Mount Erebus, currently active on the border of the continent itself. This description already suggests possible differences in basic features of snow chemistry in the two regions: stronger marine influences in Antarctica, and stronger continental influence in Greenland.

Chemical composition of preindustrial polar snows

The chemical composition of preindustrial polar precipitation is nearly the same in both hemispheres, but differs markedly in particular features from mid-latitude precipitation. (Herron 1982; Delmas and others 1982, Herron and Herron 1983, Legrand and Delmas 1984.) Currently polar snow is extremely pure, with an ion budget (the sum of all ionic species) generally lower than 10 μEq.l^{-1}. For low altitude snow in other latitudes this value is typically an order of magnitude higher. Polar snow is always slightly acidic, for two reasons. Firstly, primary aerosol contributions (alkaline sea-salt and crustal dust) are weak in comparison to secondary (gas-derived) acidic aerosol. Secondly ammonia, coming mainly from continental sources, is not a significant neutralizing agent in these remote areas.

Figure 2 compares ion balances for geographically similar sites in Greenland and Antarctica. It is worth noting that no actual ion balance has yet been established for Greenland snow because of the lack of free acidity measurements (Figure 2, caption). However, these two ion balances demonstrate that, for polar sites at similar altitudes and with low accumulation rates, chemical composition of snow is nearly the same in both hemispheres. Though various ion balances have been reported for polar snows, the basic composition shown in Figure 2 is fundamentally unvarying except at most coastal locations. This uniformity suggests strongly that a background precipitation chemistry, typical of polar areas, can be defined for both hemispheres. Several authors have recently described the physical properties of a background polar aerosol common to north and south high latitude regions.

The low primary aerosol contribution may be explained by the high altitude (above 2,500 m) of the central polar ice sheets. Precipitation forms above the boundary layer in an atmosphere which, even over marine areas, has a very low sea salt content; in both Antarctica and Greenland Na concentrations are lower than 1.5 μEq.L^{-1}. (Na is a much more conservative reference element than Cl for assessing the marine contribution). Altitude, rather than distance from the sea, is the determining parameter for the amount of sea salt transported inwards over the ice sheets. Long range transport of continental

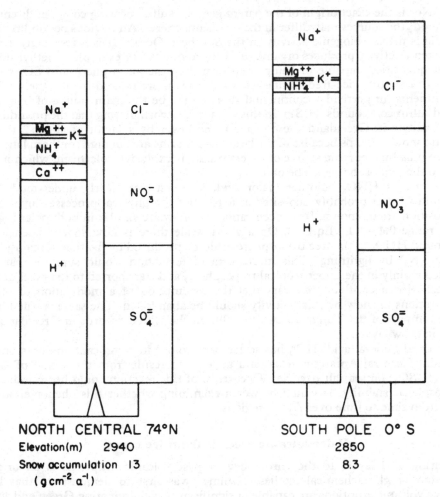

NORTH CENTRAL 74°N   SOUTH POLE 0° S

Elevation(m)   2940   2850

Snow accumulation 13   8.3

$(g\,cm^{-2}\,a^{-1})$

Fig 2. Ion balance of Greenland and Antarctic snows. As no H^+ data are available for North Central, the H^+ calculated here corresponds to an exact anion-cation balance. Na^+, Mg^{++} and K^+ have been calculated from Cl values. The ionic budget at North Central, $7\ \mu EqL^{-1}$, gives the scale.

dust also takes place above the boundary layer. Due to the remoteness of Antarctica, this contribution is extremely low on the Antarctic plateau, as indicated by an Al content rarely higher than 2 PPB. At the South Pole the insoluble part of snow impurities represents less than 10% of the mass budget. The size distribution of microparticles in Antarctic snow clearly indicates that they originate mainly from the surrounding continents and not from local sources. In Greenland snows the large continental areas of the northern high latitudes give rise to higher background levels (2 to 10 PPB).

The key compound in polar precipitation and aerosol is sulfate, in particular excess sulfate (H_2SO_4) which in central Antarctica represents more than 90% of total sulfate. This acid, ubiquitous throughout the global troposphere, is gas-derived and formed by atmospheric oxidation of SO_2. The location and processes of this important chemical transformation have not been determined precisely for the special case of polar excess sulfate. Nor is the exact origin of the parent gaseous sulfur-bearing compounds entirely understood; the sources may differ in the two hemispheres. An obvious possibility in the Antarctic is marine biogenic activity in the Southern Ocean. It is increasingly evident that biogenic activity produces organic sulfur compounds (for example dimethyl sulfide) in great quantities, capable of being transformed in the atmosphere into SO_2 and/or H_2SO_4. In the north, ice-free marine areas are much more limited than in the south, so the continents, in particular continental shores, may be the main source of biogenical reduced sulfur compounds (H_2S?). In this debate, it is worth noting that the preindustrial level of sulfate in Greenland snow seems to be lower by a factor of about 2 than in Antarctic snow, perhaps because of the larger free marine area in high southern latitudes. An additional but sporadic source of excess sulfate is explosive volcanism, which injects SO_2 into the stratosphere (see below).

The case of HNO_3, another major acid, is much less clearly understood. This compound is most probably deposited as a gas, but the physical processes involved in its deposition are unknown. Its concentrations in snow are similar in both polar regions (in the range $0.4 - 1$ $\mu Eq.L^{-1}$). On a global scale there is now some evidence that background HNO_3 in the free troposphere could be produced by fixation of atmospheric nitrogen (N_2) by lightning. This mechanism of formation would suppose that it is produced mainly in the upper tropical troposphere, and transported to the poles at high altitudes. Moreover it now appears that the assumption of a modulation of HNO_3 concentrations in snow by solar activity should be abandoned. The same is true for the assumed impact of the Tunguska event (1908) on the HNO_3 records in Greenland and Antarctic snow layers.

The third mineral acid, HCl, has so far been found in significant amounts only in well-defined central Antarctic areas, and appears to result from the action of excess sulfate (H_2SO_4) on sea salt particles. Occurence of this reaction could be an indication of atmospheric stability. It would be worth examining whether this phenomenon also occurs from time to time over Arctic regions.

Paleovolcanic records from ice cores

Relating acid layers to the chronology of past volcanic eruptions is an original application of glaciochemical studies. Hammer was first to demonstrate that large explosive volcanic eruptions are capable of significantly contaminating Greenland snows through the deposition of sulfuric acid over 1 to 2 years (Hammer 1977; Hammer and others 1980). The same effect has been found more recently in Antarctica (Delmas and others 1985). Ash emissions dust layers in snow occur only close to eruptions, whereas volcanic sulfur-bearing gases have a global (or at least hemispheric) effect. To be clearly identified, sulfate from a volcanic source must emerge clearly from the natural background of sulfate scattered in ice and snow. Levels detected as due to eruptions generally overcome background scatter by factors of 2 to 4, exceptional ones by factors of up to 10.

Most eruptions recorded in the Antarctic ice sheet appear to be linked to the stratospheric transport of sulfuric acid (the major constituent of the stratospheric Junge layer). This conclusion is less obvious for Greenland, where the ice cap shares a belt of latitude with major volcanic areas (Figure 1a). Table 1 compares signals in Greenland and Antarctic firn cores for all major eruptions of global concern catalogued over the last 200 years. Among the 80 eruptions recorded, nearly one in five left an evident signal in the Greenland ice (10%), in Antarctica (14%), or in both (5%); see Table 1. The

Table 1. Major volcanic eruptions of global concern over the last 200 years that have given a signal in polar snows. VEIs (volcanic explosivity indexes) are from Newhall and Self (1982). Quality of signals: *** excellently recorded; ** recorded; * weakly recorded; 0 definitely not recorded.

VOLCANO	LATITUDE N	S	YEAR	VEI	GREENLAND	ANTARCTICA
Lakagigar	64		1783	4	***	0
Pogromni	55		1795	4	***	0
Tambora		8	1815	7	***	***
Galungung		7	1822	5?	0	***
Cosiguina	13	1835		5	*	**
Hekla	64		1845	4		
Armagura		18	1846	3?	**	
Makjan	0		1861	3		**
Krakatau		6	1883	6	**	**
Tarawera		38	1886	5		**
Thompson Isld		54	1895?	?	0	**
Santa Maria	15		1902	6		***
Novarupta (Katmai)	58		1912	6	***	0
Hekla	64		1947	4	*	0
Agung		8	1963	4	*	***

well-known eruption of Tambora, Indonesia, in 1815 appears to be the most universally recorded volcanic event of the last two centuries.

One of the major interests of this comparison is to show that the number of events recorded in Antarctica equals or even exceeds those of Greenland, despite the substantial imbalance in the distribution of volcanos over the world. For instance forty major eruptions (VEI greater than four) of the last 200 years were located poleward of 20° N, only five poleward of 20° S. Methods used to obtain volcanic records from firn cores have been different for Antarctic and Greenland: nevertheless this observation suggests that the signal-to-background ratio (R) of volcanic eruptions is greater in the Antarctic. A possible explanation lies in the influence of snow accumulation rates (A) on R. A recent detailed study of Agung H_2SO_4 fallout in Antarctic snow layers strongly suggests that R is reduced as A grows. As values of A are generally greater in Greenland than in Antarctica, it can be deduced that Antarctic ice cores should be more suitable than those of Greenland for reconstructing a paleovolcanic history of global concern. Finally Antarctica also appears to be better protected than Greenland from the tropospheric transport of volcanic products from nearby moderate eruptions, of limited interest for global climate studies.

In the future it will be most interesting to compare long-term volcanic records in deep ice cores of both ice sheets in order to catalogue and date past cataclysmic eruptions, of Tambora magnitude, which have affected the global atmosphere.

Pollution records

Most anthropogenic products in polar snows are not directly identifiable. Increasing human influence in these regions is better revealed by gradual increasing concentrations from layers deposited in the last century to recently fallen snows. An increase must be observed over several decades before it can be clearly assessed and separated from natural background scatter.

Over 90% of atmospheric pollutants are emitted in the northern hemisphere, and inter-hemispheric exchanges in the troposphere are limited. Consequently effects of global pollution should be easier to detect in Greenland than in the Antarctic.

The first clear evidence for long-range transport of an industrial impurity (lead) to Greenland was discovered 20 years ago. Recent concentrations of Pb were shown to be at least 2 orders of magnitude higher than in layers representing 2,000 years before the present (0 AD) (Wolff and Peel 1985). Human impact was already significant before the industrial revolution. Deposition in snow during the last 100 years has increased by a factor of 5, while emissions over the same period have grown exponentially (Figure 3). Present day concentrations of Pb in north polar regions are in the range 0.15 to 0.25 ppb, with a marked winter maximum (as for most pollutants) corresponding to the intrusion of mid-latitude polluted air masses.

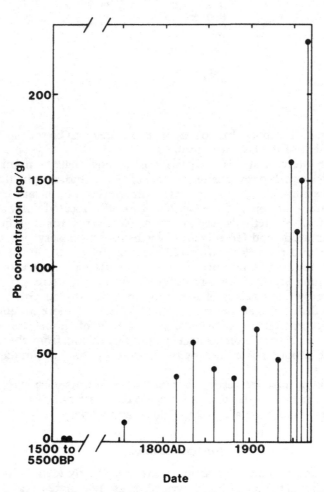

Fig. 3. The increase of anthropogenic lead pollution in Greenland snow (from Wolff and Peel 1985).

In fact the determination of pollution level of polar snows by heavy metals is a very difficult task, because of contamination during sampling and analytical work. Background levels in polar snows have not yet been clearly defined. Many published values should be treated with caution. Table 2 gives the most probable levels currently available for some of these metals in polar snows. Concentrations are 20 to 60 times higher in Greenland than in Antarctica.

Table 2. A comparison of pollution of northern and southern polar snows by heavy metals (units: 10^{-12} gg^{-1} or ppt. Data compiled by Wolff and Peel 1985).

	Pb	Cd	Cu	Zn	Ag
Greenland(G)	200	5	97	200	10
Antarctica (A)	5	0.26	1.8	3.3	0.5
G/A	40	20	50	60	20

Although the pollution level of the north polar region has been widely investigated and seems now to be clearly established, the case of Antarctic snows is still uncertain (Boutron and Patterson 1983). As the behaviour of trace metals in the atmospheric environment is not entirely understood, natural enrichment factors (in comparison with average crustal concentrations) are not accurately known. The increase of these metals in recent Antarctic snows, as generally observed, could very well be a pure natural phenomenon. Only lead can be considered to show any anthropogenic influence in Antarctica at the present time (Figure 4).

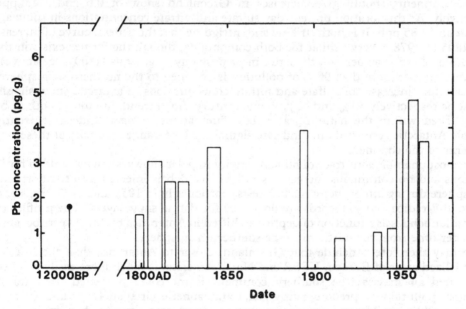

Fig 4. Concentrations of lead in Antarctic precipitation (From Wolff and Peel 1985).

Other major atmospheric pollutants that have reached Greenland are sulfate and nitrate, the two most important components of acid rain. Amounts of these compounds naturally present in the atmosphere are high in comparison with trace metals, as the impact of anthropogenic emission, even though enormous, is relatively less significant. Only long term records over more than 50 years allow detection of a possible gradual increase in concentration over year-to-year fluctuations. Sulfate and nitrate concentrations have increased respectively by factors of 3 and 2 in snow from southern Greenland

184   DELMAS

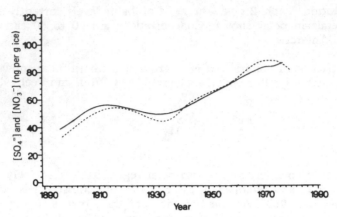

Fig 5. Trends of sulfate (stippled) and nitrate (solid) deposition at Dye III (Greenland) (From Neftel and others 1985).

since the end of the last century (Neftel and others 1985). Some authors have calculated also that an increase of 1 Tg (1 Teragramme $= 10^{12}$ g) of sulfur and nitrogen produced on the continents should give increases in Greenland snow of 0.6 and 1.43 ppb respectively. At this location present-day sulfate and nitrate concentrations in snow are equal at about 85 ppb. It is both striking and intriguing that the general curve of increase from 1895 to 1978 is very similar for both compounds, though the former exists in the polar atmosphere as an aerosol, the latter most probably as gaseous HNO_3 (see above).

Assuming that more than 90% of pollution is confined to the northern hemisphere, it appears that increases of sulfate and nitrate concentrations in antarctic snow should at most be respectively 30% and 40% in one century. Such trends are too smooth to be distinguished within the natural scatter and fluctuations of data. Indeed, currently available Antarctic records do not indicate significant long-range transport of these two pollutants to the continent.

The most unambiguous trace of human activity in polar snows, showing sudden rapid variations and no continuous increase, is radioactive debris injected into the northern stratosphere during atmospheric nuclear tests, particularly in 1954 and 1962. ^{137}Cs and ^{90}Sr (total β-radioactivity) records are now used for dating snow layers. The profiles of these radionuclides as a function of depth are shifted in Antarctica by 1 to 2 years because of transfer time from the northern to the southern hemisphere.

We may therefore conclude that Greenland, closer to Antarctica than the heavily industrialized areas of Europe, USSR and North America, is much more easily reached by polluted air masses; the southern continent is moreover protected from trans-latitudinal pollution by predominantly zonal atmospheric circulation. Antarctica may therefore still be considered a virgin area where, a few elements excepted, anthropogenic influences on snow chemistry are very difficult to establish.

Climatic changes and snow chemistry

Major changes in distribution of deserts, marine and atmospheric circulation, marine and land biogenic activity etc, occurring in periods of climate change, have induced significant modifications in the composition of polar aerosol. Chemical analysis of deep polar ice cores has already provided fundamental insight into the global environment when earth's pattern of climate differed from that of today.

Data at present available from Greenland are based on analysis of two deep ice cores, one from Camp Century in the northwest and the other from Dye III in the south. Four deep ice cores from the Antarctic—Byrd Station, D 10, Vostok and Dome C—are currently being examined, the two latter stations representing central areas.

All results (Cragin and others 1977; De Angelis and others 1984; Lorius and others 1985) lead to the conclusion that drastic climatic changes occurred on earth during the last ice age, in particular toward its end 16,000–25,000 BP, the last glacial maximum products in the snow was considerably higher during the LGM than now, which means that the atmosphere was at this time heavily loaded with dust and sea salt (Table 3).

Table 3. LGM-to-Holocene ratios for concentrations of selected elements or compounds in Greenland and Antarctic ice cores. * Cold-to-warm period ratios; ** alkaline ice during the LGM. Na_M is marine sodium. Ratios calculated from data of Cragin and others (1977), De Angelis and others (1984); Finkel and Langway (1985), Legrand (1985) and Lorius and others (1984).

	Na	K	Mg	Ca	Si	Al	SO_4	NH_4	H	NO_3	Cl	H_2SO_4	Na_m
Camp Century	3.0	8.8	9.7	32	31	11	3.5	–	**	–	–	–	–
Dye III	–	–	–	–	–	–	4.3	–	**	1.5	1.7	–	–
Byrd Station	1.7	1.4	3.1	2.4	6.9	7.3	1.5	–	–	–	–	–	–
	2.1	1.6	2.1	–	–	8	1.6	1.09	1.5	1.3	2.4	1.3	2.1
D 10	3.4	–	–	–	–	11	–	–	1.4	–	–	1.9	3.1
Dome C	6.7	10	8	20	–	27	2.5	2.7	0.9	2.0	6.3	1.2	5.6
Vostok	–	–	–	–	–	37	–	–	1.1	–	–	–	5.0

Enhancement of continental dust deposition during the LGM is due to two factors—extension of desert areas on the surrounding continents and stronger winds. Aeolian erosion and long-range transport at altitudes above 3,000 m were greater in comparison to the present. The source regions of this higher continental contribution may however have been different in the two hemispheres; it is mostly alkaline (due to $CaCO_3$?) in Greenland, and lacking in carbonaceous matter in Antarctica. Continental alkaline matter was so high at Camp Century snows from the LGM show an overall alkalinity. On the other hand acidic precipitation prevailed in Antarctica. The continental shelves exposed to wind erosion after the general lowering of sea level (by about 150 m) at this time were possible sources of $CaCO_3$ particles. Neutralization of this material during atmospheric transport could explain its disappearance in the southern hemisphere, where transport pathways were longer. Finally there is no unquestionable demonstration of enhanced or sustained volcanic activity to account for the global increase of dust during the LGM, even though sulfate levels were relatively high at the time.

Apart from stronger zonal transport, greater storminess in the subantarctic may explain the stronger concentrations of typical marine elements (such as Na, Cl, K, Mg) in LGM snow. It can be calculated that a doubling of wind speed near the sea surface could account for the ratios given in Table 3. The same phenomenon, though much less marked, is observed also in the northern hemisphere. It is worth noting that, in the Antarctic, high levels of marine aerosol contribution ceased at the end of the LGM a few thousand years later than the terrestrial contribution, the two following in sequence after the change in climate.   Deposition of secondary aerosol in snow seems relatively little affected by temperature changes. Some effects seem to exist but have not yet been clearly defined. In fact the relatively simple chemical budget, dominated by secondary aerosol contributions, that has been elaborated for recent polar precipitation is unsuited to such complex solutions of primary and secondary aerosols. Certain contributions, at present negligible, have to be taken into account to understand LGM snow chemistry. A typical example is sulfate. As already discussed, in central polar areas excess sulfate is at present largely dominant. On the other hand all potential sources of neutral sulfate (marine and crustal) contributed very significantly to the observed increase of this element during the LGM, so that the net effect of the climatic change on the concentration of excess sulfate could in fact be relatively weak (Legrand 1985) (Table 3).

Another factor which could affect the values reported in Table 3 is the accumulation rate. Little information is currently available either on snow accumulation rates during the LGM, or on relationships between rates of accumulation and concentration (see above). Studies of past environments are in their initial stages. In Greenland it has recently been shown that several cold periods seem to have similar atmospheric chemistry features (Finkel and Langway 1985). Studies concerning periods of climatic transitions would be most interesting.

Conclusions

From this comparison of north and south polar snow chemistry, it may be deduced that complementary information on past global environments can be obtained from ice core analysis. Current research in these regions involves the chemistry of background precipitation, dominated by the contribution of gas-derived acids. During the ice ages sea salt and dust particles were the dominant components. For events of global concern such as climatic changes, comparisons of Greenland and Antarctic ice records are particularly useful in differentiating between regional and global effects. Anthropogenic pollution is still a matter of serious concern only in the northern hemisphere, mainly because it affects only the troposphere and therefore does not cross the equator.

References

BOUTRON, C. F. and PATTERSON, C. C. 1983. The occurence of lead in Antarctic recent snow, firn deposited over the last two centuries and prehistoric ice. *Geochimica et Cosmochimica Acta*, 47: 1355–68.

CRAGIN, J. H. AND OTHERS. 1977. Interhemispheric comparison fo changes in the composition of atmospheric precipitation during the last cenozoic era. In: DUNBAR, M. J. (editor). *Polar Oceans*. Calgary, Canada Arctic Institute of North America: 617–31.

DE ANGELIS, M. AND OTHERS. 1984. Ice Age data for climate modelling from an Antarctic (Dome C) ice core. In: BERGER, A. L. AND NICOLIS, C. (editors). *New Perspectives in Climate Modelling*. Amsterdam, Elsevier: 23–45.

DELMAS, R. J. AND OTHERS. 1982. Chemistry of south Polar snow. *Journal of Geophysical Research*, 87: 4313–38.

DELMAS, R. J. AND OTHERS. 1985. Volcanic deposits in Antarctic snow and ice. *Journal of Geophysical Research*, 90: 12901–20.

FINKEL, R. C. AND LANGWAY, C. C. Jr. 1985. Global and local influences on the chemical composition of snow fall at Dye 3, Greenland: the record between 10 ka BP and 40 ka BP. *Earth and Planetary Science Letters*, 73: 196–206.

HAMMER, C. U. 1977. Past volcanism revealed by Greenland ice sheet impurities. *Nature*, 270: 482–86.

HAMMER, C. U. AND OTHERS. 1980. Greenland ice sheet evidence of past glacial volcanism and its climatic impact. *Nature*, 288: 230–35.

HERRON, M. M. 1982. Impurity sources of F^-, Cl^-, NO_3^- and SO_4 in Greenland and Antarctic precipitation. *Journal of Geophysical Research*, 87: 3052–60.

HERRON, M. M. AND HERRON, S. 1983. Past atmospheric environments revealed by polar ice core studies. *Hydrological Sciences Journal*, 28(1): 139–153.

LEGRAND, M. 1985. *Chimie des neige et glace antarctiques: un reflet de l'Environnement*. Doctoral thesis, University of Grenoble.

LEGRAND, M. R. AND DELMAS, R. J. 1984. The ionic balance of Antarctic snow: a 10–year detailed record. *Atmospheric Environment*, 18: 1867–74.

LORIUS, C. AND OTHERS. 1984. Late-glacial maximum-Holocene atmospheric and ice-thickness changes from Antarctic ice-core studies. *Annals of Glaciology*, 5: 88–94.

NEFTEL, A. AND OTHERS. 1985. Sulphate and nitrate concentration in snow from South Greenland 1895–1978. *Nature*, 314: 611–13.

NEWHALL, C. G. AND SELF, S. 1982. The volcanic explosivity index (VEI): an estimate of explosive magnitude for historical volcanism. *Journal of Geophysical Research*, 87: 1231–38.

WOLFF, E. W. AND PEEL, D. A. 1985. The record of global pollution in polar snow and ice. *Nature*, 313: 535–40.

FUTURE CHEMICAL MEASUREMENTS IN THE ARCTIC ATMOSPHERE

ØYSTEIN HOV

ABSTRACT. The arctic atmosphere is remote from pollution sources; measurements there are of particular interest as indicators of pollution background concentrations for the northern hemisphere. Top priority in chemical measurement should be to improve understanding of atmospheric distribution of trace gases (CO_2, O_3, CH_4, N_2O and chlorofluorocarbons) that determine the tropospheric infrared radiation budget. Equally important is to measure trace gases (CO, NO_x and individual hydrocarbons) that indirectly determine the abundance of tropospheric ozone. Light-absorbing aerosols should be measured insofar as changes in their abundance are believed to contribute to climatic change. Ice core and snow pack samples record changes in abundance of species involved in control of climate, and the history of emissions and deposition of sulphur compounds, NO_x and trace metals, which is of value in understanding how atmospheric composition has developed. If environmental effects of air pollution within the Arctic require control strategies to be considered, arctic airmasses can be tracked back to their origins from the distribution of trace metals in aerosols, and through measurement of trace gases which originate only from known specific sources.

Contents

Why measure atmospheric composition in the Arctic?

The arctic atmosphere is an important part of the global atmosphere. Chemical changes in the global atmosphere inevitably influence the composition of arctic air (sometimes in ways which shed light on why and how the changes occur) and have climatological implications on a world scale. Chemical measurements in the Arctic should therefore as a top priority aim at improving the understanding of trace gases which, together with water vapour, directly determine the infrared radiation budget in the troposphere. The gases concerned are notably carbon dioxide (CO_2), methane (CH_4), ozone (O_3), nitrous oxide (N_2O) and such chlorofluorocarbons as CFC-11($CFCl_3$) and CFC-12(CF_2Cl_2). In a recent study, Ramanathan and others (1985) estimated that future climatic change from increases in O_3, N_2O and CH_4 would be comparable to the effect of the CO_2 increase. Equally important is the determination of nitrogen oxides (NO_x), gaseous hydrocarbons

(HC), carbon monoxide (CO) and other trace gases that indirectly determine the abundance of O_3. The arctic atmosphere is of particular interest because it is far from sources of pollution; measurements there are indicative of northern hemisphere background concentrations. Furthermore, in winter chemically reactive species reach high concentrations in the arctic atmosphere. With the changes in radiation and transport patterns that occur in spring, gaseous hydrocarbons and carbon monoxide from the arctic atmosphere may contribute in an important way to the generation of secondary species like O_3 and peroxyacetylnitrate (PAN) at mid and high latitudes in the northern hemisphere (Isaksen and others 1985).

Just as air measurements in the Arctic improve understanding of current atmospheric composition and its changes, analyses of arctic snow pack and ice cores provide a record of historical changes. Greenland's ice sheets have preserved a unique, dateable record of the past composition of the global atmosphere (Wolff and Peel 1985). Chemical analysis of the upper layers for major ions (nitrate, sulphate, ammonium), pH, and such gases as CH_4 and CO_2, give evidence of increasing pollution. Not only can general changes in northern hemisphere emissions of pollutants be detected. At Svalbard, for example, where precipitation is high due to orographic enhancement (Semb and others 1984) and winds during precipitation blow mainly from east and southeast, snow profiles provide a chemical record of precipitation that has originated in specific, known regions of the industrial northern hemisphere. There is consequently a historical record of how emissions of sulphur dioxide (SO_2) and NO_x have changed and developed with time. These data are useful for global models of atmospheric development, as well as measurements of CO_2, CH_4, major ions etc, against which models can be compared. Polar ice samples also provide information about chemical processes in the atmosphere in the past. Neftel and others (1984) measured concentration in ice cores of hydrogen peroxide (H_2O_2), a powerful oxidant believed to contribute significantly to the oxidation of SO_2 to sulphate in clouds.

It is not clear how current loadings of arctic atmospheric pollutants directly affect the arctic environment. Springtime absorption of solar radiation by arctic haze has been estimated to produce instantaneous solar heating rates of the order of 1 K/d (Valero and others 1983); acid deposition effects remain undocumented, but deposition of man-made pesticides from mid-or low latitudes is a matter for concern. If environmental effects prove strong enough to warrant control strategies, the origins of arctic air pollution will need to be known. They can be traced by analyses of trace gases and aerosols. Much research effort has been spent on examining trace element concentrations of arctic aerosols. Rahn (1981) has used manganese:vanadium (Mn/V) ratios to distinguish North American from Eurasian aerosols, and radioactive lead to identify sources of arctic aerosols. Particulate elemental carbon may also be used (Heintzenberg 1982), as well as nickel, lead and zinc (Ottar and Pacyna 1984). Measurement of specific trace gases in arctic air samples provides an even more promising method than trace element analyses, because changes in the size spectrum of aerosol particles in transit cause difficulties of interpretation, while many trace gases tend not to be altered during the five- to-ten days of travel from source regions in Eurasia or North America, thus conserving source region 'signatures'.

Khalil and Rasmussen (1984) suggested as tracers $C_2Cl_3F_2$ (F-113) and the fire-extinguishing compound CF_3Br which are used primarily in the US, and CF_2BrCl, which is used primarily in Europe; none of these gases is believed to be used on a wide scale in USSR. Analytical procedures for measuring them have been refined to the level of one part per trillion by volume. A significantly higher ratio of

$(C_2H_6 + C_3H_8):(C_2Cl_4 + C_2HCl_3)$ has been measured at Svalbard during southeasterly flow off USSR, than at Barrow during springtime (Hov and others 1984). C_2H_6 and C_3H_8 originate mainly from natural gas exploitation, while C_2Cl_4 (carbon tetrachloride) and C_2HCl_3 are industrial solvents. Comparisons between Svalbard and Barrow indicate that use of chlorinated ethenes is primarily confined to western industrialized countries.

Temporal and spatial resolution of measurements

Measurements of trace species in arctic air must include the dominant aspects of their distribution in space and with time. Raatz and Shaw (1984) used noncrustal V and Mn as chemical tracers to investigate the long-range transport of pollution-derived aerosols to the Alaskan Arctic, finding that movement occurs when mid-latitudinal and arctic atmospheric circulations are quasi-persistent. The necessary meridional shift takes place mainly along the edges of near-stationary anticyclones, where pressure gradients are relatively strong. Seasonal variation in the concentration of pollution-derived aerosol linked closely with seasonal variation in the occurrence and position of mid-latitude blocking anticyclones. Blocking occurs typically when a warm anticyclone is located in about 60°N, and a cold cyclone in 40°N. West of these pressure centres the upper-atmosphere jet stream is split into two branches. Within the western part of the blocking, enhanced meridional exchange is taking place. From the occurrence of a surge at the beginning and the arrival of polluted air at the end of the episode, Raatz and Shaw (1984) estimated that transportation from mid-latitude to the Alaskan Arctic took seven to ten days.

Variations in transport mechanism and slow changes in global emissions of some trace gases together give rise to changes in concentration distributions at arctic locations on at least three different time-scales: episodic, seasonal and long-term. Episodes typically last a few days, the average time between large changes in the synoptic weather situation. Measurement of trace species should aim at resolving the smallest of these time-scales; consequently a temporal resolution of one to two days should be satisfactory when measuring most species.

Figure 1 shows how measurements of O_3 at arctic locations vary during episodes and seasons and over several years at stations ranging from Barrow, Alaska to the South Pole. Trend measurements are based on 2–8 km tropospheric average ozonesonde observations at Resolute (75°N), Hohenpeissenberg (47°N in Bavaria, 975 m above sea level), Aspendale, Australia (38°S) and the average of the ozonesondes for North America, Europe, Japan and north temperate latitudes. Variability in episodes is clearly more marked at Barrow in springtime than at other sites, though this may not be true for all seasons. Part of the variation at Fritzpeak, west of Boulder, Colorado, may be due to nearby urban influences (Oltmans 1981).

Figure 2 gives another striking example of episodicity in arctic air, this time in sulphate aerosol concentrations at two stations in the Norwegian Arctic (Joranger and Ottar 1984). In late winter and early spring monthly mean concentrations are largely determined by a few episodes. There are usually both winter and spring maximum of surface O_3 at Barrow (Figure 1), with a summer minimum. Annual mean O_3 concentrations in the 2–8 km layer indicate significant year-to-year variability at specific locations. When averaging over all north temperate sites, however, there is indication of a more than 10% O_3 increase over the period 1970–80. Measurements at Resolute, Arctic Canada, obviously play an important role in understanding the development of O_3 concentrations in the northern hemisphere. It should be mentioned here that calculations by Fishman and others (1979)

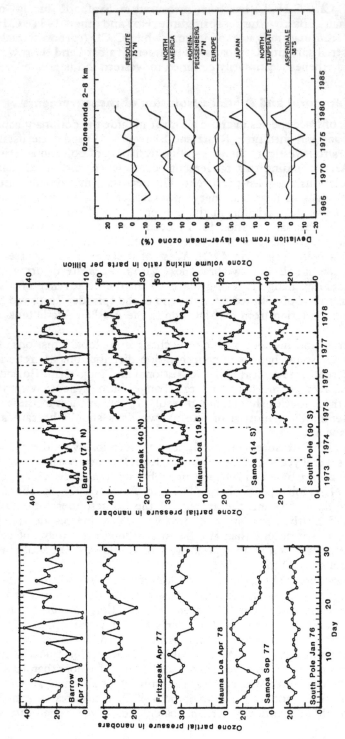

Fig 1. Left: day-to-day variations of surface ozone (Oltmans 1981). Middle: mean monthly surface ozone concentrations (Oltmans 1981). Right: variation of mean annual ozone in the 2–8 km tropospheric layer, based on ozonesonde observations at given stations, and mean variation for given regions and for the north temperate zone (Angell and Korshover 1983).

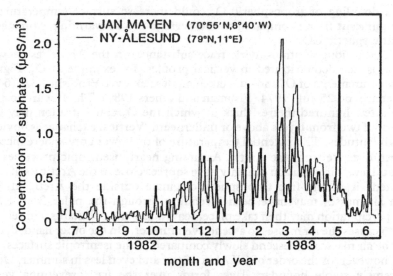

Fig 2. Concentration of sulphate in atmospheric aerosols, Jan Mayen and at Ny-Ålesund, Norwegian Arctic. Values are means for a $2+2+3$ days' sampling sequence applied each week (Joranger and Ottar 1984).

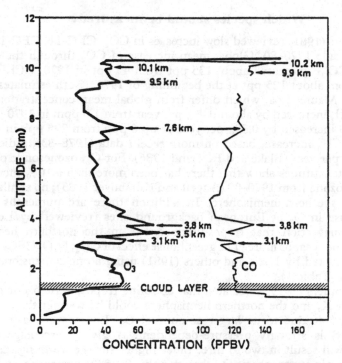

Fig 3. Vertical profiles of O_3 and CO concentrations measured during an aircraft descent over Frobisher Bay (64°N), northeast Canada, 25 July 1974. Arrows show altitudes where anomalously high/low O_3 and CO concentrations were found (Fishman and others 1980).

indicated that a doubling of tropospheric O_3 could increase surface temperatures by nearly 1 K, a significant figure compared to the average 2–3 K warming estimated for a doubling of atmospheric CO_2.

The episodic behaviour of atmospheric trace substances in the Arctic, as shown in Figures 1 and 2, is also demonstrated in vertical profiles, for example of O_3. Figure 3 shows aircraft measurements of O_3 and CO, taken on descent over Frobisher Bay (64°N), northeastern Canada, on 25 July 1974 (Fishman and others 1980). The free troposphere includes layers a few hundred metres thick in which the O_3 concentration may differ almost by a factor of two from layers above or underneath. Vertical exchange is obviously very slow at high latitudes. The potential temperature of the lower atmosphere increases significantly with distance from the Arctic. Assuming nearly isentropic processes, this means that ground level air pollution can only have sources close to the Arctic itself, while layers of polluted air in the free troposphere originate outside the Arctic in lower latitudes. Such air masses may have been ascending through the polar frontal zone. Saturation and precipitation may then take place, scavenging the pollutants and releasing latent heat which causes the air masses to ascend even further. On the other hand, diabatic cooling causes the air masses to descend slowly compared to the isentropic surfaces. This process is slow, however, of the order of 1 K/d in winter and even less in summer. During winter an extremely stable boundary layer forms over the ice, preventing vertical exchange efficiently. Further discussion of vertical exchange in arctic air is given by Iversen (1984).

Which species should be measured?

Lacis and others (1981) reviewed slow increases in CO_2, CFC-11, CFC-12, CH_4 and N_2O over the decade 1970–80. Global mean increase of CO_2 through the decade was about 12 ppm; CCl_3F increased about 135 ppt from 45 ppt in 1970, CCl_2F_2 increased about 190 ppt from about 125 ppt at the beginning of 1970. (Both estimates were from measurements at Mauna Loa, which differ from global mean concentrations less than about 10 ppt). CH_4 increased by about 0.9% per year from 1.5 ppm in 1970 to 1.65 ppm in 1980, and N_2O increased by 0.3% per year or up 6 ppb from 295 ppb in 1970. Later estimates of the CH_4 increase, based on more recent data (1978–83), indicate a mean increase of 1.1% per year (Blake and Rowland 1986). For O_3, ozonesonde observations at north temperate latitudes show that there has been more than a 10% increase in 2–8 km tropospheric ozone from 1970–80 (Angell and Korshover 1983); no similar trend can be seen for the southern hemisphere. In addition there are numerous records of significant increase in O_3 at European background sites (reviewed by Logan 1985), supporting the suggestion that tropospheric ozone in the northern hemisphere is increasing by 1% per year or more. The greenhouse effect of CH_4, N_2O, CFC-11, CFC-12 and CO_2 was calcuated by Lacis and others (1981) using a one-dimensional radiative-convective model (Table 1).

Fishman and others (1979) calculated that doubling tropospheric ozone could warm the surface by 0.9 K, and the northern hemisphere could be warmer than the southern hemisphere by 0.2 K due to its larger amount of O_3. Longwave opacity of O_3 (the greenhouse effect) is strongly pressure-dependent. A given percentage increase in tropospheric O_3 will result in two to three times higher surface warming than the same percentage increase in stratospheric O_3, though only 10–15% of atmospheric O_3 is found in the troposphere (Ramanathan 1979). Changes in O_3 concentrations around the tropopause are more effective than those in other regions of the atmosphere for causing surface temperature change (Wang and others 1980).

Table 1. Greenhouse effect of several trace gases (from Lacis and others 1981).

| Species | 1970–1980 change | | | Doubling | | |
	Start	Change	$\Delta T_{eq}(K)$[1]	Start	Change	ΔT_{eq}[1]
CO_2	325 ppm	12 ppm	0.14	300 ppm	300 ppm	2.9
CH_4	1.5 ppm	0.15 ppm	0.032	1.6 ppm	1.6 ppm	0.26
N_2O	295 ppb	6 ppb	0.016	280 ppb	280 ppb	0.65
CCl_3F	45 ppt	135 ppt	0.020	0	2 ppb	0.29
CCl_2F_2	125 ppt	190 ppt	0.034	0	2 ppb	0.36

[1] ΔT_{eq}: change in global equilibrium temperature in K.

Together with CO_2, tropospheric O_3 seems to be of considerable significance as a greenhouse gas. Arctic trace gas measurements should therefore try to contribute to the understanding of tropospheric ozone. This involves measurements of O_3 itself, and also of the gases which control it photochemically through the main production processes:

$$CO + OH \rightarrow CO_2 + H$$

$$H + O_2 + M \rightarrow HO_2 + M$$

$$HO_2 + NO \rightarrow OH + NO_2$$

$$NO_2 + hv \rightarrow NO + O$$

$$O + O_2 + M \rightarrow O_3 + M$$

$$CO + 2O_2 \rightarrow CO_2 + O_3 \text{ (net)}$$

Methane and non-methane hydrocarbons contribute to the formation of O_3 because HO_2 and organic peroxy radicals are formed when HC-molecules are broken down. The rate-determining reaction step in the decomposition of HC-molecules is usually the initial reaction:

$$CH_4 + OH \rightarrow CH_3 + H_2O$$

or, in general:

$$HC + OH \rightarrow \text{radicals, products}$$

As the abundance of CH_4 changes, there is a feedback on the overall chemistry which may change for instance the hydroxyl radical concentration or the rate of ozone formation or loss, which mainly takes place through:

$$O_3 + hv \rightarrow O_2 + O(1D), \lambda < 320 \text{ nm}$$

$$O(1D) + H_2O \rightarrow 2OH$$

$$O_3 + H_2O \rightarrow O_2 + 2OH \text{ (net)}$$

or through:

$$O_3 + HO_2 \rightarrow OH + 2O_2.$$

To understand the behaviour of tropospheric ozone, knowledge of the distribution of NO_x, CH_4, non-methane hydrocarbons and CO is required. Some of these gases (NO_x,

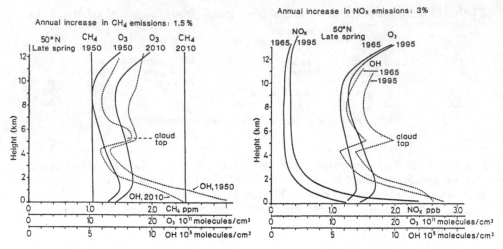

Fig 4. Calculated vertical profiles at 50°N latitude. Left: O_3, CH_4 and OH for 1950 and 2010, with a CH_4-concentration increase of 1.5% per year. Right: O_3, OH and NO_x for 1965 and 1995 with a NO_x-emission increase of 3% per year. A 2–dimensional, meridional model of the global troposphere was used (Isaksen and Hov 1986).

Table 2. List of species which should be seriously considered in future Arctic measurements.

Atmospheric species	Comments
CO_2	Climate – infrared absorbing gas
O_3	Climate – infrared absorbing gas
CH_4	Climate – infrared absorbing gas
	Important for tropospheric photochemistry
N_2O	Climate – infrared absorbing gas
CCl_3F	Climate – infrared absorbing gas
CCl_2F_2	Climate – infrared absorbing gas
$NO_x(NO_2)$	Contribute to O_3 photochemistry
NMHC	Non methane individual hydrocarbons, contribute to O_3 photochemistry
CO	Contribute to O_3 photochemistry
PAN	Good indicator of photochemical pollution processes
Light-absorbing aerosol	Climate control
Trace elements in aerosols	Determination of the origin of polluted air
$C_2Cl_3F_2$	Determination of the origin of polluted air
CF_3Br	Determination of the origin of polluted air
CF_2BrCl	Determination of the origin of polluted air
C_2Cl_4	Determination of the origin of polluted air
C_2HCl_3	Determination of the origin of polluted air

Ice core/snow pack samples	Comments
CO_2	Determination of historic concentration
CH_4	Determination of historic concentration
	Important for establishing past atmospheric composition
Ions	Historic development to determine past SO_x and NO_x depositions and emissions
H_2O_2	Important for the oxidation of SO_2 to sulphate
Trace metals	Sources of past depositions and emissions

reactive hydrocarbons) have rather short chemical lifetimes—of the order of one day—in the atmosphere, which indicates wide variations of concentration in time and space.

A two-dimensional meridional model of the global troposphere with chemistry, where complete zonal mixing is assumed in latitudinal bands of 10° width, has been used to calculate how tropospheric composition may change as a result of increases in the global emissions of CH_4 or NO_x (Isaksen and others 1985; Isaksen and Hov, in preparation). Figure 4 shows calculated vertical profiles of O_3, CH_4 and OH for 1950 and 2010 at 50°N, spring, for a case with an assumed CH_4-concentration increase of 1.5% per year, and NO_x, O_3 and OH for 1965 and 1995 for an assumed annual NO_x emission increase of 3% per year. In the first case, the OH-radical concentration is lower in 2010 than in 1950, indicating that there is a more-than-proportional increase in CH_4 concentrations compared to the emissions. Ozone is seen to increase throughout the troposphere, markedly at high altitudes (about 0.7% per year for CH_4-emission increase of 1.5% per year) where the climatological impact may be most significant. The increase in emissions of NO_x is seen to influence O_3 mainly in the atmospheric boundary layer and in the lower free troposphere. This reflects the much shorter atmospheric lifetime of NO_x compared with that of CH_4 (5–10 y).

Table 2 lists species which should be included in future arctic measurements. At ground sites measurements should be taken about every day. Aircraft measurements should resolve layers of a few hundred metres thickness in the vertical.

Acknowledgement

Ivar Isaksen, Stuart Penkett and Arne Semb have contributed to this paper.

References

ANGELL, J. K. AND KORSHOVER, J. 1983. Global variation in total ozone and layer-mean ozone: an update through 1981. *Journal of Climatology and Applied Meteorology*, 22: 1611–27.
BLAKE, D. R. AND ROWLAND, F. S. 1986. World wide increases in tropospheric methane, 1978–83. *Journal of Atmospheric Chemistry*, in press.
FISHMAN, J. AND OTHERS. 1979. Tropospheric ozone and climate. *Nature*, 282: 818–20.
FISHMAN, J. AND OTHERS. 1980. Simultaneous presence of O_3 and CO bands in the troposphere. *Tellus*, 32:456–63.
HEINTZENBERG, J. 1982. Size-segregated measurements of particulate elemental carbon and aerosol light absorption at remote Arctic locations. *Atmospheric Environment*, 16: 2461–69.
HOV, Ø. AND OTHERS. 1984. Organic gases in the Norwegian Arctic. *Geophysical Research Letters*, 11: 425–28.
ISAKSEN, I. S. A. AND OTHERS. 1985. Model analysis of the measured concentration of organic gases in the Norwegian Arctic. *Journal of Atmospheric Chemistry*, 3: 3–27.
ISAKSEN, I. S. A. AND HOV, Ø. 1986. Calculations of trends in the tropospheric concentration of O_3, OH, CO, CH_4 and NO_x. *Tellus*, in press.
IVERSEN, T. 1984. On the atmospheric transport of pollution to the Arctic. *Geophysical Research Letters*, 11: 457–60.
JORANGER, E. AND OTTAR, B. 1984. Air pollution studies in the Norwegian Arctic. *Geophysical Research Letters*, 11: 365–68.
KHALIL, M. A. K. AND RASMUSSEN, R. A. 1984. Trace gases in the Arctic: indices of air pollution. In HARRIS, J. (editor). *Geophysical monitoring for climatic change*, vol 12. Boulder, Colorado, NOAA GMCC (1983 annual issue).
LACIS, A. AND OTHERS. 1981. Greenhouse effect of atmospheric trace gases, 1970–1980. *Geophysical Research Letters*, 10: 1035–38.
LOGAN, J. A. 1985. Tropospheric ozone: seasonal behaviour, trends and anthropogenic influence. *Journal of Geophysical Research*, 90D:10463–82. NEFTEL, A. AND OTHERS. 1984. Measurements of hydrogen peroxide in polar ice samples. *Nature*, 311: 43–5.
OLTMANS, S. J. 1981. Surface ozone measurements in clean air. *Journal of Geophysical Research*, 86: 1174–80.
OTTAR, B. AND PACYNA, J. M. 1984. Sources of Ni, Pb and Zn during the Arctic episode in March 1983. *Geophysical Research Letters*, 11: 441–44.
RAATZ, W. F. AND SHAW, G. E. 1984. Long-range tropospheric transport of pollution aerosols into the Alaskan Arctic. *Journal of Climate and Applied Meteorology*, 23: 1052–64.

RAHN, K. A. 1981. The Mn/V ratio as a tracer of large scale sources of pollution aerosol for the Arctic. *Atmospheric Environment*, 15: 1457–64.

RAMANATHAN, V. 1979. Climatic effects of ozone change: a review. *COSPAR Symposium Series* 8 London and New York, Pergamon Press: 223–38.

RAMANATHAN, V. AND OTHERS 1985. Trace gas trends and their potential role in climate change. *Journal of Geophysical Research*, 90D: 5547–66.

SEMB, A. AND OTHERS. 1984. Major ions in Spitsbergen snow samples. *Geophysical Research Letters*, 11: 445–48.

VALERO, F. P. J. AND OTHERS. 1983. Radiative effects of the Arctic haze. *Geophysical Research Letters*, 10: 1184–87.

WANG, W. C. AND OTHERS. 1980. Climatic effects due to halogenated compounds in the earth's atmosphere. *Journal of Atmospheric Science*, 37: 333–38.

WOLFF, E. W. AND PEEL, D. A. 1985. The record of global pollution in polar snow and ice. *Nature*, 313: 535–40.

PART 3.

HEALTH AND ECOLOGICAL ISSUES

Back: W. A. Turner; J. C. Hansen; J. Harte; A. C. Upton
Front: M. M. Henderson; J. Middaugh (Convener); W. C. Hanson; C. D. Stutzman

PART 3. INTRODUCTION

Atmospheric pollution in lower latitudes has proven implications for human health, and for the well-being of natural ecosystems; what effects on human health and ecology, if any, might we expect from the recent increase in arctic atmospheric pollution? The seven papers in this section covered both the several kinds of pollutants that enter the Arctic in air masses, and the more personal pollutants that man introduces voluntarily into the home in the shape of cigarettes and fuels. Maureen Hendersen introduced the subject, basing her contribution on the impending report (now published) of the National Research Council Committee on the Epidemiology of Air Pollution. Though not specifically addressing the arctic situation, her paper provides a background for considering health aspects of all kinds of air pollution, in the field and in the home. Upton's paper summarized the effects on health of various forms of radiation transmitted through the atmosphere, providing a useful guide to ionizing and non-ionizing forms, again in a general rather than a specifically arctic content.

Wayne Hanson concentrated on the results of two decades' study of the effects of radionuclides on arctic ecosystems, especially the radionuclides that entered the Arctic atmospherically from fallout following atomic tests between 1952 and 1980. Surprising concentrations of ^{137}Cs and other radioactive elements had been reported in Arctic foodchains, transferred from lichens through caribou to man, and imposing on the Inuit body burdens up to 200 times greater than those of populations in lower latitudes. Exposure rates reached a maximum in 1964–66, but have now dropped to levels approximately equalling background levels. The cancer risks imposed on Alaskan natives by radioactive fallout during the nuclear tests were assessed by C. D. Stutzman and Donna M. Nelson. Though patterns of cancer epidemiology among these populations differ from those further south, differences cannot be explained by the different levels of fallout experienced; changes in lifestyle during and since the period of testing are more likely to be responsible.

Indoor air pollution was discussed in the paper by J. D. Spengler and W. A. Turner of Harvard University, who pointed out that concentrations of atmospheric particles and gases build up in homes, offices and public buildings, forming an important component of the total airborne pollution to which people are exposed. There is no public policy on these potential sources of respiratory hazards, and research has been spasmodic and uncoordinated. Relevance for Arctic populations, who frequently live in crowded, ill-ventilated conditions, is clear. J. C. Hansen of Aarhus University discussed levels of mercury, lead, calcium and selenium detected in blood and hair samples from modern Greenland Inuit. Mercury and selenium concentrations were high. Blood lead levels were similar to those of Europeans, indicating long-distance atmospheric transport of lead; cadmium was also high, reflecting smoking habits. Comparisons were made with hair and sealskin samples from 500-year-old mummified Inuit, which showed lower mercury and lead levels, and with late 19th century human hair samples which were lowest of all in mercury, cadmium and selenium. Contemporary levels of some heavy metals, due to marine and atmospheric pollution, give cause for health concern.

J. Harte considered ecological implications of arctic and sub-arctic air pollution, drawing attention to the unusual properties of arctic forests, tundra and aquatic ecosystems and their consequent vulnerability to pollution. Possible effects of acid rain on salmon fisheries, and of atmospheric warming on tundra soils, are considered as case-histories; the need for further ecological research is emphasized, before even the magnitude of the problem can be gauged.

EPIDEMIOLOGY OF HEALTH EFFECTS OF AIR POLLUTION

MAUREEN M. HENDERSON

This contribution was invited because I recently served on a National Research Council Committee which considered and wrote a report on the Epidemiology of Air Pollution. The report was to be released on September 1 1985, and I intended to discuss its conclusions and apply them to the problem before us. The report had not been officially released, so I spoke instead about issues discussed during the committee's deliberations that were likely to be of interest to this group, and are pertinent to future studies in Alaska.

This group is addressing one of the most important issues debated by that National Research Council Committee: is today's ambient air pollution damaging human health, and will ambient pollution be more or less harmful in the future?

The role of epidemiology in getting answers to those questions was equally important to us and may be important for those in this conference who will act upon one or more of our recommendatons.

We asked ourselves at the outset whether one comprehensive longitudinal study could be designed to answer the most important health effects questions. The answer was an emphatic 'no'. There was an equally emphatic 'no' to the question of a study or studies to establish safety. Evidence of safety is not just the converse of evidence of risk, and safety is much harder to demonstrate than risk. Whatever the size and rigor of population studies with negative results, they are virtually useless in assuring safety in other populations and circumstances.

How about scientific disciplines? Which would be most useful in answering contemporary air pollution questions? There was complete agreement that no one discipline is likely to be successful on its own. Research on air pollution and health needs both interdisciplinary team efforts, in which atmospheric, clinical, epidemiological, and biostatistical scientists develop research plans and implement research operations in a truly integrated manner, and also resarch within separate disciplines but built on a foundation of complete information, of up-to-the-minute knowledge and technology in all the other relevant fields, toxicological, clinical, and epidemiological.

Several committee members began with doubts and reservations about the existence of serious ambient air pollution threats. After reviewing the updated evidence from controlled exposure, toxicological and epidemiological literatures and ongoing research, committee members agreed that there is enough evidence to warrant concern that current pollution still causes bad health effects in some segments of the US populaton. The effects in question include a variety of acute and chronic respiratory conditions that have been clearly linked with previous types and amounts of air pollution. Whether or not they are viewed as serious by society depends not only on their own magnitude, but upon the relative magnitude of contemporary diseases from all other causes.

Even when air pollution was gross, as in the 1930s and '40s, it was hard to measure any impact on health because of the high rate of background infectious diseases and diseases caused by cigarette use. Given that tobacco-related background disease still exists and that current air pollution levels are low, we expect and have to design studies that are sensitive to small risks of disease. Both absolute and relative risks can be expected to be small. In fact, most of the risks will be too small to be a measurable threat to individuals. Their potential importance lies in their cumulative impact. When general air pollution is of concern, even minute individual increases in adverse health effects can aggregate into meaningful public health problems because of the large mass of people exposed. It is important to measure those public health effects. It is equally important and much rarer to measure and learn about public health benefits which result from air pollution control. Although national monitoring data show substantial descreases in most urban air pollutants, there are some pollutants, eg nitrous oxide and ozone, that have proved relatively hard to control and consequently may still pose disproportionate hazards. There are also some notable changes in dispersion. We have been hearing about some of those today. We have also been hearing about other new phenomena in the guise of identified local concentrations, and the emergence of new pollutants, like aerosols, sulphates and woodsmoke, for which there are no standards.

Although there are good data-based reasons for believing that air pollution can multiply a smoker's risk of developing respiratory deisease, there is no quantitiative assessment of the amount of cigarette-smoking-caused disease that would be avoided by air pollution control. For example, asbestos does not cause lung cancer in non-smokers; we have no knowledge of how many other pollutants work in the same way.

Now having mentioned some of the committee's reasons for supporting the idea that research on air pollution and health effects could be productive, I would like to mention some of the health effects of interest and some of the difficulties encountered in measuring them.

Very few hard facts are known about human exposure curves for common air pollutants. Setters of standards have to extrapolate from acute to chronic human health effects and from high-to-low dose human reactions. There is equal, if not more, ignorance about the detailed pathophysiology of the most pertinent chronic respiratory diseases. Even worse from the point of view of research design, most if not all health effects of interest have a number of other causes. In many instances, air pollution only aggravates or exacerbates pre-exisitng disease.

The committee was concerned with air pollutions's impact on human health, not just on its association with clinical disease. This concept encompasses growth and development and lung function as well as disease. It has a number of research implications. It immediately broadens and increases the target populations suitable for air pollution research and includes some very sensitive sub-populations such as infants and the elderly. *Unfortunately*, it also immediately increases the number of other factors that can influence the health measurement of interest. Most of these variables confound research findings because they are unevenly distributed among those exposed and not exposed to air pollution. A strict account has to be taken of all these influential factors before the relationship with air pollution can be quantitatively assessed. In this broader context, examples of confounding factors include birthweight, social class and physical activity as well as pre-existing health cnditions and smoking habits. Their identification, and the design of studies to reduce their effects and get good estimates of remaining effects, are some of the most difficult tasks in this type of observational research.

The committee fully accepted that breathed air can affect any of the body systems, but it was expedient to concentrate on the respiratory system which is in direct contact with pollutants. The absence of respiratory effects would not rule out effects on other systems but would instinctively make them less likely. While there was general agreement to focus on respiratory effects, the committee fully realized that in the long run there may be effects, even major public health effects, on other body systems. This is an important consideration because it has great pertinence for appropriate choice of study design. For example, a prospective cohort study allows the detection of a range of outcomes. The non-respiratory effects of air pollution deemed worthy of further study included the effects of lead on neurobehavioral development and on blood pressure, and the effects of carbon monoxide on ischemic heart disease. The committee felt that possible neurotoxic mutagenic or carcinogenic effects of community exposure to benzene and other volatile organic substances were much less worthy of study but could not be ignored completely.

Acute respiratory effects of interest include an increase in the frequency of asthmatic attacks, transient deficits in pulmonary funcion and increases in the frequency of upper and lower respiratory diseases in both children and adults. We badly need the results of new research to clarify the basis of individual variation in response to air pollution exposure, and new research to determine the relationship between acute affects and long-term function. These results are necessary to set the stage to measure the acute effects of future air pollution.

Chronic respiratory effects that merit study include chronic obstructive pulmonary disease, a few other conditions in which air flow in the lungs is limited, and adverse changes in the rate of growth or reduction in lung function. As mentioned earlier, the very sparse evidence of possible chronic respiratory effects of air pollution are due as much to a lack of knowledge of the natural history of the disease as to a shortage of technological and methodological expertise.

Although the bulk of lung cancer is caused by cigarette use, there is room for assessment of the reduction in lung cancer incidence that could come from successful air pollution control, because of its synergistic action.

The ways in which acute or chronic effects have to be measured add to the research difficulties and the uncertainties of the data. Biological measures and all kinds of records are used. The former include pulmonary function tests and biological markers. The latter include questionnaires, death certificates, hospital admissions and attendances, and days lost from work or school. The committee paid particular attention to records of morbidity rather than mortality, and to biological measures rather than records.

Although spirometry is the most commonly used method of measuring lung function, other techniques are being developed and offer some promise in population-based research. Some of these newer techniques measure functions other than air flow and include bronchial hyperactivity testing and tests for particle depostion, macrophage function and dust clearance.

Biologic markers are of intense interest at the present time. They have to be very carefully classified for purposes of this type of research. *Markers of Exposure* show that there has been exposure to a specific agent. They may allow some estimate of absorbed dose and dose to target tissue. *Markers of Effect* show early results of interaction between specific agents and cellular or molecular targets in the host. *Markers of Susceptibility* reflect some characteristics of the host that increase likelihood of disease for any given exposure; an enzyme deficiency is an example. Effect and exposure markers overlap and some can be used in one or other context by different investigators.

Future research will depend more and more often on effect markers and other measures of subclinical disease. There are some good and practical reasons. First, very large numbers of people now have to be studied to measure statistically significant changes in clinical events. The necessary numbers will increase as pollution levels fall. Valid markers would reduce the needed numbers of individuals. In addition, markers should appear much more quickly after exposure than clinical disease. This cuts down both time for follow-up and extraneous cases from other incidental causes.

Much more research and development is needed before biologic markers will be ready for use in population studies of the effects of air pollution. Their accuracy in measuring what they are intended to measure, their value in predicting health impairment or premature death and their applicability to various demograph groups all remain to be fully established.

Sputum cytology has been used in studying non-neoplastic pulmonary disease with some encouraging results. Biochemical tests on sputum, such as measurements of anti-proteases, also have some early promise. Some important biochemical and cellular features of nasal mucosa can also be investigated with nasal brushing, a new non-invasive technique.

It has been suggested that measurable elements in the protease-antiprotease balance could be valuable early markers of emphysema. There is some belief that the protease-antiprotease balance protects the connective tissue matrix of the alveolar wall of healthy lungs and an excess of proleopytic activity leads to destruction of the matrix as emphysema develops.

Recent advances in the understanding of the role of factors measurable in plasma such as ceruloplasmin and catalase, which protect the lung from injury by oxidents and free radicals, might lead to markers of susceptibility. Changes in lung collagen have also been proposed as markers of chronic pulmonary disease, and the amino acid hydroxyproline as a possible marker of the status of lung connective tissue.

Genotoxic markers could be of use in air pollution research, both to measure the extent of exposure and to identify individuals who are particularly susceptible to genetic damage by specific pollutants. Much more work is need to clarify the relationship between genotoxic markers and specific end points before they can be useful in epidemiological research per se.

The committee viewed air pollution research as having four equally difficult components:

(1). The specification of the research problem and the formulation of focused research questions. For example, the following four questions require very different research approaches:

Does breathed air cause a measurable public health problem?
Can breathed air affect the health of individuals or groups of individuals?
Can specific pollutants found in ambient air affect the health of individuals or groups of individuals?
Is ambient air harmless?

(2). The Assessment of Health Effects — I have already addressed that component.

(3). Exposure Assessment. I will say a little about the exposure assessment but expect others to speak more about it later in this meeting.

(4). Epidemiolgoical Tools and Strategies. I am going to say very little about the technology per se.

Exposure Assessment

The committee viewed ambient air as air breathed in 24 hours. In other words, one person's ambient air comes from a number of indoor and outdoor sources and is inhaled during a range of activities. This concept requires consideration of indoor as well as outdoor exposure and it increases the potential for big individual differences in measurements of 24–hour exposure. Whenever this concept is adopted, there is likely to be more variation between the exposure of individuals within a community than will be measured from one community to the next.

Breathed air was seen as a dynamic mixture of ingredients, many able to potentiate or neutralize each other's effects. Some have the added capability of mixing together to produce new pollutants, which can be harmful in themselves or can also potentiate or neutralize others. Masking, additive, and synergistic effects of pollutants were seen as critical areas for future research. Past epidemiologic studies have assumed that measured exposure equated with dose. Recent advances in toxicology and molecular epidemiology have led to a distinction between measured concentration in the environment (exposure), the amount of a substance or its metabolites in the tissues (internal dose), and the amount of a substance that interacts with a particular target tissue (biologically effective dose). The last two measures are still in the process of development.

Present wisdom divides pollution into three catagories by source and type:

Outdoor, such as sulphates, ozone and lead.

Outdoor and indoor; fine particles, nitrous oxides and carbon monoxide

Indoor, such as volatile organic compounds, radon and its progeny, formaldehyde and woodsmoke.

In reviewing the physical, chemical and biologic characteristics of pollutants, the committee noted that some are 'fresh' (from local sources); some are 'aged' (from distant sources). The committee identified both 'older' and emerging pollutants of concern. The older are acid aerosols, ozone, nitrogen dioxide, carbon monoxide, lead, and radon and its progeny. Emerging problems are volatile organic compounds and products of incomplete combustion.

In considering the discipline of epidemiology, the committee emphasized its role in the formulation and evaluation of preventive strategies and its ability to detect hazards under actual conditions of exposure. It also underlined epidemiology's unique ability to determine the impact of actual exposure on public health, and the contribution to public health problems made by each of a number of co-existing causes of morbidity.

Finally, it viewed epidemiologic research as playing a potential role in public safety through its association with the determination of approximate safe exposures to particular pollutants.

Genotoxic markers could be of use in other air pollution research before they are ready for use in epidemiologic studies of air pollution. For example, they could be developed for use to measure the extent of exposure and identify individuals who are particularly susceptible to genetic damage by specific endpoints.

Reference

Committee on the epidemiology of air pollution, National Research Council. 1985. *Epidemiology of air pollution*. Washington DC, National Academy Press.

BIOLOGICAL EFFECTS OF LOW-LEVEL IONIZING AND NON-IONIZING RADIATION

ARTHUR C. UPTON

ABSTRACT. Early in this century it was recognized that large doses of ionizing radiation could injure almost any tissue in the body, but small doses were generally thought to be harmless. By the middle of the century however it came to be suspected that even the smallest doses of ionizing radiation to the gonads might increase the risk of hereditary disease in subsequently-conceived offspring. Since then the hypothesis that carcinogenic and teratogenic effects also have no threshold has been adopted for purposes of radiological protection. It is estimated nevertheless that the risks that may be associated with natural background levels of ionizing irradiation are too small to be detectable. Hence validation of such risk estimates will depend on further elucidation of the dose-effect relationships and mechanisms of the effects in question, through studies at higher dose levels. In contrast to the situation with ionizing radiation, exposure to natural background levels of ultraviolet radiation has been implicated definitively in the etiology of skin cancers in fair-skinned individuals. Persons with inherited defects in DNA repair capacity are particularly susceptible. Non-ionizing radiations of other types can also affect health at high dose levels, but whether they can cause injury at low levels of exposure is not known.

Contents

Introduction

Everyone is bathed continuously with small amounts of ionizing and nonionizing radiation from natural and man-made sources. Although it is evident that the skin is often injured by sunlight, it is questionable whether the small amounts of other radiation to which people are regularly exposed causes any harmful effects. This article reviews the kinds of biological effects that may be caused by low-level radiation, and our knowledge of their frequency and severity.

Types of radiation injury

For purposes of radiological protection, radiation injuries are customarily divided into two types: stochastic effects, which are conceived to result from damage to a single cell and assumed to have no threshold (although one cannot be excluded), and non-stochastic effects, which result only from damage to many cells and have appreciable thresholds (ICRP 1977, 1984).

Stochastic effects include: (1) damage to chromosomes and genes, leading to chromosome aberrations and gene mutations, (2) disturbances in the growth and development of the embryo or fetus, including teratogenic effects, and 3) carcinogenic effects. These effects have no distinguishing features by which they can be recognized as having resulted from radiation, as opposed to other causes; thus their relationship to irradiation can be inferred only statistically. At levels of radiation too low for the increased frequency of effects to be detected, their probability can be estimated only by extrapolation from observations at higher dose levels, based on assumptions about the relevant mechanisms and dose-effect relationships. To estimate the biological effects of low-level irradiation, therefore, one must seek to understand the nature of radiation and its actions on living cells.

Radiation and its interaction with matter

Ionizing and nonionizing radiations are composed of electromagnetic waves (Figure 1). Ionizing radiations also include atomic particles such as neutrons, protons, electrons, and alpha particles. Nonionizing radiations cause their biological effects primarily by exciting and/or polarizing atoms and molecules of the cells in which they are absorbed. Ionizing radiations impart enough localized energy to disrupt atoms and molecules in their paths. In either case, molecular changes give rise to biochemical lesions, which lead ultimately to cellular injuries of various kinds.

Different types of radiation vary markedly in penetrating power. UV and alpha radiations, for example, may not penetrate through the skin, whereas high-energy x-rays may traverse the body without being absorbed.

The distribution of ionization events along the path of an ionizing radiation depends on the energy, mass, and charge of the radiation. With neutrons and charged particles, the ionizations are distributed more densely than with x-rays or gamma rays. Since the probability of injury depends on the concentration of molecular damage, charged particles generally penetrate only a short distance in tissue, with the result they pose less hazard than x-rays or gamma rays when impinging on the body from an external source.

Once taken into the body, radioactive elements tend to localize more in some organs than in others. Radioiodine, for example, concentrates in the thyroid gland, whereas radium and strontium are deposited primarily in the skeleton. Radioelements also vary in their rates of removal; radioiodine is normally eliminated from the thyroid gland with a biological half-life of about 90 days, whereas strontium-90 remains in the skeleton for many years. Because of the differences among radionuclides in distribution and retention, their doses to different tissues vary accordingly.

Radiation dosage is usually expressed in terms of the amount of energy delivered by the radiation to defined areas or volumes of tissue. With microwave radiation, for example, the standard absorbed dose (SAR) is usually expressed in W/cm^2, pulsed or continuous; with ionizing radiation the dose is usually expressed in *rad* (1 rad = 100 ergs per gram of tissue) or *gray* (1 Gy = 1 joule per kg of tissue = 100 rad). The *rem* and the *sievert* are also used with different types of ionizing radiation to normalize their doses

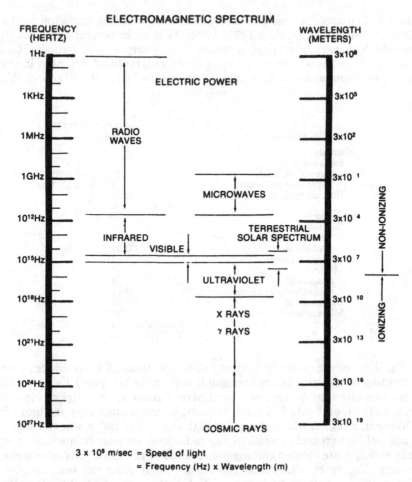

Fig 1. Radiations of the elctromagnetic spectrum (United States 1977).

in terms of biological effectiveness; ie, one *rem*, defined loosely, is that amount of any ionizing radiation which is equivalent in biological effectiveness to one rad of gamma-rays; similarly, one *sievert* is that amount which is equivalent to one *gray* of gamma rays (1 Sv = 100 rem).

Environmental sources and levels of radiation

Natural background radiation.

  Life has evolved in the presence of natural background electromagnetic (EM) radiation of a broad spectrum of frequencies, ranging from essentially steady-state electric and magnetic fields to energetic gamma rays at frequencies of 10^{23}Hz (Figure 1). The natural background ionizing radiations consist of (1) cosmic rays, which originate in outer space; (2) terrestrial radiations, which come from radium, thorium, uranium, and other radioactive minerals in the earth's crust; and (3) internal radiations, which are emitted by the potassium-40, carbon-14, and other radioactive elements normally present within

Table 1. Estimated annual doses of whole-body ionizing radiation to the US population. (From NAS/BEIR 1972, 1980). [a]Values in parentheses indicate range over which average levels from different States vary with elevation. [b]Range of variation (in parentheses) attributable largely to geographic differences in the content of radium, thorium, and uranium in the earth's crust. (See also Hanson, W. C., this volume).

Source of radiation		Average dose rates (mSy/year)
Natural		
Environmental		
Cosmic radiation		0.28 (0.28–1.30)[a]
Terrestrial radiation		0.26 (0.30–1.15)[b]
Internal radiactive isotopes		0.26
	Subtotal	0.80
Man-made		
Environmental		
Technologically enhanced		0.04
Global fallout		0.04
Nuclear power		0.002
Medical		
Diagnostic		0.78
Radiopharmaceuticals		0.14
Occupational		0.01
Miscellaneous		0.05
	Subtotal	1.06
	TOTAL	1.86

living cells. The average dose of ionizing radiation from all three souces received by a person residing at sea level is approximately 0.80 mSv per year (Table 1). However, a dose twice this size may be received by a person residing at a higher elevation (eg, in Denver, Colorado, or Sante Fe, New Mexico) where cosmic rays are more intense, or by one living in a region where there is a high content of radium in the soil.

The natural background of nonionizing radiations includes frequencies ranging from essentially steady-state electric and magnetic fields to UV radiations approaching 10^{18}Hz in frequency (Figure 1). The radiations come mostly from the sun, supplemented by terrestrial electromagnetic disturbances including thunderstorms. The dose of ultraviolet radiation varies systematically in relation to geomagnetic latitude (Figure 2), as well as in relation to climate, amount of time spend out of doors, dress, and other factors influencing exposure to sunlight. Levels of extraterrestrial radiation at radio and microwave frequencies are extremely low, averaging about 10^{-20} W/m²/Hz from a typical radio star. Similarly, the terrestrial components of the solar flux at 10 Hz do not exceed about 1 mV/M. At these levels there is little possibility of bioeffects (Adey 1981). In addition, however, there is a continuous spectrum of natural EM fields of terrestrial origin, ranging from DC to about 3kHz, some of which are related to the earth's magnetic field. In fair weather the steady atmospheric electric field is about 150 V/m, but it may increase to 10 kV/m during thunderstorms, when electrical field oscillations may be as high as 10^{-2} V/m at frequencies between 1 and 10 Hz. In the background spectrum there is also a series of resonances between the earth's surface and ionspheric particles in the upper atmosphere, at frequencies betweeen 8 and 32 Hz (Adey 1981).

Man-made radiation.

The population is exposed increasingly to ionizing and nonionizing radiations from man-made sources. The largest source of ionizing radiation is use of x-rays in medical

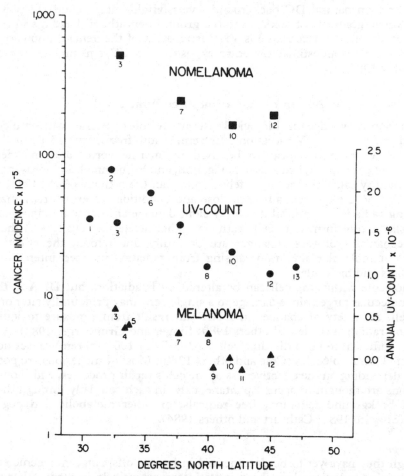

Fig 2. Variation with latitude in ultraviolet radiation levels ad age-adjusted skin cancer rates for white males in various US cities and states, 1970 (from Upton 1977). (1) Tallahassee, FL; (2) El Paso, TX; (3) Fort Worth, TX; (4) Birmingham, AL; (5) Atlanta, GA; (6) Albuquerque, NM; (7) Oakland, CA; (8) Colorado; Pittsburgh, PA; (10) Des Moines, IA; (11) Detroit, MI; (12) minneapolis, MN; (13) Bismarck, ND.

diagnosis. Doses from different types of medical and dental x-ray examinations vary widely, but the average dose to people in developed countries approximates the dose received from natural background radiation (Table 1). Smaller doses of man-made radiation are also received from such sources as radioactive minerals in crushed rock, building materials and phosphate fertilizers; radiation-emitting components of TV sets, smoke detectors and other consumer products; radioactive fallout from atomic weapons, and radiation released in the production of nuclear power (Table 1).

Background levels of non-ionizing radiation in urban and industrial environments have also icreased exponentially since the advent of electric power and EM communication systems. Ambient 60 Hz fields of 1–10V/m, with levels as high as 250 v/m in in vicinity of electric blankets, are typical in American homes. Users of hair dryers, electric shavers and other electric hand tools may be exposed to 60 Hz magnetic fields of 5–25G, 20–75 times higher than natural DC background levels. High-voltage overhead power lines produce electric gradients of 10kV/m to the ground beneath. FM radio and television broadcasting accounts for continuous VHF irradiation of the general population, with 18 per cent receiving an estimated median exposure of 5 μW/cm^2 time-averaged power density (Adey 1981).

Action of radiation at cellular level

It is common knowledge that biological systems are able to interact with ambient levels of infrared, visible, and UV radiations. Extremely low frequency (ELF) fields in the range of 10^{-7}V/cm also appear to be used by marine vertebrates in orientation, navigation, and prey attack, by birds in navigation, and by mammals in biorhythms; such fields also modify calcium binding in feline and avian brain tissue (Adey 1981).

Depending on the type of radiation, dose and conditions of exposure, ionizing and non-ionizing radiations can inhibit cell division, damage chromosomes and genes, and cause neoplastic transformation, cell death, and other deleterious changes. Although the precise mechanisms of these changes are not fully understood, the effects are all end-results of molecular alterations arising from radiation-induced interactions with cellular atoms and molecules.

Any molecule within the cell can be altered by irradiation, but DNA is the most sensitive molecular target, since damage to a single gene may profoundly alter or kill the affected cell. A variety of changes in DNA can result from exposure to ionizing or non-ionizing radiation (Cole and others 1980; Singer and Grunberger 1983). A dose of x-radiation sufficient to kill a dividing cell (eg 1–2 Sv or 100–200 rem) causes dozens of lesions in its DNA molecules (Cole and others 1980). Most of the lesions are potentially reparable, depending on the effectiveness of the cell's repair processes, and thousands of DNA lesions are thought to occur 'spontaneously' in each cell daily through the effects of natural background radiation, free radicals, or other metabolic and degradative processes (Shapiro 1981; Cathcart and others 1984).

Damage to genes.
Although they have yet to be documented in human offspring, mutagenic effects of UV and ionizing radiations have been investigated extensively in many other types of organisms. Hence the wealth of information from other species, as well as from human somatic cells, provides a basis for estimating the heritable effects of ionizing irradiation in man.

· In human lymphocytes irradiated *in vitro*, frequency of mutations at the HGPRT locus amounts to about 6 x 10^{-6} per Gy (Grosovsky and Little 1985). In mouse germ cells (spermatogonia and oocytes) exposed to ionizing radiation *in vivo*, the frequency of radiation-induced mutations per genetic locus approximates 1 x 10^{-5} per Sv (corresponding to an increase of about 100 percent per Sv), depending on the conditions of irradiation (NAS/BEIR 1980). An increase of this order of magnitude is too small to detect with existing methods in the children of A-bomb survivors, in view of the limited numbers of such children and the comparatively small average dose received by their parents. The fact that no genetic effects of A-bomb irradiation have been detected in this population is thus not unexpected (UNSCEAR 1977), BEIR 1980).

On the basis of present knowledge, the dose of ionizing radiation required to double mutation frequency in the human population approximates to 0.2 − 2.5 Sv (20–250 rem). From this estimate it is inferred that only a small percentage (0.5 − 3.0%) of genetically related human diseases is attributable to natural background irradiation (Table 2).

Table 2. Estimates of genetic detriment, per million offspring, attributable to natural background ionizing radiation. (Modified from NAS/BEIR 1972, 1980 and UNSCEAR 1977). [a]Equivalent to 0.1 rem per year, or 3 rem per parental generation (30 years), average dose from natural background ionizing radiation.

Type of genetic detriment	Natural Incidence	Contribution from natural background radiation[a]	
		First Generation	Equilibrium Generations
Dominant traits + diseases	10,000	3–50	45–300
Chromosomal + recessive traits + diseases	10,000	<30	<120
Recognized abortions			
Aneuploidy + polyploidy	35,000	33	33
XO	9,000	9	9
Unbalanced rearrangements	11,000	216	276
Congenital anomalies	15,000		
Anomalies expressed after birth; constitutional + degenerative diseases	10,000	3–300	30–3000
Total (rounded)	100,000	300–900	500–3500

Mutagenic effects of EM radiation at frequencies below 10^{15} Hz have yet to be demonstrated unequivocally in the absence of thermal effects.

Damage to chromosomes.

Through breakage of chromosome fibres and interference with the normal segregation of chromosomes to daughter cells at the time of cell division, ionizing radiation can alter the number and structure of chromosomes in the cell. The majority of such changes cause the affected cells to die at mitosis.

In irradiated cells, frequency of chromosome aberrations increases roughly in proportion to the dose of ionizing radiation within the low-to-intermediate dose range, constituting a crude biological indicator of exposure. The rate in human blood lymphocytes irradiated in culture approximates 0.1 aberration per cell per Sv (100 rem). Such aberrations increase in frequency in radiation workers and others exposed accidentally or therapeutically to large doses of ionizing radiation (Lloyd and others 1980).

Only a small percentage of all chromosome aberrations is attributable to natural background ionizing radiation. The majority result from other causes, including certain viruses, chemicals, and drugs. Although chromosome aberrations are mostly deleterious to the cells in which they occur, the ultimate impact, if any, of an increase in their frequency on the health of an affected individual cannot be predicted from present knowledge.

Cytogenetic effects in mammalian cells have been produced at high dose levels with EM radiations of frequencies below 10^{15} Hz (Hol amd Schneider 1970, Leach 1976) but not clearly in the absence of thermal effects.

Effects on cell survival.

Any cell may be killed if exposed to a large enough dose of ionizing radiation (eg hundreds of Gy); but a dose of only 1–2 Gy (100–200 rem) is usually sufficient to interfere with cell proliferation (UNSCEAR 1982). In general, only a few types of non-proliferating cells (eg lymphocytes and oocytes) are as susceptible to killing by ionizing radiation as are proliferating cells.

The percentage of cells retaining the ability to multiply tends to decrease exponentially with increasing dose, the dose-survival curve characteristically decreasing more steeply with high-LET radiation than with low-LET radiation in the low dose region. If a given dose of ionizing radiation is delivered in small increments over an extended period, its effect on cell survival is usually decreased, owing to repair of intracellular damage and proliferation of surviving cells between successive exposures.

Cytotoxic effects have not been observed with EM radiations at frequencies below 10^{15} Hz, in the absence of thermal effects.

Effects on tissues

Pathologic changes induced by radiation in tissues include secondary and indirect effects, some of which evolve slowly, as well as direct effects. Thus, degenerative changes, scarring, tissue break-down and cancer may not be manifest until years or decades after irradiation, although killing of cells may be detectable promptly. Injury from ionizing radiation generally appears most rapidly in tissues having a high rate of cell proliferation (UNSCEAR 1982).

Nonstochastic effects of irradiation generally occur only at dose levels well above present protection standards. Hence they are of little concern in relation to low-level irradiation of the population (Adey 1981; UNSCEAR 1983; Petersen 1983; ICRP 1984). With few exceptions, therefore, they will not be discussed in this report.

Blood-forming and lymphoid tissues.

In blood-forming cells, microscopic changes are detectable within minutes after a dose of 1 Sv (100 rem) of ionizing radiation, and larger doses to the whole body may deplete such cells drastically. These changes are not seen, however, at doses and dose rates compatible with present radiological protection standards (UNSCEAR 1982; ICRP 1984).

Microwave radiation has been observed to affect hematologic tissues and immune reactions in the medium-to-high dose range, but such effects are yet to be demonstrated conclusively in the low dose range (Adey 1981; Petersen 1983).

Reproductive organs.

Acute exposure of both testes to as little as 0.15 Sv (15 rad) of ionizing radiation can cause temporary hypospermia, and an acute dose in excess of 4 Sv (400 rem) can cause lasting sterility in some men (UNSCEAR 1982). Simlarly, an acute dose of 1.5 − 2.0 Sv (150–200 rem) to both ovaries can cause temporary sterility, and a dose in excess of 2.0 − 3.0 Sv (200–300 rem) permanent sterility in some women (UNSCEAR 1982).

Microwave radiation impairs reproductive function in laboratory animals at relatively high dose levels, and in mice exposed chronically at only 250–500 $\mu W/cm^2$; however, the biological significance and public health implications of the latter effects remain to be determined (Adey 1981).

Lens of the eye.

Irradiation of the lens at moderate-to-high dose levels can cause lens opacities or cataracts, which may not become evident until months or years after exposure. The threshold of x-rays for a vision-impairing cataract varies from 5 Gy (500 rad) in a single exposure to as much as 14 Gy (1,400 rad) in multiple exposures over a period of months (UNSCEAR 1982). The corresponding thresholds for neutrons are somewhat lower (UNSCEAR 1982). Infrared, UV, and microwave radiations also have been observed to cause cataracts in humans and laboratory animals, but only at relatively high dose levels (Adey 1981; Curtis and Nichols 1983; Petersen 1983).

Nervous system.

The mature nervous system appears to be relatively resistant to damage by ionizing and nonionizing radiation, but the brain is less resistant to ionizing radiation during prenatal development, as discussed below. Various neurological, psychometric, neuro-endocrine, and behavioral changes have ben reported following postnatal irradiation in the low-to-moderate dose range; however, the health significance of the changes attributed to ionizing (UNSCEAR 1982) and nonionizing (Adey 1981) radiation remains to be established.

Effects on embryonic growth and development

Embryonic, fetal, and juvenile tissues are relatively radiosensitive. During organo-genesis, acute exposure to as little as 0.25 Sv (25 rem) of ionizing radiation has caused birth defects in experimental animals (UNSCEAR 1977). After larger doses, comparable effects have been observed in children; eg mental retardation and reduced head size in atomic-bomb survivors who were irradiated between the 10th and 17 week of prenatal development (UNSCEAR 1977; Otake and Schull 1984).

Prenatal mortality and developmental disturbances have been reported in laboratory animals exposed to microwave radiation, but with dose-effect relationships as yet poorly defined (Adey 1981; Petersen 1983).

The critical stage of organogenesis most susceptible to radiation injury is brief, so that frequency of malformations is far lower if a given dose is received over a period of days or weeks than if it is received entirely within the few hours during the critical period. Whether developmental defects can result from irradiation at the low rates associated with natural background is uncertain (NAS/BEIR 1980; Adey 1981; Petersen 1983).

Effects on incidence of cancer

The incidence of many (but not all) types of cancer has been increased by ionizing irradiation in atomic-bomb survivors, patients exposed for medical purposes, and various groups of occupationally-exposed workers (NAS/BEIR 1980). The induced cancers have not been evident, however, until years or decades after exposure, and have had no distinguishing features by which they could be recognized to have resulted from radiation as opposed to other causes. In any given case, therefore, the cancer could not be linked to previous irradiation except on a statistical basis.

Likewise, with increasing intensity and duration of exposure to solar UV radiation, the incidence of skin cancers is increased, at least in fair-skinned individuals (Figure 2).

With few exceptions, effects of ionizing radiation on incidence of cancer have been evident only at moderate-to-high doses (0.5–2.0 Sv or 50–200 rem) or dose rates, and in no instance have the effects been observed over a wide enough range of doses and dose rates to define the shape of the dose-incidence curve. Hence the carcinogenic risks of low-level ionizing irradiation can be estimated only by extrapolation from observations at higher doses and dose rates, based on assumptions about the relevant mechanisms and dose-incidence relationships (UNSCEAR 1977; NAS/BEIR 1980). With UV radiation, on the other hand, effects on the incidence of skin cancer are evident at natural background exposure levels (Figure 2).

Although information about the dose-incidence relationship is fragmentary for most types of cancer, substantial data are available for leukemia and cancer of the female breast. For leukemia, the combined incidence of all types (except the chronic lymphatic type) is increased during the first 25 years after irradiation, by about 1–2 additional cases of leukemia per year per 10,000 persons at risk per SV (100 rem) to the bone marrow. The relationship between the cumulative incidence and the dose is compatible with a linear-quadratic function, in keeping with observations on experimental animals (UN-SCEAR 1977; NAS/BEIR 1980). Interpretation of the relationship is complicated, however, by the fact the different types of leukemia vary in the magnitude of the excess for a given dose, age at irradiation, and time after exposure (NAS/BEIR 1980). Incidence of chronic lymphatic leukemia, moreover, has not been detectably affected by irradiation in any of the populations studied to date.

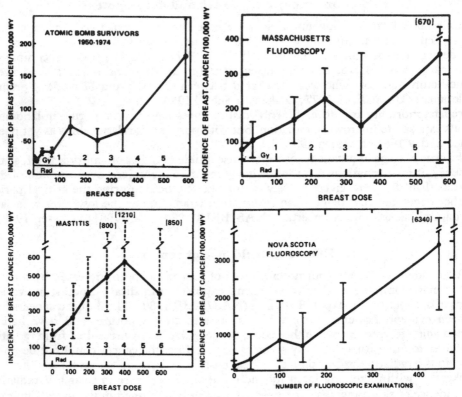

Fig 3. Incidence of cancer of the female breast as a function of dose in A-bomb survivors, women treated with x-rays for acute post- partum mastitis, and women subjected to multiple fluoroscopic examinations of the chest during treatment for pulmonary tuberculosis with artificial pneumothorax (Boice and others 1979).

For cancer of the female breast (Figure 3) the relation between dose and incidence (adjusted for age at irradiation and duration of follow-up) appears to be essentially the same in women exposed acutely (atomic-bomb survivors and radiotherapy patients) as in women exposed chronically (patients given multiple fluoroscopic examinations of the chest during treatment for pulmonary tuberculosis, and radium dial painters) (NAS/BEIR 1980; Baverstock and others 1981). Depending on age at the time of irradiation, the cumulative excess of breast cancers corresponds to about 6 cases per 10,000 women per year per Sv (100 rem) (NAS/BEIR 1980). The fact that small, widely-spaced exposures appear fully additive in their cumulative carcinogenic effects implies that the risk of breast cancer is proportional to the total absorbed dose of ionizing radiation, irrespective of the dose rate.

Additional evidence that carcinogenic effects may result from small doses includes: (1) increased incidence of thyroid tumors in persons exposed to 0.06–0.2 Sv (6–20 rem) to the thyroid gland in infancy or childhood (NAS/BEIR 1980); and (2) an association between prenatal diagnostic x-irradiation and childhood cancer, which suggests that an acute exposure *in utero* to a dose of only 0.01–0.05 Sv (1–5 rem) may increase the subsequent risk of leukemia and other cancers in childhood by as much as 40–50 percent (NAS/BEIR 1972, 1980, Monson and MacMahon 1984).

Although frequency of many types of cancer has increased in populations exposed to ionizing radiation, the data are not adequate in most instances to indicate whether there are risks at low doses. On the assumption of a linear non-threshold dose-incidence relationship, however, the total excess of all cancers combined is estimated to approximate 0.6–1.8 cases per thousand persons per year per Sv (or 6–18 cases per million per year per rem), starting 2–10 years after whole-body irradiation and continuing for the remainder of life (UNSCEAR 1977; NAS/BEIR 1980). This increase corresponds to a cumulative lifetime risk of roughly 20–100 additional cancers of various types per thousand persons per Sv (200–1,000 per million per rem) (Table 3), or a 10–60 percent increase per Sv in the natural lifetime risk of cancer (0.1–0.6 percent per rem).

Table 3. Estimated cumulative lifetime risks of cancer from low- level ionizing radiation. (From NAS/BEIR 1972, 1980; UNSCEAR 1977; Jablon and Bailar 1980).

Site of cancer	Risk per million person-rem	
	Fatal cancers	Incident cancers
Breast (women only)	50	50–200
Thyroid	10	20–150
Lung	25–50	25–100
Bone marrow (leukemia)	15–40	20–60
Brain Stomach Liver Colon Salivary glands	10–15 (each)	15–25
Bone Esophagus Small intestine Urinary bladder Pancreas Lymphatic tissue	2–5 (each)	5–10
Skin	1	15–20
Total (both sexes)	100–250	300–400

The increased numbers of cancers attributable to low-level ionizing radiation at doses of 1–50 mSv (0.1–5 rem) are so small that epidemiological investigations undertaken to verify them are not feasible; ie the study populations would have to include hundreds of thousands of exposed persons (Land 1980). For this reason, efforts to refine existing risk assessments must include research into the mechanisms of radiation carcinogenesis through experiments in laboratory animals and other model systems, in order to advance out understanding of the scientific basis for extrapolation into the low dose domain.

Table 4. Estimated contribution of ionizing radiation exposure to the burden of fatal cancer in the US population. (From Jablon and Bailar 1980). Total number of cancer deaths in the US annually from all causes is about 400,000.

Source of radiation	Annual collective dose (Person-Sy)	Lifetime cancer mortality commitment (Number of fatal cancers)	(Percentage of all cancers)
Natural background	20,000	5,000	1.2
Healing arts	17,000	4,250	1.1
(diagnostic x-rays)	(14,800)	(3,670)	
Nuclear weapons fallout	13,000	375	0.09
Technologically enhanced natural radiation (mining, milling etc)	10,000	250	0.06
Nuclear energy	360	9	0.002
Consumer products	60	1.5	0.0004
Total	393,450	< 10,000	< 2.5

Table 5. Estimated contribution of various causal factors to the US cancer burden. (From Doll and Peto 1981). [a]Allowing for possibly protective effects of antioxidants and other preservatives in food. [b]Ultraviolet radiation causes a much larger proportion of non-fatal cancers (up to 30% of all cancers, depending on ethnic mix and latitude), because of the relative frequency of non-fatal basal cell and squamous cell carcinomas on sunlight-exposed skin.

Factor or class of factors	Contribution to the cancer burden	
	'Best' Estimate	'Acceptable' Estimates
	(percentage of all cancer deaths)	
Tobacco	30	25–40
Alcohol	3	2–4
Diet	35	10–70
Food additives	<1	−5[a]−2
Reproductive and sexual behavior	7	1−13
Occupation	4	2−8
Pollution	2	<1−5
Industrial products	<1	<1−2
Medicines and medical procedures	1	0.5−3
Ionizing and ultraviolet radiation[b]	3	2−4
Infection	10?	1−?
Unknown	?	?

From existing risk estimates, it may be inferred that no more than 1–3 percent of all cancers in the general population are attributable to natural background ionizing radiation (Tables 4 and 5) but that up to 20 percent of lung cancers in non-smokers may be attributable to inhalation of radon and other naturally occurring radioelements in air. The numbers of cancers attributable to ionizing irradiation in radiation workers, estimted similarly, amount to less than one percent of the natural incidence in such persons (Table 6), from which the loss of life expectancy in radiation work may be inferred to be no greater than that in other 'safe' occupations (Sinclair 1981).

Although the above risk estimates were formulated for populations, and not for individuals, risks for individuals must also be assessed for purposes of compensation. A report (Rall and others 1985) outlinging methods for determining the attributable risk

Table 6. Estimates of cancer burden attributable to occupational exposure in various types of work involving ionizing radiation, US, 1975 (from Jablon and Bailar 1980). [a]Entire fuel cycle included. [b]Entire fuel cycle not included.

Type of work	No of persons exposed annually	Average annual dose per person (mrem)	Annual collective dose (thousands of Person-rems)	Lifetime cancer (number of fatal cancers)	Mortality commitment (Percentage of natural incidence)
Healing arts	500,000	80–160	40–80	10–20	0.01–0.02
Manufacturing and industrial	7,000,000	7	50	12.5	0.001
Nuclear energy[a]	62,000	840	52	13	0.1
Research	100,000	120	12	3	0.02
Naval reactors[b]	36,000	220	18	2	0.03

of cancer in an irradiated individual has recently been published in response to a mandate from the US Congress.

As concerns the contribution of naturally occurring UV radiation to the cancer burden (Table 5), the data imply that most skin cancers in the US are attributable to sunlight (Curtis and Nichols 1983).

EM radiations with frequencies below 10^{15} Hz have not been shown conclusively to exert carcinogenic effects in laboratory animals or humans in the absence of thermal effects (Curtis and Nichols 1983, Petersen 1983), although co-carcinogenic effects have been observed at high dose levels under certain experimental conditions (eg Balcer-Kubiczek and Harrison 1985).

Acknowledgements

Preparation of this report was supported in part by Grants ES 00260 and CA 13343 from the US Public Health Service and Grant 8–0248–302 from the American Cancer Society.

References

ADEY, W. R. 1981. Tissue interactions with nonionizing electromagnetic fields. *Physiological Review*, 61: 435–514.

BALCER-KUBICZEK, E. K. AND HARRISON, G. H. 1985. Evidence for microwave carcinogenesis *in vitro*. *Carcinogenesis*, 6: 859–64.

BAVERSTOCK, K. F. AND OTHERS. 1981. Risks of radiation at low dose rates. *Lancet*, 1: 430–33.

BOICE, J. D. JR AND OTHERS. 1979. Risk of breast cancer following low-dose radiation exposure. *Radiology*, 131: 589–97.

CATHCART, R. AND OTHERS. 1984. Thymine glycol and thymidine glycol in human and rat urine: a possible assay for oxidative DNA damage. *Proceedings of the National Acedemy of Sciences USA*, 81: 5633–37.

COLE, A. AND OTHERS. 1980. Mechanisms of cell injury. In MEYN, R. E. AND WITHERS, H. R. (editors). *Radiation biology in cancer research*. New York, Raven Press: 33–58.

CURTIS, R. AND NICHOLS, M. 1983. Nonionizing radiation. In ROM, W. N. AND OTHERS (editors) *Environmental and occupational medicine*. Boston, Little, Brown: 693–705.

DOLL, R. AND PETO, R. 1981. The causes of cancer: quantitative estimates of avoidable risks of cancer in the United States today. *Journal of the National Cancer Institute*, 66: 1192–1308.

GROSOVSKY, A. J. AND LITTLE, J. B. 1985. Evidence for linear response for the induction of mutations in human cells by x-ray exposure below 10 rads. *Proceedings of the National Academy of Sciences USA*, 82: 2092–95.

HOLM, D. A. AND SCHNEIDER, L. K. 1970. The effects of non-threshold radio frequency radiation on human lymphocytes *in vitro*. *Experientia*, 26: 992– 94.

INTERNATIONAL COMMISSION ON RADIOLOGICAL PROTECTION. 1977. Recommendations of the International Commission on Radiological Protection. *Annals of the ICRP*, 1(3). Oxford, Pergamon Press. (ICRP Publication 26.)

INTERNATIONAL COMMISSION ON RADIOLOGICAL PROTECTION. 1984. *Nonstochastic effects of radiation*. Oxford, Pergamon Press. (ICRP Publication 41)

JABLON, S. AND BAILAR, J. 1980. The contribution of ionizing radiation to cancer mortality in the United States. *Preventative Medicine*, 9: 219–26.

LAND, C. E. 1980. Estimating cancer risks from low doses of ionizing radiation. *Science*, 209: 1197–1203.

LEACH, W. M. 1976. On the induction of chromosomal aberrations by 2450 MHz microwave radiation. *Journal of Cellular Biology*, 70: 387.

LLOYD, D. C. AND OTHERS. 1980. The incidence of unstable chromosome aberrations in peripherical blood lymphocytes from unirradiated and occupationally exposed people. *Mutation Research*, 72: 523–32.

MONSON, R. R. AND MACMAHON, B. 1984. Prenatal x-ray exposure and cancer in children. In BOICE, J. D. JR AND FRAUMENT, J. F. JR (editors). *Radiation carcinogenesis: epidemiology and biological sgnificance*. New York, Raven Press: 97–105.

NATIONAL ACADEMY OF SCIENCES. ADVISORY COMMITTEE ON THE BIOLOGICAL EFFECTS OF IONIZING RADIATION (BEIR). 1972, 1980. *The effects on populations of exposure to low levels of ionizing radiation*. Washington, DC, National Academy of Sciences, National Research Council.

OTAKE, M. AND SCHULL, W. 1984. *In utero* exposure to A-bomb radiation and mental retardation: a reassessment. *British Journal of Radiology*, 57: 409– 14.

PETERSEN, R. C. 1983. Bioeffects of microwaves: a review of current knowledge. *Journal of Occupational Medicine*, 25: 103–110.

RALL, J. E. AND OTHERS. 1985. *Report of the National Institutes of Health ad hoc working group to develop radioepidemiological tables*. Washington, DC, US Department of Health and Human Services. (NIH Publication No 85–2748.)

SHAPIRO, R. 1981. Damage to DNA caused by hydrolysis. In SEEBERG, E. AND KLEPPE, K. (editors). *Chromosome damage and repair*. New York, Plenum Press: 3–18.

SINCLAIR, W. K. 1981. Effects of low-level radiation and comparative risk. *Radiology*, 138: 1–9.

SINGER, B. AND GRUNBERGER, D. 1983. *Molecular biology of mutagens and carcinogens*. New York, Plenum Press.

UNITED NATIONS SCIENTIFIC COMMITTEE ON THE EFFECTS OF ATOMIC RADIATION (UNSCEAR). 1977, 1982. *Sources and effects of ionizing radiation*. Report to the General Assembly, with annexes. New York, United Nations.

UNITED STATES. 1977. National Institute of Environmental Health Sciences. *Human health and the environment — some research needs. Report of the Second Task Force on Research Planning in Environmental Health Science*. Washington, DC, US Government Printing Office.

UNITED STATES. 1979. Department of Health, Education, and Welfare. *Interagency Task Force on the health effects of ionizing radiation. Report of the Working Group on Science*. US Government Printing Office, Washington, DC.

UPTON, A. C. 1977. Radiation effects. In HIATT, H. H. AND OTHERS (editors) *Origins of human cancer*. Cold Spring Harbor, Cold Spring Harbor Laboratory: 477–500.

ECOLOGICAL PROCESSES IN THE CYCLING OF RADIONUCLIDES WITHIN ARCTIC ECOSYSTEMS

WAYNE C. HANSON

ABSTRACT. Worldwide fallout radionuclides in arctic ecosystems was investigated ecologically by circumpolar nations during 1959–80. Several of the radionuclides are isotopes of elements which currently contribute to arctic haze; they thus serve as effective tracers of biogeochemical processes. Investigations demonstrated the effective concentration of several radionuclides, particularly strontium-90 (an alkaline earth metal) and cesium-137 (a light alkali metal) which are chemical analogs of calcium and potassium, two very important stable elements in biotic systems. Transfer of ^{137}Cs through the lichen-caribou/reindeer-man food chain characteristic of circumpolar nations, resulted in body burdens in Inuit that were 20 to 200 times greater than those in human populations of temperate latitudes. Radiation exposures from ^{90}Sr, ^{137}Cs and other natural and worldwide fallout radionuclides, were two to three times greater than for most other world populations. These results demonstrate the concentration capabilities of arctic ecosystems for several groups of chemical elements that have counterparts in arctic haze. These elements, therefore, provide the basis for considering the ecological implications of current situations.

Contents

Introduction

Current interest in local and long-range atmospheric transport of industrial pollutants to arctic environments strongly parallels similar interest in worldwide deposition of radioactive debris from nuclear explosions during 1952–80. Following simultaneous and independent observations of elevated concentrations of certain radioactive elements in successive links of the atmosphere-lichen-caribou/reindeer-man food chain of northern Alaska and Scandinavia during the early 1960s (Lidén 1961; Palmer and others 1963), international interest prompted several circumpolar research programs to define the issues and concerns. The foremost concern was to evaluate quickly the radiation exposure to Inuit [Eskimos], Indians, and Lapps because of their strong reliance on caribou and reindeer for food and livelihood. Also important was the definition of mechanisms by which this unforeseen problem arose, to avoid similar situations in the future and define routes, rates and amounts of worldwide fallout in other ecosystems.

One of the most extensive arctic research efforts during 1962–80 was supported by the US Atomic Energy Commission as an extension of baseline investigations in the Cape Thompson region of north-western Alaska (Wilimovsky and Wolfe 1966). Ecological research focussed on defining the transfer of radionuclides in arctic ecosystems. This paper reviews these results, and suggests that similar mechanisms may apply to the transfer of atmospheric pollutants to arctic ecosystems.

Worldwide fallout

Types and sources

Radioactive fallout consists of fission products, unexpended fissile material such as uranium-235 and plutonium-239, and a variety of activation products resulting from neutron capture by materials contained in the nuclear devices or in the environment of explosions (Edvarson and others 1959; Whicker and Schultz 1982). The type and composition of the nuclear device markedly affects the kinds of radioactivity produced, while the location and size of detonation determine the quantity of radioactivity released to the biosphere. Thus the factors affecting fallout have varied in much the same manner as those currently considered in arctic air pollution investigations.

Fallout from nuclear explosions has been classified as *local* or *worldwide*. Local fallout consists of relatively large (more than 35 μm diameter) particles that normally fall to earth within 24 hours of the explosion and are mainly composed of shorter-lived radionuclides with physical half-lives of a few hours to one week. Tropospheric worldwide fallout originates in the troposphere (earth's surface to stratosphere, 9,000–18,000 m [30,000–60,000 ft] above sea level) and is transported solely within it. The fallout remains airborne for more than 24 hours and is deposited by gravity, wind currents and precipitation. Major deposition occurs along a band near the latitude of the detonation and is composed mainly of intermediate half-life (1–8 weeks) radionuclides. Iodine-131 has been one of the most important radionuclides in this category. Areas near the Nevada Test Site were occasionally contaminated with this type of fallout from US continental tests (Pendleton and others 1963; Martell 1964; Beck and Krey 1983), and Japanese populations received radioiodine deposited by fallout from Soviet nuclear tests (Yamagata and Iwashima 1962).

Most of the worldwide fallout produced to date has been termed stratospheric because it originates from relatively high-yield nuclear devices that propel the radioactive debris into the stratosphere, somewhere between 9–18,000 m and 50,000 m [30,000–60,000 and 160,000 ft] above sea level (Carter and Moghissi 1977). The stratosphere is relatively stable, cloud-free, and has an increasing temperature gradient; it serves as a vast reservoir for fallout with physical half-lives for some radionuclides of several years. Mean residence time of debris in this reservoir ranges from one to five years. The longer residence time of stratospheric fallout has several important implications. It results in considerable mixing of debris from different nuclear tests, fosters broader areas of deposition, and allows radioactive decay to modify the spectrum of radionuclides of interest in the biosphere.

Deposition mechanisms

Gaps in the upper boundary of the troposphere (tropopause), at about 40° to 50°N and 40° to 50°S, allow stratospheric fallout to be injected into the troposphere and deposited in those latitudes (Collins 1963; Peirson 1971; Volchok 1964). These are temperate regions with high rainfall, and deposition is accelerated (Miyake and others 1961; Van der Westhuizen 1969); as a result northern states of the conterminous US and southern Canada have received most of the fallout on the North American continent. Northern Alaska has received about a quarter as much fallout per unit area as the northern conterminous states (Hanson 1966). Within Alaska, more fallout has apparently been deposited in the south because of higher rainfall and snowfall, which are important mechanisms for scavenging atmospheric pollutants.

Radionuclides of interest

Investigations of worldwide fallout in Alaska have shown that although several fallout constituents could be routinely measured in biota, the radionuclides of major interest in arctic ecosystems are the fission products cesium-137 and strontium-90 (Hanson 1967, 1982; Hanson and Thomas 1982). Cesium-137 has received most attention because it is relatively easy to measure, has a relatively long physical half-life (30 years), is concentrated at successive levels of the food webs, and because of its chemical behavior as an analog of potassium (Miettinen 1966; Lidén and Gustafsson 1966; Nevstrueva and others 1967). Strontium-90 is also important because it too has a relatively long physical half-life (28 years), is concentrated in bone, and is similar to calcium in biotic systems. Both radionuclides contribute to increased radiation exposures of northern Alaskan natives.

Concentration processes in arctic ecosystems

Vegetation

Direct absorption of radionulides is the most important process by which fallout is concentrated by arctic flora, although several parameters may determine the subsequent deposition. Lichens, because of their physiological characteristics, are particularly efficient accumulators of fallout radionuclides and other atmospheric pollutants (Gorham 1959; Rickard and others 1965; Salo and Miettinen 1964). Concentrations of ^{137}Cs in lichens are about five times those in soil of the same community. Lichens function as a reservoir of radionuclides and determine the radionuclide levels in caribou and their consumers (Figure 1). Measurements of ^{137}Cs deposition resulting from precipitation, compared with ^{137}Cs content of lichens, indicate that 95% of radionuclide deposited was retained for long periods involving tens of years (Svensson and Lidén 1965, Hanson and Eberhardt 1969 and 1973).

Herbivores

Reindeer and caribou (*Rangifer tarandus*) browse extensively on lichens, particularly in winter. Thus they ingest significant amounts of fallout radionuclides and other atmospheric pollutants, of which ^{90}Sr and ^{137}Cs are the most important biologically. Several hundred analyses of fallout materials in reindeer and caribou tissues reflect seasonal changes in diet. Maximum radionuclide concentrations occur at the end of winter, following several months of lichen feeding; minimum concentrations occur at the end of summer, when the animals leave summer pastures of fresh green foods (Hanson and Palmer 1965). Slower turnover of ^{90}Sr in bone also reflects the metabolic differences between ^{90}Sr and ^{137}Cs.

Cesium-137 concentrations in caribou meat harvested by Anaktuvuk Pass hunters during spring migration were four times the ^{137}Cs concentration in lichens sampled from the caribou winter range. After a summer spent grazing on green sedges and other fresh vegetation, ^{137}Cs concentrations were minimal in autumn. This is apparently due to small amounts of fallout ^{137}Cs in the summer forage, substantial amounts of potassium, and increased summer body water turnover rates.

The importance of lichens as vectors of fallout radionuclides was further illustrated

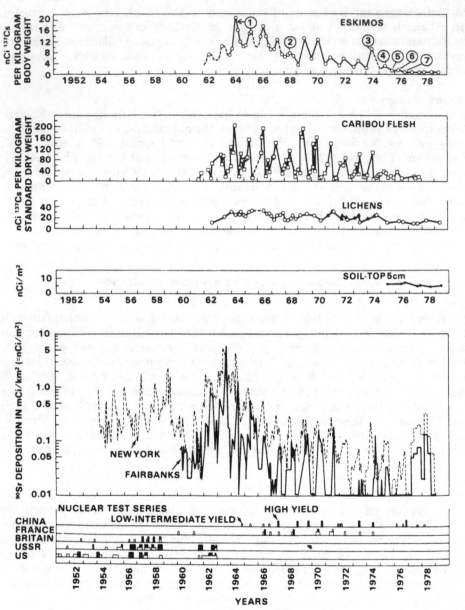

Fig 1. Fallout cesium-137 deposition in New York City and Fairbanks, Alaska, and subsequent concentrations in soils and biota of northern Alaska ecosystems during 1962–79. A previous period of fallout deposition and ecosystem cycling in the lichen-caribous-Inuit food chain during 1952–58 is inferred from measured deposition in New York City, the relation to similar data from Fairbanks, and ^{137}Cs concentrations in Alaska lichens during 1959–62. (Hanson 1982.)

by a tenfold difference in ^{137}Cs and ^{90}Sr concentrations in caribou tissues compared with moose (*Alces alces*) collected from the same environment at the same time during 1963 and 1964 (Hanson 1966). Moose, which browse willows, birch and other trees and shrubs, showed no seasonal variation in ^{137}Cs concentrations comparable to those observed in caribou and reindeer.

Carnivores, including man

During the 1960s Hanson and others (1964) and Palmer and others (1963) reported appreciable ^{137}Cs levels in Inuit and Indian populations of northern Alaska. During the summers of 1962–64, ^{137}Cs and other gamma-emitting radionuclides were measured in about 2,500 natives from the major ethnic groups of northern Alaska, and in samples of their food. Radionuclides observed in humans included sodium-22, iron-55, cobalt-60, cesium-134, polonium-210, and (more recently) plutonium/americium. Nearly all of these represent a concentration by the lichen-caribou/reindeer-man food chain. It can be inferred from our observations of Alaskan reindeer and caribou that iodine-131 would have been present in native peoples in small concentrations during 1962–65 (Hanson and others 1963a, 1963b).

These radionuclides represent a broad spectrum of chemical elements ranging from light alkali metals (cesium) and alkaline earth metals (strontium) to heavy metals (iron, cobalt, and polonium) and actinides (plutonium). As such they are effective tracers of ecological processes that may substantially concentrate atmopheric pollutants, the importance of which has only recently been recognized. In the case of ^{137}Cs, concentrations in northern Alaskan native populations during 1962–64 were 20 to 200 times those of temperate zone populations. This difference was almost entirely due to cycling of the radionuclide through the lichen-caribou-man food chain.

Within the various cultural groups of northern Alaska, Hanson and Palmer (1964) reported the following ranking of ^{137}Cs body burdens in 952 native residents during the summers of 1962 and 1963:

Inland Inuit 100
Kobuk and Noatak Valley Inuit 25–30
Coastal Inuit 6–10
Athapascan Indians 5

This ranking was directly proportional to the importance of caribou in the diets of the people, as estimated from about 800 interviews. Further studies of native residents of Anaktuvuk Pass, Ambler, and Arctic Village showed that during the two years of study there was considerable variation in dietary habits due to environmental factors, but that in general the villages maintained an overall ranking of 1:1:0.5 which relates directly to caribou consumption.

In most cases, amounts of ^{137}Cs in native Alaskans have ranged from one to two times the concentration in dietary caribou flesh. Maximum amounts were found particularly in adult men between the ages of 20–50 years, due to their greater proportion of muscle and greater consumption of caribou. Minors (15–20 years) possessed about 70% as much as adults, and children (3–14 years) 50% or less than adults. During the past 10 years the concentrations have declined substantially, coinciding with decreased numbers of atmospheric nuclear weapons tests and reduced worldwide fallout transmission through the lichen-caribou-man food chain (Hanson 1982).

Strontium-90 has consistently been considered the most critical long-lived fission product in terms of potential dose to human bone marrow. During the major periods of radioactive fallout, considerable attention and funds were devoted to dietary studies in the US and other worldwide temperate zones. Implications of ^{90}Sr in Alaskan food webs, and its ultimate deposition in native populations, are more difficult to define because it is a bone-seeker and requires more complicated measurement techniques. Hanson and Thomas (1982) attempted to define the ecological relationships of ^{90}Sr by analysing various components of soil, vegetation, animals, and processed food samples collected in northern Alaska from 1959–79; by computing ^{90}Sr ingestion by caribou consumers using accurate tracer kinetics equations; and by using the data as input to two tested predictive models of ^{90}Sr in humans.

Strontium-90 concentrations in successive trophic levels of the lichen-caribou-wolf food chain increased by about sevenfold for caribou bone:lichens, and 260 for wolf bone:caribou muscle. Ratios in Inuit residents of Anaktuvuk Pass have ranged from 1.2 in peak fallout years of 1964–66 to 0.25 in 1979, reflecting their lesser consumption of caribou flesh in recent years. Modelling of ^{90}Sr in the lichen-caribou-man food chain has predicted that maximum concentrations occurred in 1974 in skeletons of people born in 1954, based upon several assumptions and our knowledge of the system.

Radiological health considerations

No discussion of the ecological implications of arctic atmospheric pollutants would be complete without an evaluation of the consequences to the human populations of northern Alaska. Of the several types and sources of atmospheric pollutants that have been and are currently of concern, none is better documented and understood than worldwide fallout. While there are several radionuclides and ecological settings that could yield valuable information, the ecosystems and radionuclides of greatest concern from a radiological health standpoint have been comprehensively investigated.

Radiation exposures of northern Alaskan natives that resulted from worldwide fallout transmitted through the lichen-caribou-man food chain reached a maximum during 1964–66 and averaged 100–140 mrad per year from ^{137}Cs; 100 mrad per year from natural ^{210}Po; 10–15 mrad per year from ^{90}Sr; and a few mrad per year from ^{55}Fe to give a total exposure of about 250 mrad per year. Exposure rates have subsequently declined to about 120 mrad per year. These exposures are additional to the approximately 100 mrad per year typically received from natural sources by human populations in most areas of the world.

Conclusions

Two decades of research into cycling of worldwide fallout in arctic ecosystems of northern Alaska have illustrated the abilities of ecosystems to concentrate certain elements. These elements represent groups that have analogs in the pollutants of the arctic atmosphere. They therefore serve as tracers with defined routes, rates, and amounts.

Although arctic ecosystems are less complex than those of temperate regions, conditions of temperature, logistics, large geographical area, and existing data bases complicate investigations. Although the transmission of worldwide fallout radionuclides through the lichen-caribou-man food chain of northern Alaska is understood, experience in simulation modelling of the arctic food chain components has shown that there remains much to be learned about processes, conversion factors and constants within the relatively simple ecosystems involved. However, many of the transfer coefficients for materials have been defined and would serve as an effective base for investigating current topics (Hanson and Eberhardt 1973).

Radiation ecology studies have been extended to initial studies of the effects of sulfur dioxide upon lichen communities of northern Alaska (Moser and others 1980). Results indicate that SO_2 concentrations ranging from 0.5–1.0 ppm would destroy the photosynthetic capabilities of test species, *Cladonia stellaris*, depending upon thallus moisture during exposure, within a matter of hours. Even three years after exposure the lichens could not photosynthesize normally.

The results of mineral cycling through arctic food webs and sensitivity of basic components suggest that an assessment of the ecological effects of acid rain, arctic haze and other atmospheric pollutants needs a higher research priority than is currently perceived. Current federal research that emphasizes the development of computer models, investigates origins of acid rain, and studies mechanisms through which emissions are converted to sulfuric and nitric acids, needs re-ordering to achieve essential definitions where ecology and health requirements are pressing.

References

BECK, H. L. AND KREY, P. W. 1983. Radiation exposures in Utah from Nevada nuclear tests. *Science*, 220: 19–24.

CARTER, M. W. AND MOGHISSI, A. A. 1977. Three decades of nuclear testing. *Health Physics*, 33: 55–71.

EDVARSON, K. AND OTHERS. 1959. Fractionation phenomena in nuclear weapons debris. *Nature*, 184: 1771–74.

GORHAM, E. 1959. A comparison of lower and higher plants as accumulators of radioactive fallout. *Canadian Journal of Botany*, 37: 327–29.

HANSON, W. C. 1966. Fallout radionulides in Alaskan food chains. *American Journal of Veterinary Research*, 27: 359–66.

HANSON, W. C. 1967. Radioecological concentration processes characterizing arctic ecosystems. In ÅBERG, B. AND HUNGATE, F. P. (editors). *Radioecological concentration processes*: 183–91. Oxford, Pergamon Press.

HANSON, W. C. 1982. ^{137}Cs concentrations in northern Alaskan Eskimos, 1962– 79: effects of ecological, cultural and political factors. *Health Physics*, 42: 433–47.

HANSON, W. C. AND EBERHARDT, L. L. 1969. Effective half-times of radionuclides in Alaskan lichens and Eskimos. In NELSON, D. J. AND EVANS, F. C. (editors). *Symposium on radioecology*. Washington, US Atomic Energy Commission: 627–34. (USAEC Technical Information Division Report CONF-670503).

HANSON, W. C. AND EBERHARDT, L. L. 1973. Cycling and compartmentalizing of radionuclides in northern Alaskan lichen communities. In NELSON, D. J. (editor). *Proceedings 3rd national symposium on radioecology*. Washington, US Atomic Energy Commission: 1: 71–75. (USAEC Technical Information Division Report CONF-701501).

HANSON, W. C. AND PALMER, H. E. 1964. The accumulation of fallout cesium-137 in norhtern Alaskan natives. *Transactions of the North American Wildlife and Nature Resources Conference*, 29: 215–25.

HANSON, W. C. AND PALMER, H. E. 1965. Seasonal cycle of 137-Cs in some Alaskan natives and animals. *Health Physics*, 11: 1401–06.

HANSON, W. C. AND THOMAS, J. W. 1982. Prediction of ^{90}Sr body burdens and radiation dose in Anaktuvuk Pass, Alaska Eskimos due to fallout. *Health Physics*, 43: 323–33.

HANSON, W. C. AND OTHERS. 1964. Radioactivity in northern Alaska Eskimos and their foods, summer 1962. *Health Physics*, 10: 421–29.

HANSON, W. C. AND OTHERS 1963a. Iodine-131 in the thyroids of North American deer and caribou: comparison after nuclear tests. *Science*, 140: 801–02.

HANSON, W. C. AND OTHERS. 1963b. Thyroidal radioiodine concentrations in North American deer following 1961–1963 nuclear weapons tests. *Health Physics*, 9: 1235–39.

LIDÉN, K. 1961. Cesium-137 burdens in Swedish Laplanders and reindeer. *Acta Radiologica*, 56: 237–40.

LIDÉN, K. AND GUSTAFSSON, M. 1967. Relationships and seasonal variation of ^{137}Cs in lichens, reindeer and man in northern Sweden 1961–1965. In ÅBERG, B. and HUNGATE, F. P. (editors). *Radioecological concentration processes*: 193–208. Oxford, Pergamon Press.

MARTELL, E. A. 1964. Iodine-131 fallout from underground tests. *Science*, 143: 126–29.

MIETTINEN, J. K. 1966. Radioactive food chains in arctic regions. In *Third UN international conference on peaceful uses of atomic energy*, vol 14: 122– 27. New York, United Nations.

MIYAKE, Y. AND OTHERS. 1962. Seasonal variation of radioactive fallout. *Journal of Geophysical Research*, 67: 189–93.

MOSER, T. AND OTHERS. 1980. Effects of a long-term field sulfur dioxide fumigation on Arctic caribou forage lichens. *Canadian Journal of Botany*, 58: 2235–40.

NEVSTRUEVA, M. A. AND OTHERS. 1967. The nature of ^{137}Cs and ^{90}Sr transport over the lichen-reindeer-man food chain. In ÅBERG, B. and HUNGATE, F. P. (editors). *Radioecological concentration processes*: 209–15. Oxford, Pergamon Press.

PALMER, H. E. AND OTHERS. 1963. Cesium-137 in Alaskan Eskimos. *Science*, 142: 64–66.

PEIRSON, D. H. 1971. Worldwide deposition of long-lived fission products from nuclear explosions. *Nature*, 234: 79–80.

PENDLETON, R. C. AND OTHERS. 1963. Iodine-131 in Utah during July and August 1962. *Science*, 141: 640–42.

RICKARD, W. H. AND OTHERS. 1965. Gamma-emitting radionuclides in Alaskan tundra vegetation 1959, 1960, 1961. *Ecology*, 46: 352–56.

SALO, A. AND MIETTINEN, J. K. 1964. Strontium-90 and caesium-137 in arctic vegetation during 1961. *Nature*, 201: 1177–79.

SVENSSON, G. K. AND LIDÉN, K. 1965. The transport of ^{137}Cs from lichen to animal and man. *Health Physics*, 11: 1393–1400.

VAN DER WESTHUIZEN, M. 1969. Radioactive nuclear bomb fallout: a relationship between deposition, air concentration and rainfall. *Atmospheric Environment*, 3: 241–48.

VOLCHOK, H. L. 1964. Strontium-90: estimation of worldwide deposition. *Science*, 145: 1451–52.

WHICKER, F. W. AND SCHULTZ, V. 1982. *Radioecology: nuclear energy and the environment*. 1: 103–16. Boca Raton, FL, CRC Press.

WILIMOVSKY, N. J. AND WOLFE, J. N. (editors). 1966. *Environment of the Cape Thompson Region, Northwest Alaska*. Springfield VA, US Atomic Energy Commission. (USAEC Publ PNE-481).

YAMAGATA, N. and IWASHIMA, K. 1962. Environmental contamination with iodine- 131 in Japan. *Nature*, 193: 892.

CANCER INCIDENCE AND RISK IN ALASKAN NATIVES EXPOSED TO RADIOACTIVE FALLOUT

CHARLES D. STUTZMAN AND DONNA M. NELSON

ABSTRACT. Cancer incidence in northern Alaskan villages exposed to radioactive fallout from nuclear weapons testing in the late 1950s and early 1960s was assessed using data from the Alaskan Native Tumor Registry. Previous studies have shown that cancer incidence in Alaskan natives differs from that in residents of the rest of the United States: rates of cancer of the nasopharynx and liver are higher in Alaskan native men and rates of cancer of the nasopharynx, gallbladder, cervix, and kidney are higher in Alaskan native women. Leukemia, breast cancer and bone sarcoma are the cancers most likely to result from fallout exposure in the Arctic, but the incidence of these cancers in the North Slope villages appeared to be lower than in either the entire Inuit population or the US population.

The fallout radionuclides of potential health concern are cesium-137 and strontium-90, because of their abundance, long half-life, and chemical characteristics that facilitate transport through and concentration in the food chain and accumulaton in sensitive tissues of the body. Radionuclide body burdens were determined in North Slope Inuit 25 years ago, because of their possible exposure to radioactive fallout via the lichen-caribou-man pathway. Cancer risk estimates have been calculated using highest average dose measurements from residents of Anaktuvuk Pass, under the assumption that peak exposure levels of the mid 1960s remained steady over the following 20 years. Worst-case estimates of expected cancer excess were calculated for leukemia, breast cancer and bone sarcoma. The maximum expected numbers of cases per year due to cesium-137 and strontium-90 exposure were too small to be detected by epidemiology. Other possible explanations for the different cancer incidence found in Alaskan Inuit include westernization of dietary and lifestyle patterns, and certain biological factors such as the Epstein Barr and Hepatitis B viruses.

Contents

Introduction

The possible link between cancer incidence in the Inuit villages of the North Slope region of Alaska, and exposure to radioactive fallout from nuclear weapons' testing approximately 25 years ago, has recently been cause for concern now that a significant latent period for cancer induction has passed. We have addressed this concern by calculating the cancer incidence that might result from fallout exposure (using estimates

229

of dose and population exposed in Alaska), evaluating incidence and histological types of cancer in exposed villages, and examining other possible explanations for cancer incidence observed.

Most atmospheric nuclear weapons tests, 379 in all, were carried out in the northern hemisphere by the USA and USSR between 1945 and 1963. In the 10 years following the 1963 Limited Test Ban Treaty 29 atmospheric tests were conducted by France and 14 by the People's Republic of China; neither was party to the treaty (Whicker and Schultz 1982). Subsurface nuclear weapons testing by the major powers has continued since 1963. However, the peaks periods of biosphere contamination, extensively monitored throughout the world, were the 1950s and the early 1960s.

Radioactive fallout can be local or global, depending on its spatial and temporal distribution. Local fallout, consisting of particles generally over 35 microns in diameter, returns to earth within about a day and up to several hundred miles from the detonation site. Global fallout consists of smaller particles that rise higher into the atmosphere, and become widely dispersed over the earth's surface. Low-yield detonations yield mainly tropospheric fallout, which returns to the surface within a month or so from detonation; deposition depends on weather conditions, usually occuring along a band in the same latitude as the detonation site. High-yield detonations push radioactive material into the stratosphere, and this material returns to earth over months or years. Deposition is a function of latitude, and deposits occur in higher concentrations in the temperate zones. Since most tests have taken place in the northern hemisphere, most fallout has occurred in the north temperate zone. Most global fallout from nuclear weapons testing has been from the stratosphere.

Fallout is composed of 200 or more radionuclides, only a few of which pose long-term health hazards. To be significant, the radionuclides must be abundant, with relatively long half-lives and chemical characteristics that result in concentration through food chains, accumulating considerable radioactivity in sensitive body tissues. Two major radionuclides concentrated in the Arctic and sub-Arctic through the lichen-caribou-man food chain are cesium-137 and strontium-90. Lichens absorb and retain particles, including radioactive fallout material. In winter Alaskan caribou *Rangifer tarandus* feed almost exclusively on lichens in the Kobuk and Koyukuk River drainage areas. In spring they migrate to the northwestern Arctic to feed mainly on seed plants, in which radionuclide concentrations are 2–10 times lower. In the 1950s and 1960s Alaskan natives hunted caribou (their main meat source) on migration through the Brooks Range in autumn and spring. Cesium-137 concentrations in caribou were high in spring after the winter lichen-feeding, and lower in autumn after the summer grazing.

Cesium-137 has a half-life of 30 years. Biological elimination half-time is about 135 days in adult males, about 85 days in adult females, and from about 60 days in older children to 12 days in infants (NRCPM 1979). Metabolism of cesium-137 is similar to that of potassium. It is distributed fairly uniformly throughout the body, although concentrations are generally higher in muscle than in bone or fat. Since cesium emits penetrating high-energy gamma radiation, organs most subject to carcinogenesis are the relatively sensitive bone marrow and breast (NRC/NAS 1980).

Strontium-90, with a physical half-life of 28 years, is also absorbed by lichens and concentrates in caribou bone after ingestion; its effective half-life in bone is 18 years. Metabolism of strontium is similar to that of calcium; hence its affinity for bone. Emitting high-energy beta radiation, it can irradiate both calcified bone and adjacent bone marrow (NCRPM 1979). Bone sarcoma and leukemia are likely to be induced by internal strontium exposure.

Iodine-131, another fallout radionuclide, has a physical half-life of 8 days and a slightly shorter biological half-life. It emits beta particles and concentrates in the thyroid gland, one of the organs most sensitive to cancer induction by radiation (NRC/NAS 1980).

Because of their possible long-term health effects, these three are the primary radionuclides of concern. Other fallout nuclides detected in arctic ecosystems appear not to be significant long-term health hazards for Alaskan natives because concentration processes do not occur that would result in signficant accumulation in humans, their physical half-life is too short, or they do not exist in sufficient quantities.

Dose assessment

During the 1960s radionuclide levels were measured in lichens, caribou, and Inuit. Most important were cesium-137 body burden measurements carried out in many north Alaskan villages, particularly Anaktuvuk Pass, where caribou was the primary component of the diet. The highest average cesium-137 body burdens (over 1,000 nCi) were recorded in Anaktuvuk Pass residents in summer 1964. From 1962–67, residents of Kotzebue, Noatak, Selawik, Noorvik, Kiana, Shungnak, Kobuk and Ambler had intermediate levels (150–650 nCi); those of the north coastal villages Barrow and Point Hope had the lowest levels (3–150 nCi) (Hanson and others 1964; Hanson and Palmer 1965; Rechen and others 1967). Maximum levels in Anaktuvuk Pass occurred about two years after the peak period of atmospheric testing in 1962, a time lag consistent with stratospheric fallout deposition and subsequent concentration of cesium through the lichen-caribou-man food chain. Body burden measurements in the early 1970s showed cesium levels in Akaktuvik Pass residents to have decreased since 1964 (Hedlund 1976). The decrease is slow because cesium-137, with a half-live of 30 years, is retained in lichens for 10 years or more. The highest Inuit cesium levels were in summer, the result of eating caribou which had fed on lichens throughout the winter.

Strontium-90 body burdens in Anaktuvuk Pass residents were estimated rather than directly measured; concentrations were measured in samples of caribou meat from local Inuit hunters from 1964–66, from which adult male Inuit body burdens were estimated to be about 900 pCi. This was similar to body burdens of residents of New York and San Francisco during the same period (Hanson and Thomas 1982), whose elevated levels were ascribed to consumption of dairy products and vegetables. Strontium-90 ingestion rates and estimated body burdens for adult females and for children were, respectively, 50% and 20% of those for adult males. During the 1960s caribou meat provided 80–95% of the strontium-90 body burdens of northern Alaskan Inuit. The steadily decreasing body burdens since 1966 resulted more from a decreased dependence on caribou as a food source, rather than from decreasing levels of strontium in caribou meat (Hanson and Thomas 1982).

Although iodine-131 was not directly measured in the thyroid glands of Alaskan natives, measurements and dose calculations were made on thyroids of deer, elk, caribou and reindeer from Alaska and other states during and after the peak 1962–63 nuclear testing period. Those from Alaskan herbivores carried the lowest burden, suggesting either that arctic deposition of iodine was lower, or that iodine did not enter arctic food chains as readily as those in other states. A major source of exposure to iodine-131, as for strontium-90 (see above), is dairy products; these are not part of the traditional Inuit diet.

Naturally-occurring radionuclides measured in Anaktuvuk Pass residents, lichens, and caribou included lead-210 and polonium-210, the solid decay products of radon-222. The concentration process for this 'natural fallout' is similar to that of cesium-137 and strontium-90. Concentrations of polonium-210 in caribou flesh were about 10 times greater than lead-210 concentrations. Measurements of polonium-210 in urine samples from Anaktuvuk Pass residents in the early 1960s showed levels 200 times higher than those measured in other states, and corresponded to about 10% of the maximum permissible body burden for polonium-210 (Beasley and Palmer 1966).

Cancer risk estimates

Cancer risk estimates were based on information from the National Research Council Committee on the Biological Effects of Ionizing Radiations, 1980 (BEIR III Report). They are expressed as ranges rather than as precise values because of uncertainty of assessing cancer risk from radiation. There are three important points regarding these risk estimates and expected cancer incidence rates derived from them:

(1). The highest average dose measurements were used in calculating expected cancer incidence. These occurred in Anaktuvuk Pass residents, who in the early and middle 1960s still largely depended on caribou as their food source. The actual number of persons who received these maximum doses was probably less than 100—the total Alaskan native population of Anaktuvuk Pass in 1970 was 97 (US Bureau of the Census 1971). In 1980 the population of the northern Alaskan villages where measurements were taken was 5,715 (US Bureau of the Census 1982). (The total Alaskan native population was 64,047 in 1980 and 50,819 in 1970.) Thus the percentage of Alaskan natives with additional radiation exposure through the lichen-caribou-man food chain is small.

(2). Expected cancer incidence rates were calculated assuming that peak exposure levels of the middle 1960s remained the same over the next 20 years. Body burdens in fact steadily decreased during that period. Average cesium-137 body burdens in Anaktuvuk Pass residents measured in 1979 result in a dose of 8 mrem per year (Hanson 1982), over 20 times lower than peak levels in 1964. Since 1970 strontium-90 body burdens have decreased by about 9% per year (Hanson and Thomas 1982).

(3). When risk estimates for a particular cancer site were based on more than one risk model in the BEIR III Report, the highest risk estimates were used to calculate the upper limit of the expected cancer incidence rate.

These three factors result in worst-case estimates of expected excess cancer incidence rates; most-likely-case estimtes may be 10 to 1,000 times lower. Expected excess cancer rates were estimated for cancer sites that have the greatest potential for the induction of cancer by radiation in this situation.

Leukemia may result from cesium or strontium exposure. Radiation-induced leukemia has a relatively short latent period (median seven to eight years in Japanese A-bomb survivors). Latency appears to be shorter in younger age groups and with higher doses. At lower radiation doses, risk per unit of exposure (risk-per-rem) decreases. Estimation of risk from high-dose studies (the only type available) may overestimate risk from lower doses by a factor of two to 10. Evidence also suggests a relative risk two to five times higher among the very young and the very old. Acute leukemia and chronic myelogenous leukemia are the major types associated with radiation exposure. Chronic lymphocytic leukemia has not been shown to be related to radiation (NRC/NAS 1980).

Breast cancer may result from cesium exposure. The female breast is very sensitive to induction of cancer by radiation. A conservative lower limit for the minimum latent period is five to nine years. The maximum latent period is 30 or more years. Latency appears to be independent of dose but strongly dependent on age at exposure. The lower the age at exposure, the longer the latent period tends to be. The occurrence of radiation-associated breast cancer parallels the age distribution of 'spontaneous' breast cancer, after a minimal latent period. Dose-response for breast cancer appears to be linear down to zero dose, ie risk-per-rem is similar for low and high doses. Risk does not seem to depend on dose rate, but may depend on age at exposure; a precise age pattern is not clear. Risk estimates based on Japanese A-bomb survivors show a twofold to threefold higher risk in the 10–19 year age group, compared with the those of 20–39 years and 50 + years. There is now evidence of increased risk for exposure before age 10 years. Risk in the 40–49 year exposure age group is slightly negative, for reasons that are not apparent. These variations in risk of breast cancer with age at exposure may be due to changes in tissue sensitivity to radiation carcinogenesis, resulting from variations in ovarian function at different ages (NRC/NAS 1980).

Bone cancer may result from strontium exposure. Risk estimates are based mainly on studies of exposure to alpha emitters (eg radium-226) which have a high relative biological effectiveness compared with beta and gamma emitters. This would tend to overestimate risk from strontium-90, a beta radiation emitter. Radiation-induced bone cancers have shown latent periods ranging from four to 52 years. Generally latency is directly related to duration of exposure. Short exposure periods show a peak latency of six to eight years, while continuous long-term exposures (which would result from strontium ingestion) show much longer periods. The most common types of radiation-induced bone cancers, in order of decreasing frequency, are osteosarcoma, fibrosarcoma, and chondrosarcoma. No cases have occurred in radium dial painters at doses much below 900 rads. No increase in bone sarcoma was noted in Japanese A-bomb survivors (NRC/NAS 1980).

Thyroid cancer may result from iodine exposure. Radiation-induced types of thyroid cancer are papillary carcinoma and follicular carcinoma; anaplastic carcinoma of the thyroid has not been associated with radiation. The minimum latent period is about 10 years. The peak latent period, if one actually exists, is probably from 15–25 years. External gamma radiation has a higher carcinogenic risk than internal beta radiation, such as occurs with iodine-131, perhaps partly because the iodine resides mainly in the colloid of the thyroid follicle and gives a variable beta dose to the sensitive cellular component of the follicle. Risk from iodine-131 is also lower than that from shorter-lived radioactive iodine isotopes (which are a local rather than global fallout problem), probably because iodine-131, with a longer half-life, gives a lower dose rate and may allow for cellular recovery or repair).

Minimal or occult microscopic thyroid cancer, found at necropsy in 30% of the Japanese population and 15 % of the American population, is thought to have no malignant potential and is not known to be induced by radiation. It should not therefore be included with clinical disease when radiation risk estimates are developed or applied.

Radiation-induced benign thyroid adenomas are three times more common than malignant carcinomas. Other nonmalignant radiation effects on the thyroid gland are associated with higher doses than those that induce cancer: acute thyroiditis—20,000 rads; and hypothyroidism (thyroid ablation)—2,000 rads external or 5,000 rads internal irradiation (NRC/NAS 1980).

The range of expected cancers that may result annually from fallout exposure in Alaska is given in Table 1 and compared with age-adjusted rates for the United States. These

Table 1. Cancer risk in Alaskan natives due to radioactive fallout from atmospheric nuclear weapons tests, based on worst-case estimate, assuming a 20-year exposure at the maximum dose rate. [1]1,000 nCi cesium-137 gives 143 mrem/yr whole body and average skeletal dose. 1.0 pCi strontium-90 per gram of calcium in bone (pCi/gm Ca) gives 4.5 mrem/yr skeletal dose. [3]Age-adjusted to the 1970 census population.

Radionuclide	Cancer or tumor type	Highest average body burden or tissue concentration	Corresponding dose rate	Risk coefficient (cases per rem per year per million persons)	Expected annual excess cancer rate (per 100,000) for a 20-year exposure	Annual age-adjusted[3] cancer rate (per 100,000) for United States from SEER program 1973–77
Cesium-137	Leukemia	1330 nCi	190 mrem/yr[1]	0.01 to 2.2	0.004 to 0.8	9.8
Cesium-137	Breast cancer	1330 nCi	190 mrem/yr[1]	0.60 to 6.1	0.23 to 2.3	85.4
Strontium-90	Bone cancer	2.8 pCi/gm Ca	12.5 mrem/yr[2]	0.09 to 0.76	0.002 to 0.019	0.8
Strontium-90	Leukemia	2.8 pCi/gm Ca	12.5 mrem/yr[2]	0.01 to 2.2	0.0002 to 0.055	9.8
Iodine-131	Thyroid cancer	unknown	unknown	4	—	4.0
Iodine-131	Benign thyroid adenomas	unknown	unknown	12	—	—

Table 2. Maximum expected annual number of cancer cases due to cesium-137, and strontium-90 body burdens for northern Alaskan villages.

Village 1980 native population	Anaktuvuk Pass 191	Kotzebue 1,574	Barrow 1,720	Point Hope 434	River villages 1,796	Total 5,715
Cancer or tumor type						
Leukemia	0.002	0.01	0.01	0.004	0.02	0.049
Breast cancer	0.004	0.04	0.04	0.01	0.04	0.13
Bone sarcoma	0.00004	0.0003	0.0003	0.0001	0.0003	0.001

are worst-case estimates of cancer risk, and the percentage of Alaskan natives to which they might apply is probably extremely small. Expected cancer rates from iodine exposure were not estimated because reliable dose estimates are not available.

Mean US levels of natural background radiation, including cosmic radiation, external gamma radiation from naturally-occurring material in the earth's crust, and radiation from naturally occurring radionuclides found in the body, range from 100 to 250 mrem per year (Klement and others 1972). Certain areas of the world have unusually high natural radiation levels, averaging as high as 3,000 mrem per year and ranging up to 12,000 mrem per year (Eichholz 1976). Epidemiology in some of these regions has not shown increased cancer incidence attributable to these exposures, although the number of people exposed was generally only a few thousand. Doses received by Japanese A-bomb survivors ranged from zero to over 400,000 mrem. There is little evidence of excess cancer at doses below 10,000 mrem, and most of the excess cases in this group occurred at doses over 50,000 mrem (Beebe and others 1978).

Table 2 shows maximum-expected numbers of cancer cases per year due to cesium-137 and strontium-90 body burdens for the northern Alaskan villages. With a population totalling 5,715 in 1980, and assuming the worst-case estimate, a maximum of 3.6 cases of cancer would have developed in 20 years. These numbers are so small that an increase could not be detected by epidemiological study.

Cancer epidemiology in Alaskan natives

Incidence of cancer seen in Alaskan natives differs from that in US Caucasians. A cancer-incidence survey among Alaskan natives, including Inuit, Indians and Aleuts, 1974–78, demonstrated significantly higher risks of cancer of the nasopharynx and liver in men, and cancer of the nasopharynx, gall bladder, cervix and kidney in women. Significantly lower risks were found for leukemia, lymphoma, and cancer of the prostate and bladder in men, and lower risks in women were seen for lymphoma and cancer of the uterus (Lanier and others 1982). No increase could be demonstrated in cancers expected from arctic fallout exposure, such as acute and chronic myelogenous leukemia, and cancer of the breast, bone or thyroid.

Using data from the Alaskan Native Tumor Registry, cancer incidence in northern Alaskan villages was determined (Table 3). Annual rates (cases per 100,000) of pertinent cancer types were calculated for 1969–83 based on the 1970 Inuit population. Rates for the entire Inuit population for 1969–83, and the US Surveillance Epidemiology and End Results (SEER) rates for 1973–77 are shown for comparison. Incidence of acute and myelogenous leukemia and of breast, thyroid and bone tumors appear lower for Inuit in the exposed villages than for the whole Alaskan Inuit population and the US SEER study population. Caution is needed in these comparisons because of the small Inuit population. It appears also that incidence of liver, stomach, and gall bladder cancer is higher in Inuit than in US Caucasians.

Cancer incidence patterns in Alaskan natives cannot be explained adequately by exposure to radioactive fallout. Marked changes in lifestyle occurring during and since the period of atmospheric testing, are more likely to be implicated.

During the 1950s and early 1960s, subsistence hunting and fishing provided the economic bases for the northern Alaskan native villages. Increasing outside influences, notably oil industry developments on the North Slope during the 1970s, have caused a gradual shift toward capital-based economies. Changes in lifestyle and accompanying economic problems were evident in the Inupiat of Anaktuvuk Pass. Before 1961 the Inupiat were nomads, subsisting on the caribou herds. In 1961 the State of Alaska built

1970 Inuit population.

Table 3. Annual cancer rates[1] (cases per 100,000) for 1969–83. Rates based on 1970 Inuit population. [2]Higher exposure villages; Anaktuvuk Pass, Kotzebue, Noatuk, Noorvik, Kiana, Shungnak, Kobuk, Ambler, Selawik. [3]Lower exposure villages; Barrow, Point Hope.

Cancer types	Acute leukemia	Lymphoma	Myelogenous leukemia	Breast	Thyroid	Bone tumor	Liver	Stomach	Gall bladder	Colo-rectal
Higher exposure villages[2] n=3,115	0	2	0	17	2	0	4	17	9	26
Lower exposure villages[3] n=2,265	6	6	3	6	0	0	3	15	12	33
Alaska Inuit n=28,186	3	2	1	20	3	1	7	10	6	33
US SEER rate, n=10% of total US population	10	12	4	85	4	1	2	10	2	59

a school in Anaktuvuk Pass. With their children attending school daily, families could no longer follow the migrating caribou herds, but were limited to hunting them as they passed near the village. This did not always provide sufficient food. Fuel was limited to the willows growing near Anaktuvuk Pass, which were soon depleted. For the first time the Inupiat required money to purchase food and fuel. However, there were few local jobs and men were sometimes forced to the cities in search of employment, causing a disruption of family units and further deviation from the tradiational way of life.

In the absence of opportunities to earn a sufficient income, these changes in lifestyle caused considerable economic hardship. Welfare and food stamps were introduced and foods such as carbonated beverages and pre-packed frozen dinners became a staple part of Inupiat diet (Morgan 1974). Thus 'the Inupiat had found a precarious balance between two worlds—a balance that depended, unfortunately, on the whims of the caribou herds and the welfare agencies' (Morgan 1974).

The shift to a capital-based economy may have contributed to the increasing incidence of certain cancers, for example colo-rectal cancer. In the early 1960s squamous cell carcinoma of the lung and upper respiratory tract was rare in Alaskan natives. Now it nearly equals the US incidence and is most likely attributable to tobacco smoking and alcohol use, introduced in the 1940s (Lanier and others 1976). Elevated rates of nasopharynx and liver cancers follow similar trends; both are increasingly predominant in the Inuit-Aleut population, showing familiar clustering, and are associated with the Epstein Barr and Hepatitis B viruses, respectively (Lanier and others 1882). Thus many of the cancer incidence patterns seen in the Alaskan Inuit may be related to the westernization of their lifestyle, or to certain genetic or biological factors.

Conclusions

Studies over the past 25 years have identified and measured fallout radionuclides of potential significant health importance in Alaska. Cesium-137 is the main concern because of its accumulation in the lichen-caribou-man food chain, although strontium-90 and iodine-131 are important also. Cesium levels were measured by whole-body counting throughout the areas where caribou was a significant food source, using techniques that were sensitive and precise enought to detect cesium body burdens of potential concern. Strontium-90 levels in humans were assessed from dietary information and measurements in caribou flesh, while iodine-131 levels were measured in caribou only, but no significant exposure pathway exists for humans.

Cancer risk due to levels of fallout radionuclides in Alaskan natives is very low, consistent with the observation that the maximum annual dose rates from measured body burdens of cesium were comparable to dose rates from natural background radiation in some regions of the US. Cancers that might be related to radioactive fallout exposure have not increased in northern Alaskan villages. On the basis of current knowledge of radiation carcinogenesis, the radiation doses received and the populations potentially exposed are too small to expect epidemiology to detect any effect. Changing cancer trends seen in Alaskan natives in recent years are more likely due to other factors, such as changing dietary and lifestyle patterns, associated viral exposure and hereditary factors.

References

BEEBE, G. W. AND OTHERS. 1978. Studies of the mortality of A-bomb survivors: 6. Mortality and radiation dose, 1950–1974. *Radiation Research*, 75: 138.
BEESLEY T. M., AND PALMER, H. E. 1966 Lead-210 and polonium-210 in biological samples from Alaska. *Science*, 152: 1062–64.

EICHHOLZ, G. G. 1976. *Environmental aspects of nuclear power*. Ann Arbor, Ann Arbor Science Publishers.

HANSON, W. C. 1982. [137]Cs concentrations in northern Alaskan Eskimos, 1962– 79: effects of ecological, cultural and political factors. *Health Physics*, 42: 433–47.

HANSON, W. C. AND PALMER, H. E. 1965. Seasonal cycle of ^{137}Cs in some Alaskan natives and animals. *Health Physics*, 11: 1401–06.

HANSON, W. C. AND THOMAS, J. M. 1982. Prediction of ^{90}Sr body burdens and radiation dose in Anaktuvuk Pass Alaska Eskimos due to fallout. *Health Physics*, 43: 323–33.

HANSON, W. C. AND OTHERS. 1964. Radioactivity in northern Alaskan Eskimos and their foods, summer 1962. *Health Physics*, 10: 421–29.

HEDLUND, J. D. 1976. Radiocesium in native residents of Anaktuvuk Pass, Alaska, 1970–74. *Health Physics*, 30: 247–49.

KLEMENT, A. W. JR. AND OTHERS. 1972. *Estimates of ionizing radiation doses in the United States: 1960–2000*. Rockville, MD, US Environmental Protection Agency.

LANIER, A. P. AND OTHERS. 1976. Cancer incidence in Alaskan natives. *International Journal of Cancer*, 18: 409–12.

LANIER, A. P. AND OTHERS. 1982. Cancer in Alaskan natives: 1974–78. *National Cancer Institute Monograph*, 62: 79–81.

MORGAN, L. 1974. Anaktuvik Pass. *And the Land Provides*. New York, Doubleday: 1–43.

NCRPM. 1979. *Management of persons accidentally contaminated with radionuclides.* . Washington, National Council on Radiation Protection and Measurements. (NCRP Report 659).

NRC/NSA. 1980. *The effects on populations of exposure to low levels of ionizing radiation*: 1980. Washington, National Academy Press. (National Research Council/National Science Academy Report).

RECHEN, H. J. L. AND OTHERS. 1968. Cesium-137 concentrations in Alaskans during the spring of 1967. *Radiologic Health Data Report*, 9: 705–17.

US BUREAU OF THE CENSUS. 1971. *General population characteristics, Alaska*: 1970. Washington, US Government Printing Office.

US BUREAU OF THE CENSUS. 1982. *General population characteristics, Alaska*: 1980. Washington, US Government Printing Office.

US DEPARTMENT OF HEALTH AND HUMAN SERVICES. 1981. *Surveillance, epidemiology, and end results: incidence and mortality, 1973–77*. Bethesda, MD, National Cancer Institute. (*National Cancer Institute Monograph* 57).

WHICKER, F. W. AND SHULTZ, V. 1982. *Radioecology: Nuclear energy and the environment. Vol I*. Boca Raton, CRC Press.

INDOOR AIR POLLUTION AND PUBLIC HEALTH CONCERNS

JOHN D. SPENGLER AND WILLIAM A. TURNER

ABSTRACT. Concentrations of atmospheric particles, carbon monoxide, nitrogen dioxide, radon, and aldehydes and other organic compounds build up in indoor atmospheres; bacteria, fungal spores and aero-allergens are often present. Surveys of private homes, offices and public buildings have shown that atmospheric contaminants indoors form an important component of the total airborne pollution to which people are exposed, especially in winter when most time is spent indoors and ventilation is reduced. The significance of indoor pollution for public health has not been evaluated. There is no public policy on indoor contaminants in non-occupational settings; government responsiblity is diffuse and ill-defined, and there has been little integrated or coordinated research. New domestic and industrial products, changes in home fuel from gas and oil to coal and wood, new kinds of housing, reduced ventilation and increased awareness of environmental health demand resolution of the question: how much of a public health problem is indoor air pollution?

Contents

Introduction

For the last eight years we have been involved in a long-term health study on respiratory effects of air pollutants (Ferris and others 1979). The effects we are studying are small compared with the well-documented major impacts of cigarette smoking, auto accidents, poor diet and other hazards in the public health field, but this is not to say that they are unimportant. Even a small percentage of change in a widespread effect can have a significant impact on society. For example, acute respiratory diseases on average account for 4.5 workdays lost per person per year in the adult population. Subtle changes in respiratory disease rates, due perhaps to reduced air exchange in indoor environments or to outdoor air pollution, could therefore have large economic and public health effects.

In all public health research it is often difficult to establish clear-cut relationships between environmental pollutants and health effects. Influences of confounding and often co-varying factors may mask the factors of primary interest that we are trying to study. We rely on instrumentation to provide measures of exposure, but just how the measured quantities relate to actual doses that subjects receive is not well-quantified, whether relating to rats in cages or humans in New York City. Many epidemiological

239

studies rely on measurement of ambient air quality as quantitative indicators of dose, but individuals typically spend only 10% of their time outdoors (Chapin 1974).

This is a major reason why we should be concerned about indoor air pollution. Both direct and indirect health effects may be associated with pollutants produced from combustion, volatilization or suspension of gases and particles, occurring indoors as well as outdoors. Just as there are uniquely outdoor contaminants such as ozone, there are indoor contaminants that do not have outdoor counterparts. Several are established health hazards, for example, asbestos, radon, and a variety of fungal spores. Other contaminants can cause indirect health effects, such as behavioral changes, increasing irritation and general discomfort.

To quantify exposure accurately, we need to know how indoor concentrations contribute to the overall pollutant exposure of our population. Pollutants such as CO, NO_2, and particulates are regulated outdoors and have established National Air Quality Standards. We are spending billions of dollars to control outdoor sources when we are not even sure of their contribution to personal exposures (Spengler and Soczek 1984). Quantifying the contribution of indoor and outdoor sources will enable us to improve the efficiency by which we regulate sources to reduce the exposure of our population.

We are just beginning to understand the dimensions of the problem. The behavior of pollutants in air is complex. Outdoor sources contribute to outdoor concentrations, some of which penetrate indoor environments in as yet undefined amounts. Indoor sources also contribute to indoor concentrations. Both indoor and outdoor sources are diluted and removed by various decay mechanisms.

Through their daily activity patterns, individuals integrate over time their exposures to indoor and outdoor air pollution. Individual activity patterns thus determine exposure, and the relative importance of one source versus another. While we may measure or model exposure in this manner, ideally we would like to know actual dose to specific organs. However, in thousands of subjects, direct or indirect dose measures would be very difficult to ascertain. Therefore, in our study involving 22,000 people, we find it is possible only to improve upon a measure of exposure. Our aim is to understand what factors influence exposure, and use these as better measures than have previously been used in air pollution epidemiology.

Indoor pollutants

Concern for energy conservation is probably the most important factor in determining ventilation, and thus the quality of air indoors. Literature on indoor air pollutants of the 1950s and 1960s never mentioned the implications of reduced air exchange. There are really two issues that have brought indoor air quality to the forefront in the last few years: (1) new sources of contamination now being found in work places and homes; and (2) reduced ventilation in houses and offices, resulting from such energy-conserving measures as chimney blocking and window sealing.

Various gases and particles occur at higher levels indoors than out (Table 1). Contaminants such as carbon monoxide and nitrogen dioxide result from combustion. Formaldehyde and organic vapors evaporate from solvents, resins and glues. Radon and radon-progeny are decay products from trace amounts of radioactive materials in granite, soils and water. Water vapor is not generally considered an indoor contaminant, but high concentrations favor house mites, molds and fungi, which may produce allergenic responses in man, and excess moisture can damage property. Due to its high reactivity ozone is generally more prevalent outdoors, but high indoor concentrations are reported close to ion generating sources and in aircraft cabins. Acrolein is one of many potentially

Table 1. Indoor pollutants, emission sources and concentrations (adapted from National Research Council 1981). Column 3 shows typical ranges of indoor concentrations in the presence of indoor emission sources. Taken from Spengler and Sexton 1983. LV = limited and variable (limited measurements, high variation) during cooking. NA = not applicable. *** Annual average. +One-hour average in homes with gas stoves.

Pollutant concentrations	Major emission sources concentration	Typical indoor	Indoor/outdoor
Origins predominantly outdoors			
Sulfur oxides (gases, particles)	Fuel combustion, smelters	0–15 µg/m³	< 1
Ozone	Photochemical reactions	0–10 ppb	< 1
Pollens	Trees, grass, weeds, plants	LV*	< 1
Lead, manganese	Automobiles	LV	< 1
Calcium, chlorine, silicon, cadmium	Suspension of soils, industrial emissions	NA**	< 1
Organic substances	Petrochemical solvents, natural sources, vaporization of unburned fuels	NA	< 1
Origins indoors or outdoors			
Nitric oxide, nitrogen dioxide	Fuel burning	10–120 µg/m³***; 200–700 µg/m³+	> 1; > 1
Carbon monoxide	Fuel burning	5–50 ppm	> 1
Carbon dioxide	Metabolic activity, combustion	2000–3000 ppm	> 1
Particles	Resuspension, condensation of vapors, combustion products	10–1000 µg/m³	1
Water vapor	Biological activity, combustion evaporation	NA	> 1
Organic substances	Volatilization, combustion, paint, metabolic action, pesticides	NA	> 1
Spores	Fungi, molds	NA	> 1
Origins predominantly indoors			
Radon	Building construction materials (concrete, stone), water	0.01–4 pCi/liter	> 1
Formaldehyde	Particalboard, insulation, furnishings, tobacco smoke	0.01–0.5 ppm	> 1
Asbestos, mineral, and synthetic fibers	Fire retardant materials, insulation	0–1 fiber ml 1	> 1
Organic substances	Adhesives, solvents, cooking, cosmetics	LV	> 1
Ammonia	Metabolic activity, cleaning products	NA	> 1
Polycyclic hydrocarbons, arsenic, nicotine, acrolein etc	Tobacco smoke	LV	> 1
Mercury	Fungicides, paints, spills in dental-care facilities or labs, thermometer breakage	LV	> 1
Aerosols	Consumer products	NA	≫ 1
Microorganisms	People, animals, plants	LV	> 1
Allergens	House dust, animal dander, insect parts	LV	> 1

Table 2. RSP concentrations for indoor/outdoor samples collected over a year at 55 homes (Spengler and others 1981b).

Location	No of homes	No of samples	Mean concentration (μg m^{-3})	Standard deviation of home means
Outdoor	55	1676	21.1	11.9
Indoor – no smokers	35	1186	24.4	11.6
Indoor – 1 smoker	15	494	36.5	14.5
Indoor – 2+ smokers	5	153	70.4	42.9

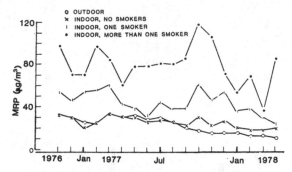

Fig 1. Monthly mean mass respirable particulate concentrations (ug/m³) in six cities (Spengler and others 1981b).

harmful gases and vapors given off by cigarette smoke. Particles are often classed together and measured as mass concentration. Particulate matter includes dust, powders, fibres, spray-can aerosols, organic materials from incomplete indoor combustion, and spores, aero-allergens and other livng organisms.

The impact of indoor pollutants on exposure can be illustrated from our studies on particles and nitrogen dioxide, both of which can have substantial indoor sources (NRC 1981). In an extensive indoor air quality survey of homes in six US cities (Spengler and others 1981) we collected particles of respirable size using a small pump and cyclone pre-separator, which gathered on filters 50% of the particles with aerodynamic diameters less than 3.5 micrometers (Turner and others 1979). Twentyfour-hour samples of respirable-sized particulate (RSP) were collected every sixth day for up to two years in over 80 homes. Results are summarized in Table 2 and Figure 1. Mean indoor concentrations of particles were higher and more variable than concentrations outdoors. Table 2 compares readings from tobacco-smoking and non-smoking homes: smoking clearly contributes significantly to indoor respirable particulate (RSP) concentrations. Questionnaire data from our survey of 10,000 children from the six cities revealed how extensive cigarette smoke exposure may be in the general population. The percentage of children living with one or more smoking adults ranged from 60% to 75%; however, even in non-smoking homes (Figure 2) indoor RSP concentrations were usually higher than outdoors, due to a variety of causes (Ju and Spengler 1981). Thus, with RSP, we encountered problems in attempting to investigate the health effects of ambient air because variation in indoor concentrations among homes was greater than inter-city variation in outdoor concentrations.

In personal exposure studies we asked subjects, most of them adults, to carry personal monitors and record their activities; in Topeka, Kansas, 46 non-smoking adults carried

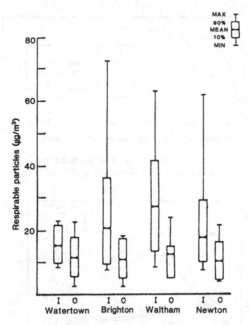

Fig 2. Indoor and outdoor concentrations of respirable particles in four private homes (Ju and Spengler 1981).

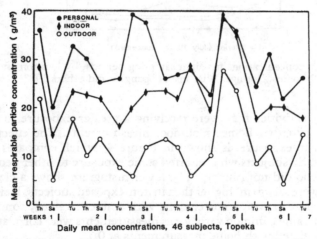

Fig 3. Indoor, outdoor, and personal RSP concentrations, Topeka, Kansas (Spengler and Tosteson 1981).

RSP monitors during the active 12–hour periods of the day (Spengler and Tosteson 1981), husband-and-wife teams participating in an 18–day study. At the same time we monitored RSP at a central outdoor site. The object was to discover how reliably such a central measurement reflected personal exposures within a community. Figure 3 shows mean daily outdoor, in-home and personal RSP; indoor concentrations are higher than outdoor, and personal exposures are higher than both. This indicates that somewhere in

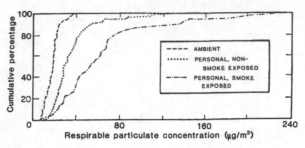

Fig 4. Percentage distribution of personal RSP concentrations: non- smoke exposed and smoke exposed samples, Topeka, Kansas (Spengler and Tosteson 1981).

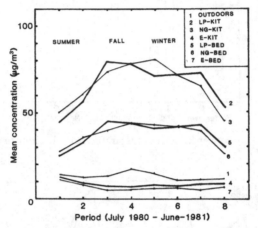

Fig 5. Mean NO$_2$ concentrations for eight sampling periods, by location and type of cooking fuel; 137 homes, Portage, Wisconsin (Spengler and others 1983).

these people's daily activities they were receiving a greater exposure to particles than could be explained by either home or outdoor measurements. The effect of one likely source, passive smoke exposure, is shown in Figure 4, which compares RSP concentrations experienced by subjects who reported some exposure to other peoples' smoking (lower) and those who did not (upper). The lower histogram shows a clear shift to the right; mean values were 20 μg/m^3 higher than in non-exposed subjects. Neither husbands nor wives showed clear exposure correlations with ambient values; concentrations for both showed considerable common variance as measurements were correlated at 0.5. The wifes' exposure correlated with home measurements at 0.7.

While indoor concentrations do not completely explain personal exposures, this and similar studies in Tennessee (Spengler and others 1981a) and Vermont (Sexton and others 1983) have indicated the importance of home measurements in understanding total personal exposure. The next phase will involve direct measurement of indoor particulates in 1,800 homes to obtain an improved measure of exposure for children. We shall also measure air exchange and attempt to use measurements of a nicotine derivative as a marker of cigarette smoke.

Nitrogen dioxide (NO$_2$), originating from indoor combustion, is another possible health hazard. Use of gas cooking fuel is associated with increased respiratory illness and

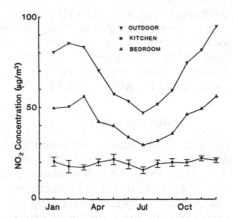

Fig 6. Mean NO_2 concentrations at homes with gas cooking stoves throughout the year, Topeka, Kansas. Outdoor (* 2 × standard errors of the mean), kitchen and bedroom locations (Letz and others 1984).

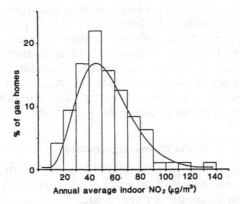

Fig 7. Frequency distribution, indoor annual average NO_2 concentrations; gas-cooking homes, Portage, Wisconsin: with fitted distribution curve (Spengler and others 1983).

even deficits in pulmonary functions (Melia and others 1979; Goldstein and others 1979; Florey and others 1979; Speizer and others 1980; Comstock and others 1981), though other studies have not demonstrated these effects (Keller and others 1979; Mitchell and others 1974). In two extensive field studies we have measured indoor and outdoor NO_2 at several hundred homes in Portage, Wisconsin (Spengler and others 1983) and Topeka, Kansas (Letz and others 1984). Both communities had relatively low ambient NO_2 concentration (10–20 ug/m³). Week-long samples were collected by passive diffusion tubes in kitchens and bedrooms, and outside each home; eight measurements were taken at approximately six-week intervals.

Figure 5 shows mean values of indoor NO_2 concentrations from 137 gas-cooking homes; effects of reduced air infiltration, and perhaps more cooking, appear in higher winter concentrations. Figure 6 shows similar curves for 55 gas-cooking Topeka homes. Figure 7 shows the frequency distribution of indoor mean annual NO_2 concentrations in Portage homes. It is important to note that the variation in air exchange rates, cooking fuel use, and perhaps other factors lead to a generally broad distribution of NO_2 concentrations. In Portage, where ambient level was 15 ug/m³, gas cooking fuel added on average 45 ug/m³ to indoor concentrations; 3% of these homes exceed the National Ambient Air Quality Standard limit of 100 ug/m³. There is a strong implication that, from combined outdoor and indoor sources, a substantial portion of the US population is exposed to levels of NO_3 that exceed NAAQS standard.

Future research

Epidemiologic research on indoor pollutants is just beginning. There have been very few systematic investigations of exposures to passive smoke, aero-allergens, radon, and other common indoor contaminants. Even the few studies reported to date have not adequately characterized exposures with direct measurements. To improve understanding of pollutant behavior we need simpler, lighter-weight and cheaper measuring instruments. Passive collectors providing time-integrated measures are useful, but not always best for particular applications. For some pollutants, exposure to short-term peak concentrations, measurable only by continuous monitoring, may have greater biological relevance.

Improved measurements, made in conjunction with health research, are needed. We shall ultimately be able to provide well-defined exposure-response relationships, which will in turn lead to the design of truly effective control strategies. We shall also be able to address questions as to how new housing should be constructed to reduce energy consumption. Specifically we shall know what products and materials to exclude, whether to use air-to-air heat exchangers or other venting systems, and whether air cleaners are needed to remove vapors and particles are needed. There is much still to be done to characterize the healhfulness of our homes, offices and public buildings.

Acknowledgements

This work was supported in part by NIEHS grant ES-01108, and EPRI grant RP-1001. The editorial assistance of Mary Lou Soczek, and typing and editorial assistance of Allison Maskell, are greatly appreciated.

References

CHAPIN, F. S. 1974. *Human activity patterns in the city*. New York, Wiley-Interscience.
COMSTOCK, G. W. AND OTHERS. 1981. Respiratory effects of household exposures to tobacco smoke and gas cooking. *merican Review of Respiratory Disease*, 124: 143–48.
FERRIS, B. G. AND OTHERS. 1979. Effects of sulfur oxides and respirable particulates on human health: methodology and demography of population in study. *American Review of Respiratory Disease*, 120(4): 767–779.
FLOREY, C. AND OTHERS. 1979. The relationship between respiratory illness in schoolchildren and the use of gas for cooking. *International Journal of Epidemiology*, 8: 347–53.
GOLDSTEIN AND OTHERS. 1979. The relationship between respiratory illness in primary schoolchildren and the use of gas for cooking. II. Factors affecting nitrogen dioxide levels in the home. *International Journal of Epidemiology*, 8: 339–45.
JU, C. AND SPENGLER, J. D. 1981. Room to room variations in concentration of respirable particles in residences. *Environmental Science and Technology*, 15.
KELLER, M. D. AND OTHERS. 1979. Respiratory illness in households using gas or electrivity for cooking. *Environmental Research*, 19: 495–515.
LETZ, R., MILLER, AND OTHERS. 1984. Relationships of measured NO_2 concentrations at discrete sampling locations in residences. In *Proceedings: National Symposium on Recent Advances in Pollutant Monitoring of Ambient Air and Stationary Sources, Raleigh NC, May 3–6 1983*. (USEPA- 600/9–84–001).
MELIA, R. H. W. AND OTHERS. 1979. The relationship between respiratory illness in primary schoolchildren and the use of gas for cooking. I. Results from a national survey. *International Journal of Epidemiology*, 8: 333–38.

MITCHELL, R. I. AND OTHERS. 1974. Household survey of the incidence of respiratory disease in relation to environmental pollutants. In *Proceedings, Recent Advances in the Assessment of the Health Effects of Environmental Pollutants*.

NATIONAL RESEARCH COUNCIL. 1981. *Indoor pollutants*. Washington, DC, National Academy Press.

SEXTON, K. AND OTHERS. 1983. Winter air quality in a valley community where residential wood combustion is a major emission source. Presented at the 76th Annual Meeting of the Air Pollution Control Association, Atlanta, GA, June 1983.

SPEIZER, F. E. AND OTHERS. 1980. Respiratory disease rates and pulmonary function in children associated with NO_2 exposure. *American Review of Respiratory Disease*, 124: 143–48.

SPENGLER, J. D. AND SEXTON, K. 1983. Indoor air pollution: a public health perspective. *Science*, 221: 9–17.

SPENGLER, J. D. AND SOCZEK, M. L. 1984. Evidence for improved ambient air quality and the need for personal exposure research. *Environmental Science and Technology*.

SPENGLER, J. D. AND TOSTESON, T. D. 1981. Personal exposures to respirable particles. Presented at Environmetrics '81, Conference of the Society for Industrial and Applied Mathematics. Alexandria, Virginia, April 8–10, 1981.

SPENGLER, J. D. AND OTHERS. 1981a. Personal exposures to respirable particles: tale of two cities—Kingston and Harriman, Tennessee. Presented at the International Symposium of Indoor Air Pollution, Health, and Energy Conservation, Amherst, Massachusetts, October 13–16, 1981. SPENGLER, J. D. AND OTHERS. 1981b. Long-term measurements of respirable sulfates and particles inside and outside homes. *Atmospheric Environment*, 15: 23–30.

SPENGLER, J. D. AND OTHERS. 1983. Nitrogen dioxide inside and outside 137 homes and implications for ambient air quality standards and health effects research. *Environmental Science and Technology*, 17: 164.

TURNER, W. A. AND OTHERS. 1979. Design and performance of a reliable personal monitoring system for respirable particulates. *Journal of the Air Pollution Control Association*, 29(7): 747–48.

EXPOSURE TO HEAVY METALS IN GREENLAND FROM NATURAL AND MAN-MADE SOURCES

JENS C. HANSEN

ABSTRACT. Since 1979 samples of blood and hair from Inuit in Greenland have been collected in order to evaluate the exposure to mercury, lead, calcium and selenium. Samples are at present taken from a total of 1,352 individuals. Mercury and selenium concentrations in blood were high and closely related to the daily consumption of local food. Blood lead concentrations were found at a level corresponding to European countries, indicating long distance atmospheric transport. Blood cadmium concentrations reflected primarily smoking habits. Comparisons between present-day and 15th century hair samples indicate that mercury and lead have increased while selenium has decreased with time. Present exposure levels may be of risk to foetal development and legal regulations should be considered to protect the Inuit population from adverse effects of marine and atmospheric pollution.

Contents

Introduction

Since 1979 biological monitoring of humans exposed to heavy metals has regularly been carried out in Greenland. This followed analyses of marine mammals and fish caught along the Greenland coast, that revealed concentrations of heavy metals mercury and cadmium which, according to dietary calculations, could be in excess of the World Health Organization (WHO) provisional tolerable weekly intake (PTWI). Lead was also present in food animals, but in concentrations lower than the PTWI standard (Johansen 1982). Mercury and cadmium monitoring involves the collection of Inuit blood and hair samples from various districts (Figure 1). Samples are currently collected from approximately 1,300 individuals, about 2.7% of the total population. Table 1 summarizes the various programmes carried out so far. Lead was also included in the monitoring programme, in accordance with the EEC Directive of 29 March 1977 on the risk of lead exposure in the non-occupationally exposed part of the population. A remote area like Greenland lacks such primary lead pollution sources as motor vehicles and industrial plant, and could provide background data. Selenium, an essential micronutrient, was later included

Fig 1. Heavy metal sampling areas, Greenland.

in the monitoring, as it has been shown to be a powerful antagonist to the toxic effects of mercury and cadmium.

An historical perspective has been made possible by comparing concentrations of heavy metals in contemporary hair samples with samples from eight 15th century mummies found in West Greenland, and from four samples collected in 1884 in East Greenland, Angmagssalik district, by the first expedition to this area.

Index media

An index medium is a tissue, body fluid or sample of excreta in which concentration of a toxicant reflects the body burden. For mercury, blood as well as hair are good indices (Hansen 1981). Sherlock and others 1984 gave the relationship between intake and blood concentration as:

Blood Hg $(\mu g/1)$ = 0.8 x daily intake (μg)

The hair:blood ratio is around 250 (Hansen 1981). Lead exposure is also generally agreed to be reflected in blood concentration, and blood samples were used in the EEC survey. Hair lead concentration has been accepted as an exposure index, but in the Greenlandic survey blood lead was considered the best choice. Blood cadmium concentrations are considered an expression of recent exposure and not of body burden, but in cases where exposure is practically constant, blood cadmium concentrations are useful in estimating exposure level. The value of hair cadmium concentrations as an index is dubious. Blood selenium concentrations are reported to reflect daily intake, but information on hair selenium is scarce. Results from the surveys among Inuit indicate that hair selenium values are not of great use.

In Greenland, with its supposedly constant exposure level to the elements under investigation, blood concentration seems a reasonable index of actual exposure. Hair concentration, even if not applicable on an individual basis, provides some information on changing exposure levels within groups.

Table 1. Heavy metal sampling programmes, Greenland and Denmark 1979–85.
References: 1) Hansen 1981; 2) Hansen and others 1983a; 3) Hansen and others 1983b; 4) Hansen and others 1984a; 5) Wulf and others 1986; 6) Hansen and others 1985; 7) Hansen and others 1984b; 8) Hansen andd Sloth-Pedersen 1986; 9) Unpublished.

Region	Sampling place District	Year of sampling	No of samples	Sex	Hg	Cd	Analyses Pb	Se	SCE	Hair	Source
West Greenland	Upernavik Umanak Godthåb Julianehåb	1979	111	M	+	+	+		+	(1)	
East Greenland	Angmagssalik	1981	178	M+F	+	+	+	+	+	+	(2, 3, 4, 5)
Denmark	Copenhagen*	1981	29	M+F		+	+	+	+		(3, 6)
West and East Greenland	Upernavik Umanak Scoresbysund Angmagssalik	1982	98x2	Mothers and Babies	+	+	+				(7)
North Greenland	Thule	1983	97	M+F	+		+	+		+	(8)
North, East and West Greenland	Godthåb Upernavik Umanak Thule Scoresbysund Angmagssalik	1984–85	357x2	Mothers and Babies	+		+	+		(9)	
Total			1,325								

* Inuit living in Denmark.

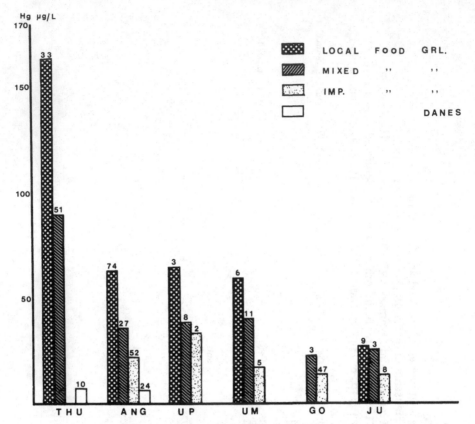

Fig 2. Mean blood concentrations and eating habits, various sampling districts.

Actual exposure levels

Mercury

As expected, mercury concentrations were high in blood and hair, and closely reflected eating habits, ie how much local food (marine mammal meat) was included in the daily diet (Figure 2). Figure 2 also shows geographical variation in mercury exposure; the north-to-south decline reflects the degree of technological development in Greenland. In the far north traditional hunting is still the main occupation, resulting in a high per capita consumption of local products; in the south modern technology and industrialization is dominant, influencing especially the younger generation towards a western lifestyle and resulting in an increased consumption of imported food.

From the daily intake blood mercury relationship (Sherlock and others 1984) and the median blood concentrations found in different regions, weekly intake can be estimated. Table 2 shows that the WHO PTWI of mercury (as methyl mercury) of 200 μg/week can be exceeded by a factor as great as 8.3. WHO has indicated a blood mercury concentration of 200 μg/1 as the lower limit for appearance of unspecific signs of toxic effects in adults; this was exceeded by 16% of the investigated adult population in North Greenland (Thule); the highest concentration recorded was 267 μg/1. According to Junghans (1983) specific signs of intoxication appear when blood mercury concentration exceed 1,000 μg/1. So in general there seems to be no immediate risk to the adult

Table 2. Estimated weekly intake of methyl mercury in Greenland regions.

Region	Estimated weekly intake μg		Estimated PTWI	
	Mean	Extreme intake	Mean	Extreme
North	896	1666	4.5	8.3
North-west	343	546	1.7	2.7
South-west	175	238	0.9	1.2
East	392	496	2.0	2.4

population. However, methyl mercury passes the placental barrier and foetal tissues are said to be more sensitive to toxicants than maternal tissues; the problem for Inuit hunting populations may thus be mercury exposure in utero. Blood samples from the umbilical cord are currently collected for analysis from six districts.

Cadmium

Johansen (1982) calculated a weekly intake of cadmium of 5 mg/person as the highest exposure level; this is up to 12.5 times the WHO PTWI of 0.4 to 0.5 mg/week. However, blood cadmium concentrations in West and East Greenland showed no influence of eating habits; smoking is the only determining factor. As with mercury, the highest exposure level should have been in North Greenland. Hansen and Sloth-Pedersen (1986) found both smoking and eating habits reflected in blood cadmium levels (Figure 3).

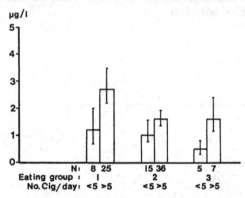

Fig 3. Blood cadmium concentrations from North Greenland in relation to eating and smoking habits. Geometric mean and 2 x Standard error of the mean indicated (Hansen and Sloth-Pedersen 1986).

From the blood cadmium data, smoking is the most important factor affecting cadmium exposure in Greenland, just as it is worldwide. Cadmium found in Inuit diets is presumably not easily absorbed; it is reflected only moderately in blood cadmium concentrations, and only when intake level is high. Dietary intake cannot be estimated from blood concentrations; no relationship been established, and cadmium exposure occurs from several sources. Preliminary results from a food survey have given median concentrations of camium in meat and organs from various food animals. By applying this to data on daily intakes of local food items, weekly intake of cadmium can be roughly estimated (Table 2). Dietary cadmium intake in Greenland exceeds the WHO PTWI standard by 12 times. Smoking seems, however, more important to the blood cadmium level, and further investigation of the absorption of cadmium from natural bindings in foodstuffs are needed to determine dietary cadmium risks.

Table 3. Estimate of weekly dietary intake of cadmium in Greenland regions.

	Estimated weekly intake μg		
Region	Mean	Extreme exposure	Estimated PTWI
North	2583	4802	5.2 − 12.0
North-west	987	2296	2.0 − 5.8
South-west	483	693	1.0 − 1.7
East	1134	1365	2.3 − 3.4

Lead

Lead burden in Greenland was thought to be low because of the absence of the most important environmental sources, vehicular traffic and industry; in addition, low lead concentrations are found in food. Surprisingly, blood lead levels recorded in West and East Greenland were similar to those found in European industrialized countries, median concentrations ranging from 8 μg/100 ml in the Thule district to 16 μg/100 ml in Angmagssalik, East Greenland. Results from East Greenland are shown in Figure 4, where four samples exceeded 35 μg/100 ml, the level set within the EEC as the upper acceptable blood lead concentration in non-occuaptionally exposed people. Reasons for this relatively high blood level are unknown. No relationships with eating habits have been found; although high alcohol consumption cannot explain lead levels, it is probably implicated. Low calcium and high protein and iron intake, characteristic of Inuit diet, may favour intestinal lead absorption.

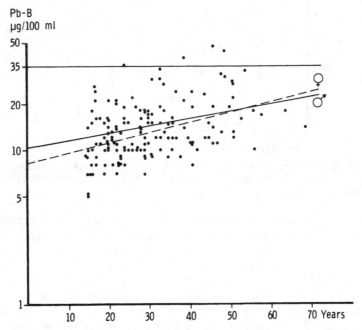

Fig 4. Blood lead concentrations from East Greenland (Angmagssalik) in relation to age.

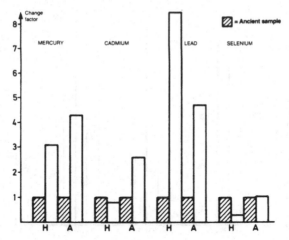

Fig 5. Relative metal concentrations in ancient and present-day hair samples from Greenland. Concentration in ancient hair = 1. H = Human hair. A = Animal hair.

Selenium

Blood selenium concentrations, determined in the districts of Angmagssalik and Thule, were high compared with those of European countries, but differed in the two districts with median values respectively of 130 μg/1 and 1,225 μg/1. The differences cannot at present be explained, but may result from interspecific variations among the food animals, and different eating patterns in the two districts. A close relationship between blood selenium concentrations and eating local food was noticed. On a molar basis, more selenium than mercury was found in human blood. According to animal experiments, selenium provides some protection against toxicity of mercury and cadmium. To assume that this happens in Greenland would be rash, as laboratory conditions very seldom reflect natural exposure. Findings in Angmagssalik demonstrated that frequencies of sister chromatid exchanges could be significantly related to blood mercury and blood cadmium concentrations, though no influence from selenium could be detected. Thus actual mercury exposure level among Greenland Inuit should be regarded as a serious problem.

Historical perspective

The discovery in 1978 of eight well-preserved mummies in Qilakitsoq, West Greenland, and their dating by C^{14} to 1475 $\pm$ 50 years, made it possible to compare analyses of hair samples and obtain an idea of metal concentrations in an ancient population living entirely off local food. The mummies wore sealskin furs from which hair was also analysed. Metal concentrations in old and modern Greenland hair samples appear in Figure 2. Mercury and lead levels have increased with time in both human and animal samples; cadmium shows increase only in seal hairs. Figure 6 shows the incorporated results of analyses of four 19th century samples collected in Angmagssalik in 1884, including the essential microelement selenium. Surprisingly, mercury, cadmium and selenium levels were lower in 19th century samples than in those from either 15th and 20th centuries; these are dietary components, and their dearth may reflect the severe famine suffered by East Greenland Inuit during the 1880s.

The discovery that lead concentrations were higher in the 19th century samples than

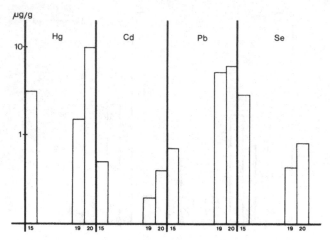

Fig 6. Metal concentrations in Greenland hair samples from the 15th, 19th and 20th centuries.

in those from the 15th century, and comparable to modern ones, indicates that lead exposure in Greenland is not primarily dietary, though diet may enhance absorption. Selenium concentrations were highest in the 15th century samples, as were those of iron and bromine. These values reflect a changing lifestyle in which local food is being replaced by western imported food.

Sources of heavy metals in the Arctic

Toxic heavy metals are ubiquitous and have always been present in man's environment in natural background concentrations. These may vary according to geographical conditions; time related fluctuations may also occur. Consequently it is difficult to distinguish between background or natural concentrations and the contribution by man-made pollution. Mercury analysis reported here indicates increased input into the marine environment, corresponding to increased technological use of this metal. The increase in cadmium concentrations in seal hair indicates that this metal too is now a significant marine pollution, though its industrial use began only in this century. Lead has shown the biggest relative increase, confirming other views that environmental lead concentrations have increased steadily in the northern hemisphere in historical time. Rahn and McCaffrey (1980) demonstrated long distance atmospheric transport of pollutants from industrialized zones in Asia, Europe, and USA to the Arctic. A long-distance transport of airborne lead particles would agree with our findings that lead is present in the Arctic environment, and gives rise to relatively high human blood concentrations, while there is no support for the idea of a dietary exposure. The actual transfer pathway of lead to man remains to be revealed.

All surveys carried out so far in Greenland show an impact of heavy metals originating from a general environmental pollution; no links with local activities (eg mining) have been seen. But Greenland today is far from being an isolated area; it has become a part of rapidly-developing western technological society. This implies progress by the Greenland society, who are now partaking in problems created by pollution. Living conditions still prevailing in Greenland make the Inuit very vulnerable to adverse effects of environmental toxicants; careful medical supervision as well as implementation of legal guidelines are needed to protect them and their culture in the future.

References

HANSEN, J. C. 1981. A survey of human exposure to mercury cadmium and lead in Greenland. *Meddelelser om Grønland, Man & Society*, 3: 1–36.

HANSEN, J. C. AND SLOTH-PEDERSEN, H. 1986. Environmental exposure to heavy metals in North Greenland. *Arctic Medical Research*, (in press).

HANSEN, J. C. AND OTHERS. 1983a. Human exposure to heavy metals in East Greenland. I. Mercury. *Science of the Total Environment*, 26: 233–43.

HANSEN, J. C. AND OTHERS. 1983b. Human exposure to heavy metals in East Greenland. II. Lead. *Science of the Total Environment*, 26: 245–54.

HANSEN, J. C. AND OTHERS. 1984a Selenium and its interrelation with mercury in wholeblood and hair in an East Greenlandic population. *Science of the Total Environment*, 38: 33–40.

HANSEN, J. C. AND OTHERS. 1984b. Concentrations of mercury, selenium and lead in blood samples from mothers and their newborn babies in four Greenlandic hunting districts. *Meddelelser om Grønland, Man & Society*, 6: 1–19.

HANSEN, J. C. AND OTHERS. 1985. Cadmium concentrations in blood samples from an East Greenlandic population. *Danish Medical Bulletin*, 32: 277–79.

JOHANSEN, P. 1982. Heavy metals in marine mammals and heavy intake in humans in Greenland. In HARVALD, B. AND HART HANSEN, J. P. (editors). *Proceedings of the 5th International Symposium Circumpolar Health, Copenhagen 1981*. Copenhagen, Nordic Council for Arctic Medical Research: 540–42. (Report Series 33).

JUNGHANS, R. P. 1983. Review of the toxicity of methyl mercury compounds with application to occupational exposures associated with laboratory uses. *Environmental Research*, 31: 1–31.

RAHN, K. A. AND McCAFFREY, R. 1980. On the origin and transport of the winter arctic aerosol. *Annals of the New York Academy of Sciences*, 338: 486–503.

SHERLOCK, J. AND OTHERS. 1984. Elevation of mercury in human blood from controlled chronic ingestion of methylmercury in fish. *Human Toxicology*, 3: 117–31.

WULF, H. C. AND OTHERS. 1986. Sister chromatid exchange (SCE) in Greenland Eskimos. Dose-response relationship between SCE and seal diet, smoking, and blood cadmium and mercury concentrations. *Science of the Total Environment*, 48: 81–94.

ECOLOGICAL IMPLICATIONS OF ARCTIC AIR POLLUTION

JOHN HARTE

ABSTRACT: The three-way coupling ecosystem behaviour, climate and air quality is nowhere else so strong as it is in the Arctic, where boreal forests, tundra, and marine and freshwater ecosystems show unusual properties arising from their extreme environmental setting. This review examines potential disruption of northern biogeochemical cycles by Arctic and sub-Arctic air pollution, including direct responses of ecosystems to pollution, exemplified by the Bristol Bay salmon fishery, and indirect responses by which haze-induced climatic changes can produce ecological effects which further alter climates.

Contents

Introduction

Earth's northern latitudes possess unique geographical and biogeochemical characteristics that shape the ways in which northern ecosystems respond to air pollution. Because of these characteristics, the three-way coupling of ecosystem behavior, climate, and air quality is probably nowhere on the planet as strong and complex as it is there. Figure 1 shows the pathways of this coupling and illustrates the scope of this review, which emphasizes potential disruption of biogeochemical cycles by Arctic and sub-Arctic air pollution, and feedback processes by which climate change due to Arctic haze can engender ecological impacts that, in turn, bring about further climatic effects.

The physical and chemical setting

Boreal forests and tundra are the dominant terrestrial life zones of northern latitudes. Along with the marine and freshwater ecosystems with which they are closely linked by runoff waters, these are the focus of our interest. Like ecosystems everywhere, these habitats are influenced by their environmental setting, and so it is no surprise that the extreme characteristics of Arctic and sub-Arctic climate lead to unusual life zone properties. Figure 2 shows aspects of the physical and chemical setting relevant to our discussion, in a progression which attempts to impose a partial causal ordering.

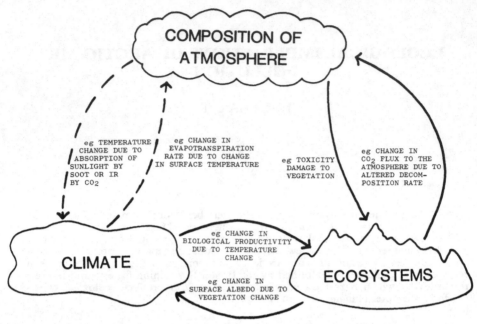

Fig 1. Linkages between air quality, climate and ecosystems. Linkages denoted by solid arrows are the primary focus of this paper.

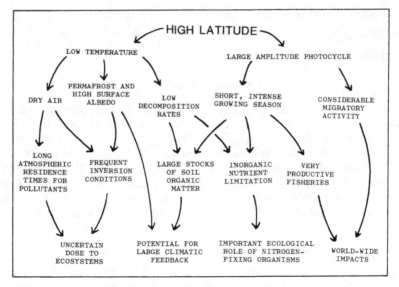

Fig 2. Environmental influences on ecosystems of the northern latitudes.

The influence of physical factors on ecosystem properties is well illustrated by the case of soil organic matter. Photosynthesis, the mechanism of energy transduction that initiates the formation and organization of all life, requires light and inorganic substances such as water, carbon dioxide, nitrogen and phosphorus. Decomposition of organic matter, which thermalizes the chemical energy stored in the process of photosynthesis and releases inorganic nutrients back to the environment, requires warmth and a sufficient supply of suitable enzymes. In northern latitudes, insolation ranges from feast to famine on a yearly cycle, with the long summer sunlit days producing photosynthesis rates in boreal forests that on an annual basis are nearly half those of tropical forests, and exceed those of savanna and grassland. Temperatures, however, remain relatively low even in summer, and decomposition is relatively suppressed. Therefore the organic content of soils, which is determined by the balance between photosynthesis (resulting in production of soil organic matter via litterfall) and decomposition (which removes soil organic matter), might be expected to be greater in northern latitudes than in the temperate zone or tropics.

Table 1. Soil carbon and nitrogen, to 1 m depth, in Earth's major terrestrial life zones (adapted from Zinke and others 1984).

No	Life zone	Area (1012m2)	Carbon density (kg m-3)	Nitrogen density (g m-3)	Carbon content (1015g)	Nitrogen content (1012g)
1	Ice	2.0	0.0	0	0	0
2	Dry tundra	0	3.1	500	0	0
3	Moist tundra	2.7	10.9	640	29	1,700
4	Wet tundra	1.5	20.7	1,250	31	1,900
5	Rain tundra	0.03	36.6	2,200	1	67
6	Boreal desert	0.03	10.0	500	0.2	14
7	Boreal dry bush	1.2	10.2	630	13	820
8	Boreal moist forest	12.8	15.5	1,030	198	13,000
9	Boreal wet forest	4.3	15.0	980	64	4,200
10	Boreal rain forest	0.31	32.2	1,510	10	470
11	Cool temperate desert	1.6	9.7	500	16	810
12	Cool temperate desert bush	4.0	9.9	780	40	3,200
13	Cool temperate steppe	9.2	13.3	1,030	120	9,500
14	Cool temperate moist forest	9.4	12.0	630	110	5,900
15	Cool temperate wet forest	1.6	17.5	930	28	1,500
16	Cool temperate rain forest	0.27	20.3	800	5	210
17	Warm temperate desert	9.4	1.4	110	13	1,000
18	Warm temperate desert bush	7.7	6.0	300	46	2,300
19	Warm temperate thorn steppe	7.7	7.6	540	59	4,100
20	Warm temperate dry forest	9.7	8.3	640	80	6,200
21	Warm temperate moist forest	9.3	9.3	650	87	6,000
22	Warm temperate wet forest	0.67	26.8	1,800	18	1,200
23	Warm temperate rain forest	0.04	27.0	700	1..2	31
24	Subtropical desert	0.56	1.4	100	0.6	56
25	Subtropical desert bush	0.55	3.0	2,300	1.6	1,200
26	Subtropical thorn woodland	1.4	5.4	380	8	540
27	Subtropical dry forest	4.9	11.5	1,070	57	5,300
28	Subtropical moist forest	9.9	9.2	990	91	9,800
29	Subtropical wet forest	0.93	9.4	2,900	8.7	2,700
30	Subtropical rain forest	0.02	15.0	600	0.3	14
31	Tropical desert	1.5	1.0	50	1.4	72
32	Tropical desert bush	0.21	1.0	50	0:0.2	11
33	Tropical thorn woodland	0.32	2.6	260	0.8	83
34	Tropical very dry forest	1.5	6.9	600	10	870
35	Tropical dry forest	6.5	10.2	890	66	5,800
36	Tropical moist forest	7.2	11.4	800	82	5,800
37	Tropical wet forest	0.30	14.5	66	4	190
38	Tropical rain forest	0.003	18.0	600	0.03	1.8
	Totals	131			1,300	97,000

Table 1 shows that this is indeed the case, with soils of the rain tundra having the highest organic carbon density and soils of the boreal moist forest having the highest total organic carbon content among all major terrestrial life zones on the planet.

Comparison of the last two columns in the table reveals also that soils of boreal moist forest and moist tundra have nitrogen-carbon ratios that are respectively 10% and 20% lower than the global average. This relative nitrogen depletion is a fairly general feature of Arctic and sub-Arctic ecosystems. Here too the physical environment is an important determining factor. The carbon supply to growing vegetation comes from the global carbon dioxide reservoir, whereas nitrogen, phosphorus, and other essential elements for plant growth come from the local soil. Thus in general terms nitrogen deficiency occurs in terrestrial ecosystems that are subject to harsh climates, including alpine as well as Arctic habitats (Haag 1974), where release of these nutrients by decomposition occurs slowly. Over the long term in such areas, nitrogen deficiency can feature in inland and coastal waters as well as in the soils. In such circumstances the ecological role of organisms capable of nitrogen fixation, for example legumes, alders, and many species of bluegreen algae and lichens, is enhanced.

The remaining items in Figure 2 should be self-explanatory. What may be less obvious is the manner in which these geographical and biogeochemical characteristics of the northern latitudes influence the direct response of ecosystems to pollution, and the indirect feedback responses by which ecosystem alteration leads to amplification or diminution of pollution-induced climate change. The following two sections describes in some detail an example of each of these classes of impact.

Pollution and Bristol Bay salmon

Figure 3 illustrates one possible sequence of linkages by means of which Arctic haze could inflict significant ecological and even economic damage. The sequence begins with the deposition of pollution, in the form of acid rain and snow or possibly as dry arosol. One class of northern latitude organisms likely to be sensitive to such pollution is the nitrogen fixers. Nitrogen fixation is a process by which molecular nitrogen is transformed into ammonia, a form of nitrogen useful to plants as fertilizers. Nitrogen-fixing organisms use an enzyme, nitrogenase, that is intolerant of low pHs; values of pH in the range of 4.5 − 5.0 are believed to impair its action. Organisms such as the nitrogen-fixing species of lichens may be particularly vulnerable to acidity because their nitrogen-fixing parts are above ground. Thus whatever buffering capacity the soil has is not available to protect the organism. So the first step of this hypothesized causal chain linking Arctic haze to extensive ecological and economic damage is impairment of the nitrogen fixation process.

With a decrease in the rate of nitrogen fixation, there will be a decrease in the rate of flow of nitrogen from tundra slopes or through boreal forest soils into northern lakes and streams. If recently-fixed nitrogen contributes significantly to nitrogen budgets of these ecosystems, then their nitrogen content will decline. The possibility that such a process could link air pollution to nitrogen availability was first raised by Schofield (1980) and Treshow (1980).

In ecosystems that are already nitrogen deficient, such as nitrogen-limited lakes, a decrease of this kind in available nitrogen will lead to a decrease in rate of photosynthesis of the phytoplankton, probably to a decrease in secondary productivity (ie of zooplankton and other organisms that feed off phytoplankton) and in turn to a decrease in productivity of such organisms as fish that consume the zooplankton. Thus the decreased flow of nitrogen to these water bodies poses the risk of further biological effects. The magnitude of these biological effects depends on the quantitative characteristics of the watersheds

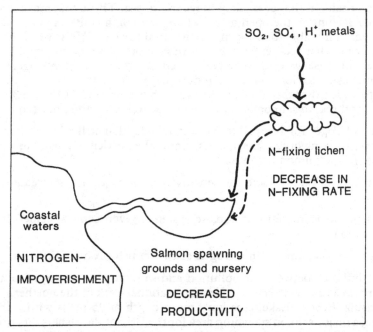

Fig 3. Schematic diagram showing how atmospheric pollutants can influence salmon productivity via the intermediate linkage of nitrogen-fixing organisms.

at risk. The threat to aquatic ecosystems will be significant only to the extent that nitrogen-fixing organisms in the watershed indeed play a large role in supplying nitrogen to the lakes and streams. This was first investigated in the Arctic by Alexander and others (1975) in the vicinity of the Colville River, which flows north from Alaska into the Beaufort Sea. They demonstrated that nitrogen-limited coastal waters were enriched by nitrogen from the Colville River, a significant source of incoming nitrogen having been fixed by blue green algae in the soils of the drainage basin.

This finding is of considerable ecological interest though not necessarily economically important, because the rich fisheries of Alaska are found, not in the Beaufort Sea, but in more southerly Alaskan waters such as Bristol Bay. Indeed, Bristol Bay is the site of the world's largest salmon fishery, with a salmon catch in 1983 (a banner year) of approximately 250 million dollars wholesale value. In addition, these waters are rich sources of sole, herring, cod, pollack, and crab and are important feeding grounds for over a dozen species of whales.

The life cycle of Bristol Bay sockeye salmon, the most important of the commercially-fished species, begins with eggs laid in small streams flowing into inland freshwater lakes bordering the bay. The emerging fry spend the first one to three years of their life in these lakes, growing to a weight of approximately 20 grams before migrating out to sea. There they spend the next two to four years, putting on four to ten kg, and return to the inland waters where they were born. Those that avoid capture by commercial fishermen return up the streams to spawn and die, thus completing their life cycle. Young salmon growing up in the lakes are preyed upon by other lake fish, including trout, dolly varden, and grayling; a fast rate of growth is important, to avoid being eaten. Thus a decrease of nitrogen input into the lakes, resulting in a decrease in plankton productivity, may affect the population of salmon that eventually move out to sea. Those that make it apparently do most of their feeding in the open ocean where upwelling controls available nitrogen, and are no longer vulnerable to nitrogen deficits from the land. However, other fish such as sole and herring, that feed extensively in near-shore waters, may be affected if the nitrogen content of these waters is strongly influenced by river-born nitrogen; a decrease in terrestrial nitrogen fixation could also reduce their productivity.

To assess the odds that the sequence of events hypothesized in Figure 3 will indeed pose a significant threat to salmon fisheries, four research objectives need to be achieved:

1. A nitrogen budget for representative watersheds, that will allow quantification of the role of nitrogen-fixing organisms in determining the nutrient status of aquatic systems that are critical to the salmon life cycle.

2. Pollution deposition rates and the sensitivity of nitrogen-fixing organsisms to that deposition.

3. Changes in nitrogen status of aquatic systems, given some specified intensity of pollutant deposition.

4. The impact of that change in nitrogen status on fish productivity.

To explore linkages between air pollution and salmon productivity, Gunther and I studied the Brooks Lake watershed, in Katmai National Park at the northern end of the Alaskan peninsula. Brooks Lake (area km², mean depth of 45 m) is within the Naknek River drainage, a major salmon spawning area for the Bristol Bay fishery; its productivity is known from earlier studies to be nitrogen limited (Goldman 1960). Begun in 1985, this research is still in progress, but initial results can be reported. From two months of lichen censusing and chemical measurements we have dedeuce that two genera of nitrogen-fixing lichens abundant in the watershed could contribute significantly to the nitrogen budget of the lake. The quantity of nitrogen fixed by these lichens each year in the watershed is of the same magnitude as that consumed each year by the growing salmon fry in Brooks Lake, and about 10% of the amount of nitrogen incorporated in primary productivity each year by lake phytoplankton. The flow of nitrogen in soil water to the lake itself, from locations in the watershed where nitrogen-fixing organisms are abundant, is now being measured. This will allow us to estimate how much of the nitrogen fixed by lichens and other organisms actually enters the lake. Other inputs of nitrogen to the lake, for example precipitation nitrate and organic nitrogen in corpses of spawning salmon, appear to be less significant than the lichen contribution. Contributions from other nitrogen-fixing organisms, such as alders and free-living bluegreen algae in th soil and lake water, have yet to be quantified and may prove to overwhelm the lichen contribution. Short-term laboratory and field exposures of lichen to various levels of acid rain have so far yielded inconclusive results; further investigations will be needed to obtain reliable dose-response information. A mathematical model of the nutrient-plankton-fish interactions in the lake will be constructed to permit synthesis of the diverse experimental information that our field and laboratory measurements are providing.

The hypothesis presented in Figure 3, though complex and speculative, is analyzable by a combination of field and laboratory research and mathematical modelling. It appears likely that the magnitude of the risk will be characterized, at least with respect to its qualitative significance, before it becomes manifest as possible ecological and economic loss.

Table 2. Some entries from the global carbon budget (Harte 1985).

Compartment	Stock of Carbon (10^{12} kilograms)
Living organisms	
marine	2
terrestrial	560
Non-living organic matter	
marine	2000
terrestrial soils	1500
Inorganic compartments	
atmosphere	735
fossil fuels	10^4
seawater	4×10^4
rock and sediment	10^7

Flows of Carbon	Magnitude of Flow (10^{12} kilograms /year)
to atmosphere from decomposition (nearly exactly balanced by a return flow in photosynthesis)	50
to atmosphere from fossil fuel combustion (1984)	5.3

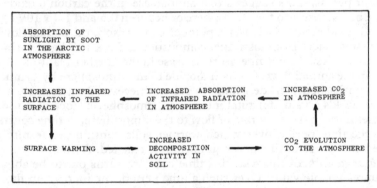

Fig 4. A bioclimatic feedback process that may be of particular concern in the northern latitudes where large stocks of organic carbon reside in the soils.

Bioclimatic feedback

The second specific case involves bioclimatic feedback and Figure 4 illustrates the feedback loop of concern. A warming of Earth's surface in northern latitudes, caused by absorption of sunlight in an atmospheric soot layer lying over the high-albedo surface, initiates the feedback. Of course any perturbation altering the temperature of that surface, including a global carbon dioxide greenhouse warming, could do the job as well.

For reasons discussed in the introduction, northern soils contain a large pool of organic carbon; Table 2 compares Earth's carbon pools, indicating the relatively large quantity of organic carbon stored in soils in general, and particularly in northern soils. As Figure 4 suggests, a speed-up in rate of decomposition of organic matter in these soils will lead to a net flux of carbon dioxide to the atmosphere. To estimate the magnitude of the effect, we must look at the data describing the effect of temperature on decomposition activity in soils of the boreal forest and the tundra. There are two categories of data: those describing the change in soil carbon content along a natural temperature gradients, and those describing the effect of a laboratory temperature manipulation on soil respiration rates.

In the first category, Van Cleve's measurements of organic carbon content of taiga soils as a function of ambient temperature in inland Alaska (Van Cleve, personal communication) lead to an estimate of a negative gradient in soil carbon of -0.3 kilograms of carbon per m^2 per °C. If this value is assumed to apply to all of the boreal forest and tundra of the northern latitudes, then every 1°C increase in surface temperature will be accompanied by roughly a 1% increase in atmospheric carbon dioxide. Because Arctic haze is not likely to give rise to more than 2°C temperature rise at the Arctic surface, this feedback effect appears to be small. Worldwide warming could of course give a larger result, if soil respiration rates worldwide are as responsive to temperature as are those of the taiga. For example, a doubling of atmospheric carbon dioxide, which wuld be expected to generate global temperature increases of 2–4°C, could deplete sufficient soil carbon to increase the atmospheric stock of carbon dioxide by 10–20% above the present value.

Next let us estimate the effect of a temperature increase on the flow of carbon from soil to the atmosphere, basing the estimate on laboratory data. Schlentner and Van Cleve (1985) manipulated temperature and moisture levels of taiga soils in the laboratory, estimating that soil respiration increased by between 1 and 2×10^{-5} kg of carbon dioxide per m^2 per hour per °C. This suggests that an increase in the carbon dioxide flow from the soils of the boreal forest to the atmosphere of between 0.5 and 1.0×10^{12} kg of carbon per year per °C could result. Aa a useful point of comparison, the annual flow of carbon dioxide to the atmosphere from fossil fuel combustion is 5.3×10^{12} kg of carbon per year (Table 2). Thus a 2°C surface temperature increase in the northern latitudes could result in an increase in the annual flow of carbon dioxide to the atmosphere by as much as 40%.

These two unabashedly crude estimates are not necessarily inconsistent with each other. The first yields a value for total increase in atmospheric stock of carbon dioxide. The second yields the increase in rate of flow to the atmosphere. A time constant should relate them—the time period over which warmer soils continue to respire additional carbon to the atmosphere until a new equilibrium is reached. Assuming soil respiration is active for four months of the year, this time constant turns out to be about 20 to 40 years, a not unreasonable value. Over such a time period, the increase in flux of carbon dioxide to atmosphere is probably not terribly serious; if the measurements of Van Cleve and his colleagues can be applied as blithely as we have done here, the feedback process of concern appears to be relatively small.

However, several important factors have not been included in our estimates, and their inclusion could change the picture considerably. Most important is the effect of moisture on soil respiration. Increased soil moisture in the sub-Arctic can exert a major positive influence on soil respiration. Although estimates of the effect of Arctic haze on precipitation rates are not available, some hint of the type of effect that could result can be obtained from global circulation model predictions of the effects on precipitation when carbon dioxide levels are doubled (Manabe and Stouffer 1980). These models indicate that surface warming induced by carbon dioxide would significantly increase precipitation rates in northern latitudes. It is likely that net soil moisture would also increase, although that would depend on the net balance between precipitation and evapotranspiration. Increased precipitation would be likely to increase the rate of pollutant deposition, thus exacerbating the direct ecological effects of Arctic air pollution but decreasing the amount of climate-altering soot in the atmosphere. More work is clearly needed to analyze these multiple feedback effects.

Another factor neglected in our simple estimates is increased primary productivity resulting from the warmer surface temperature. This could affect our estimates in two ways. First, increased photosynthesis will deplete carbon dioxide from the atmosphere, thus cancelling out some of the effect of increased soul respiration. But this correction is likely to be fairly small, because inorganic nutrient and light limitations will exert a braking effect on photosynthesis in the northern latitudes, whereas they are of relatively minor cnsequence to respiration. Secondly, increased photosynthesis will lead to additional litter fall to the soil, thus augmenting the stock of organic carbon available for respiration. This effect is also likely to be of secondary importance.

Conclusion

We have looked in some detail at two potential types of impact on ecosystems resulting from air pollution in the northern latitudes. Both examples illustrate the biogeochemical complexity of the far North, and the extent to which unique features of Arctic and sub-Arctic ecosystems, dictated by geographical considerations, influence how they might respond to air pollution.

Numerous other examples of potential ecological effects of Arctic haze could be given. Damage to lichens could deplete a major food supply for caribou. Melting of permafrost as a result of surface warming could accelerate erosion and alter nutrient profiles within soil, thus affecting vegetation. Water-logging and flooding could result from the melting of glacial ice. Deposition of trace metals and chlorinated hydrocarbons found in atmospheric pollution, and subsequent food chain biomagnification, could lead to toxic damage to organisms. Acid precipitation could lead to surface water acidification and subsequent toxic damage to fish and other organisms; in Alaska at least this is unlikely to affect the most productive waters, which are generally well-buffered. A warmer surface will likely lead to an increase in the rates of nitrogen fixation, nitrification, and denitrification in both soils and surface waters, with possibly significant consequences for nitrogen availability to vegetation. The absorption of sunlight by a sooty layer of Arctic haze in the atmosphere will lead to a decrease of sunlight at the surface and thus to decreased rates of photosynthesis.

The list could be expanded, but our purpose here was not to be encyclopedic. Far more work will be needed to identify the key ecological concerns, predict the magnitude of the likely ecological damages, and delineate the time frame over which they are likely to materialize. There may arise synergistic effects on organisms, resulting from the combined presence of toxicological and climatic stresses. And, of course, the ecological stresses resulting from Arctic haze will combine with those from other types of development in the northern latitudes, such as coastal extraction of oil or mining inland. Determining the human consequences of ecological degradation, both those of an aesthetic, cultural and political nature on the one hand and those with direct economic impact on the other, will require even more study.

At present it is not possible even to state with certainty whether direct toxic effects of Arctic haze on ecosystems, secondary climate-mediated impacts on ecosystems, or tertiary bioclimatic feedback effects are likely to be the most serious. I suspect that reliance on single-species bioassays, simulation modelling, or indeed any single-method reductionist approach, is unlikely to clarify these uncertainties; a broad biogeochemical perspective will be needed and success will require the combined efforts of field and laboratory researchers from many disciplines. Sensible mathematical modelling, particularly of the 'back of the envelope' variety, can help distinguish environmentally important dynamical mechanisms and processes from thsoe that are merely esoteric, thereby helping experimentalists select measurement objectives that will provide the greatest insights.

To make progress in estimating the types and likelihood of ecological damage that will occur, information is needed about rates of deposition of atmospheric pollutants. Unfortunately, the considerable body of information now accumulating on the composition of the Arctic atmosphere is not matched by a comparably extensive data base on deposition rates; more monitoring of deposition is therefore recommended as a high priority. Another high priority concerns the geographical focus of research. The present preoccupation with monitoring pollution and potential climate change in the Arctic, to the exclusion of the sub-Arctic, will not serve well the cause of ecological assessment. The reason is that many of the ecosystems most at risk from pollution in the northern latitudes lie within the sub-Arctic. A final recommendation is for a broadening of the scope of climate modelling. Climate modellers have tended to focus their efforts on deriving predictions about temperature changes due to Arctic haze. But the hydrological changes may be even more important to ecosystems. Changing temperatures and precipitation rates will alter soil moisture and runoff rates, both factors that exert major influences on ecosystems.

The Arctic atmosphere is becoming polluted today because it is, in a sense, a pollution commons for many of the industrialized countries of the world. The northern latitudes are also a resource commons, however, valuable economically for their rich fisheries and aesthetically for their diverse and unusual ecosystems. The pleasures of observing these ecosystems and the keen scientific challenge of probing their mysteries are, sadly, now matched by the urgency of determining the extent to which tragedy awaits this irreplaceable planetary commons.

Acknowlegements

This work was supported in part by the Hewlett Foundation and the World Resources Institute.

References

ALEXANDER, V. AND OTHERS. 1975. *Environmental studies of an Arctic estuarine system: final report.* Corvallis, US Environmental Protection Agency (EPA-660/3–75–026).

FEERY, B. W. AND OTHERS. 1973. *Primary air pollution and lichens.* Toronto, University of Toronto Press.

GOLDMAN, C. 1960. Primary productivity and limiting factors in three lakes of the Alaska Peninsula. *Ecological Monographs*, 30: 207–30.

HAAG, R. W. 1974. Nutrient limitation to plant production in two tundra communities. *Canadian Journal of Botany*, 52: 103–16.

HARTE, J. 1985. *Consider a spherical cow: a course in environmental problem solving.* Los Altos, Kaufmann.

MANABE, S. AND R. J. STOUFFER. 1980. Sensitivity of a global cimate model to an increase of CO_2 concentration in the atmosphere. *Journal of Geophysical Research*, 85: 5529–54.

SCHLENTNER, R. AND VAN CLEVE, K. 1985. Relationships between CO_2 evolution from soil, substrate temperature, and substrate moisture in four mature forest types in interior Alaska. *Canadian Journal of Forest Research*, 15: 97–106.

SCHOFIELD, E. AND HAMILTON, W. L. 1970. Probable damage to tundra biota through sulfur dioxide destruvtion of lichens. *Biological Conservation*, 2(4): 278–80.

SCHOFIELD, E. 1980. The implications of alternative national energy policies for tundra ecosystems. In HARTE, J. (editor) *Energy and the fate of ecosystems.* Washington, National Academy Press.

TRESHOW, M. 1980. Pollution effects on plant distribution. *Environmental Conservation*, 7: 279–86.

VAN CLEVE, K. AND ALEXANDER, V. 1981. Nitrogen cycling in tundra and boreal ecosystems. In CLARK, F. E. AND ROSSWALL, T. (editors) *Terrestrial Nitrogen Cycles. Ecological Bulletin (Stockholm)*, 33: 375–404.

ZINKE, P. AND OTHERS. 1984. *Worldwide organic soil carbon and nitrogen data.* Springfield, Oak Ridge National Laboratory (ORNL/TM-8857).

PARTS 4 AND 5.

INTERNATIONAL COOPERATION AND STATE RESPONSIBILITY: CONCLUSIONS

Back: R. A. Nevé; H. Sparck; L. Rey
Front: M. Bean; J. W. Winchester (Convener): J. G. Roederer

PARTS 4 AND 5. INTRODUCTION

The final session of the conference was devoted to national and international issues arising from arctic atmospheric pollution and the conclusions of the scientists who study it. It was in this session that the absence of Soviet scientists was perhaps most apparent. L. Rey, of the Comité Arctique Internationale, addressed the legal implications of arctic pollution; with its delicate ecosystems, the Arctic is a sink for long-distance pollutants which may have crossed several international boundaries, and against which there are currently no legal remedies. He outlines three possible lines of advance based on accepted international legal conventions, which would circumvent a sectored approach and provide a truly international solution. John Winchester and his colleagues provided an example of international cooperation in research, citing the work of laboratories in Japan, China and the USA that have combined successfully to monitor the progress of pollution, both natural and man-made, from mid-latitude sources to the Arctic. J. G. Roederer, of the University of Alaska-Fairbanks, stressed the need for further coordinated circumpolar work under some system of international cooperation, citing the example of non-governmental Scientific Committee for Antarctic Research (SCAR) and suggesting possible alternative approaches, including an International Year of the Arctic.

H. Sparck and D. Friday drew attention to the problems of the native Alaskan economy under the impact of influences from the south, citing examples of political and social changes during which the Bush Alaskans themselves were patronized or ignored. Recent changes in weather have significantly altered aboriginal economy, making the natives less self-supporting in ways likely to prove expensive to the US taxpayer. It is clearly in everyone's interest to determine whether these changes have been brought about by arctic haze, and if so to find remedies as quickly as possible before Alaskan culture suffers further. In a brief paper the Senior Science Advisor to the Governor of Alaska, Richard Nevé, pointed the need for scientists in atmospheric pollution, seeking State aid, to make their research needs clearly known. The following paper, also brief but no less telling, was a plea from Matthew Bean, an Alaskan native, for something to be done quickly about arctic haze, the onset of which had accompanied many undesirable ecological shifts, and changes in the native ways of life.

C. C. Wallén's conclusions, delivered in the final session and later revised, form a masterly summary of the symposium's findings.

THE IMPACT OF HUMAN AND INDUSTRIAL ACTIVITIES ON THE ARCTIC ENVIRONMENT: LEGAL CONSIDERATIONS AND INTERNATIONAL LIABILITIES

L. REY

Contents

The arctic setting

The harsh and sensitive arctic regions

The arctic regions, because of their geographical situation and climatic constraints, are under permanent stress. As such they are much more vulnerable to outside interference than many other regions. In summer and intermediate seasons the sun shines continuously, but it is low over the horizon and delivers a weak average energy because of the slant illumination and the often considerable cloud coverage. In winter, when there is permanent night, the energy balance is negative and the lands radiate most of their energy reserves into the infinite heat-sink of outer space. In all seasons the high Arctic is a real desert, with negligible precipitation. Some areas have none; their dry, barren lands are deprived of the blanketing effect of snow, and air-borne pollutants that settle on them are not washed away by the summer thaw. Temperatures drop very low and raging winds dramatically increase the chill factor.

In these conditions animals and plants live under constant stress. Having little time to adjust themselves to this changing, stringent environment, they are compelled to expedite growth, biosynthesis, reproduction and migration in a few months, sometimes a few weeks, before they initiate appropriate resistance mechanisms to challenge the winter period (Rey 1985a).

Throughout the year the Arctic exerts constant environmental pressures on all living organisms, including man who has his full share of physical strain and mental tension. Indeed, though they have adapted during nearly 25,000 years to one of the harshest climates on earth, native arctic dwellers are still fighting hard to survive on the vast frozen expanses of this last frontier. They achieve permanent residence only by living in perfect harmony with their environment, and are extremely sensitive to any disturbance in the delicate balance of their surroundings. Arctic residents are also under high physiological load; many react strongly to the shifting photoperiods, which desynchronize their

271

equilibrium, vigilance, power of concentration and mood. In these conditions it is easy to understand that pollutants in the high Arctic might have dramatic and long-lasting biological clocks, rupture their sleep patterns and impair their neurophysiological effects, since their ecological and physiological action might be potentiated by the adverse environment (Rey 1983).

A delicate issue: the socio-cultural patterns

Any project, whether urban, industrial or resource-related, is bound to interact with the environment, since environment introduces multiple hazards and constraints in all human activities, that in turn might affect its diverse sensitivities. This is particularly true for arctic regions, where it is extremely difficult to exercise an objective environmental mandate based on hard facts, and devoid of emotional, often irrational, issues. Indeed, on top of the particular sensitivity of the boreal regions to foreign environmental stress, we have to take into account that arctic native populations have developed, in the course of their long history, specific cultural patterns and traditional subsistence practices which are intimately related to their environment. Any development that disturbs this balance is bound to impact strongly on arctic residents. This is why industrial ventures in the North (mining, gas or oil-field developments, hydroprojects etc, require preliminary in-depth assessments of potential socio-cultural and socio-economic effects (Rey 1983).

The Arctic sink

Unfortunately the sensitive ecosystems of arctic regions are challenged also by pollution from distant sources, carried to circumpolar destinations by atmospheric and oceanic currents (Rey and others 1983, Rahn and McCaffey 1979, Ottar 1982, Shaw 1982). The high Arctic is thus a pollution sink where foreign wastes converge and settle permanently. This is most unfortunate, since the very low rate of biogenesis in northern soils inhibits chemical turnover. Lichens, working as permanent bio-accumulators, stockpile atmospheric pollutants for decades, releasing them abruptly to grazing caribou and, through them, to native populations. These facts have been well documented for radionuclides and organo-chlorinated compounds in Alaska, Greenland, northern Scandinavia and Spitsbergen (Hanson 1985); the same applies also to chemical inputs resulting from volcanic activity.

Pollution does not originate only at lower latitudes. Concern has recently arisen from pollution in the confined atmosphere of the arctic home, which is often almost hermetically tight. Sources include gases and particles released during heating, cooking, natural physiological activity, and the de-gassing of plastics, paints and insulating materials. Whatever the importance of such phenomena, it remains true that within the last 50 years there has also been a marked increase of outdoor pollution into arctic and antarctic areas by foreign chemicals, some of which have been traced back as far as the tropics.

An international dilemma: long-distance pollution

This is obviously a difficult issue, since we cannot relate particular targets in the high north with particular pollution sources in mid-latitudes. After a contaminated air-mass has travelled thousands of miles and crossed many different national boundaries, it is almost impossible in the present state of scientific knowledge to get an identification reliable enough to use as a legal instrument for indicting the source polluter or nation. Exceptions include only a handful of major, well-identified sources, for example the Norilsk nickel smelting Kombinat in northern Siberia or the Sudbury plant in Ontario.

Moreover we lack in international law instruments to prevent or punish long-distance transfrontier pollution, since individual responsibility cannot be easily identified. This prevents the easy settlement of environmental disputes, as is current practice between adjacent countries where only short distances are involved.

A further complication is that atmospheric transport pathways differ in quality from one geographical area to another, and change with the seasons. For example polluted air-masses generated in eastern USA and Canada and carried seaward are heavily scavenged by precipitation over the rainy North Atlantic. Atmospheric releases from Japan and the Peoples Republic of China will similarly be washed over the Northern Pacific Ocean. In contrast pollution aerosols generated in winter over Europe and western and central USSR follow a dry, cold, continental pathway, powered by Icelandic low and Siberian high pressure systems which channel them northwards over the Arctic Sea toward North America. There, almost completely unscavenged, they settle and provoke arctic haze phenomena (Rahn and McCaffey 1979, Shaw 1982).

Development of international environmental law

The *Sic utere* principle

A fundamental element of international law is that no state is ever allowed to use its territory in a way which might be detrimental to another state. This is the so-called *Sic utere* principle: 'Sic utere tuo ut alienum non laedas'. The binding element is *Opinio juris*, and it is understood that the rules apply to any state, whatever its economic, political or military power. In other words there is no supra-order in the community of nations (Johnson-Theutenberg 1976).

However, customary international law does not contain executive elements; it cannot be enforced unless it is incorporated into a court decision, settling a given environmental dispute between different states or pursuant to a treaty in force between the parties to a dispute. Along these lines, the Trail Smelter Case of 1935 has a particular significance. In a decision concerning the massive release over Washington State of sulphur-dioxide smokes generated in Canada, the Court stated:

> Under the principles of International Law, as well as of the Law of the United States, no State has the right to use or permit the use of its territory in such a manner as to cause injuries by fumes in or to the territory of another or the properties of persons therein, when the case is of serious consequence and the injury is established by clear and convincing elements.

The main guideline therefore was to reach an 'equitable decision' which could foster the development of an harmonious 'neighborly relationship'.

The same rules were applied, in 1949, in the settlement of the Corfu Channel Case, by the International Court of Justice, which claimed that:

> Every State has an obligation not to allow knowingly its territory to be used for acts contrary to the rights of other States.

Treaty Law

However, the absence of a real enforcement power in the hands of international judicial authorities drove many countries towards more reliable instruments, whereby given rules of conduct were developed within 'bilateral or multilateral conventions' based upon treaties, duly signed by the respective countries, and thus having a legally binding character.

The 'opinio juris' was this time reinforced by the written endorsement of the states that were parties to the convention. This Treaty Law was, then, designed to replace the internationally unlawful 'self-help' doctrine to which many states were obliged to resort to protect their interests—in absence of any effective protection by the community of states—under the pressure of their own development needs and the influence of their national pressure groups (Johnson-Theutenberg 1976).

The Oslo 'Convention for the Prevention of Marine Pollution by Dumping from Ships and Aircraft' (1972) as well as the Nordic Environment Convention (1974), are typical examples of this procedure. Indeed, they are of special interest for our analysis of the arctic case, since they introduce new elements in the assessment of potential pollution damages:

> Noting with deep concern the increasing pollution of the Baltic Sea Area originating from many sources such as discharges through rivers, estuaries, outfalls and pipelines, dumping and normal operations of vessels *as well as through air-borne pollutants...*

> ... Environmentally harmful operations shall mean...*air pollution*, noise, vibration, *changes in temperature*... (Nordic Environment Convention)

However, as stressed by Johnson-Theutenberg, the treaty law can be effective only if it is implemented in domestic legislation. For this reason the self-help principle is resorted to in the first instance via incorporation in national legislation. A striking case is the 'Canadian Arctic Waters Pollution Prevention Act' (1970) though which Canada unilaterally takes stringent anti-pollution measures in its arctic archipelago waters, enforced (*as an interim measure*), inter-alia, by 'pollution prevention officers', 'pending multilateral development *which would push the frontiers of international law*'. The case, in fact, is much more complex and has a wide geopolitical significance since it helps Canada to anchor its claim on the alleged 'status of internal waters' of its arctic archipelagic water bodies—a claim which has always been vigorously challenged by the USA and by a fair proportion of the international community.

OECD recommendations on transfrontier pollution

Many developments have been made in that line, particularly at the initiative of the OECD (1977, 1981) and different mechanisms have been proposed to deal with international transfrontier pollution and settle disputes between states.

International Financial Transfers (IFT) for instance, are often an effective means to prevent and control transfrontier pollution from the dual standpoint of equity and economic efficiency. In that case, however, the transfrontier pollution is clearly defined and the IFT can only officialize this situation and help solve it by appropriate strategies based upon the 'polluter pays' principle, or by reaching equitable settlement between the up-stream polluter and the exposed country, taking into account their respective development needs and environmental standards.

The Principle of Equitable Use has also often been applied to resolve disputes in the case of internationally-shared natural resources, and it has been proposed that any given nation 'be under an international legal obligation to assess the transfrontier impact of a given industrial project to its commencement'. It is self-evident that, in this particular case, an open, unrestricted exchange of technical and scientific information, as well as an extensive research and monitoring programme, are basic requirements.

The Principle of International Solidarity attempts to maintain a common concern and a correlative common liability between states involved in international environmental protection and conservation. The purpose is to induce states to seek an equitable balance of their rights and obligations in regard to the zones affected by transfrontier pollution.

The Principle of Non-discrimination is almost a consequence of the preceeding ones, since it is obvious that nothing can be achieved if states do not have equal rights of access to information and review, as well as to investigation and monitoring on site. Polluters should be subject to legal statutory provisions no less severe than those that would apply to any equivalent pollution incident occurring within their country. Similarly any state applying the 'polluter pays' principle should apply it to all polluters within the country *without distinction*, whether the pollution remains within the country or crosses its frontiers.

Efficient settlement of disputes. An important point has also been to forge ad hoc mechanisms which bring environmental conflicts to an end as soon as possible, since on-going pollution can only be damaging for all parties concerned. Thus, 'should negotiations and other means of diplomatically settling disputes concerning transfrontier pollution fail, countries should have the opportunity (if not the obligation) to submit such a dispute to a procedure of legal settlement which is *prompt, effective* and *binding*.' Mutually agreed international institutions could serve that purpose, though most conventions, for example those of Helsinki (1972), London (1973) and Paris (1974) recommend an arbitration procedure.

The denial of actio popularis. As can be seen, the development of international environmental law was most advanced by treaty law within regional conventions which clearly delineate the rights and liabilities of all parties concerned. This has been confirmed many times by the International Court of Justice which has consistently refused to take action when the applicant states were *not directly involved*, viz, in the 1973 case filed by Australia and New Zealand against France in the Pacific. Turning down the request to exert pressure on France to stop its nuclear testing, the judges ruled 'that the applicants had *no legal title* authorizing them *to act as spokesmen* for *the international community*.

It was thus quite evident that states not directly affected did not enjoy standing to bring selected cases on *general grounds* to the Court with the aim of preventing imminent or forthcoming pollution. For this reason some kind of new global ethics on environmental protection had to be developed.

De uso publico. In the wake of industrialization of frontier and offshore areas, and of space exploration, a new fundamental element appeared: the basic need to foster common concern and develop a *common responsibility* among states, extending *to those regions of the world which lie outside their exclusive sovereignty*. This principle in fact dates back to the 17th century when Grotius claimed in 'Mare liberum' that the sea was large enough for everyone's use, a principle fundamental to the 1958 Geneva Convention on the High Seas. All states may use the sea in the same way, provided they: (a) do not unjustifiably interfere with the rights of other states to use the same areas and (b) demonstrate reasonable regard to everyone's interest, including those of land-locked states.

The International Convention on the Law of the Sea opened the door to a remarkable awareness of the 'common heritage of mankind', an entirely new concept in international law which reached world recognition in Stockholm in 1972.

UN Conference on the Human Environment, Stockholm 1972

This major meeting recognized the 'Interdependence of man and his environment' and emphasized the risk of seeing uncontrolled human activities result in such deleterious effects as impairment of environmental balance, harm to living resources, reduction of amenities and hazards to human health.

The custodian principle

The conference put industrial development and natural resources management into historical perspective and underlined the importance of moral and human issues. In the light of the long-term requirements of mankind it became evident that Man was no longer the 'Owner of the Earth' but a temporary tenant, 'a Custodian of the World' on whom rising generations could later impose global responsibility. A point had been reached 'where actions throughout the World had to be shaped with a more prudent care for their environmental consequences'.

It was obvious that any development, whether urban, industrial or agricultural, had to be planned well ahead, and its potential impact on the environment duly assessed. This implied a close collaboration among states which became compulsory partners within a finite world with limited resources and diverse sensitivities. The Conference passed unanimously 26 general principles and 109 recommendations for global actions. Most were new in their phrasing, but they were based on general principles of law. For instance Principle 21 is directly in line with the *Sic utere* principle, though it refers initially to this new international code of ethics which was adopted after World War II to help delineate and control the rights and duties of nations:

> States have, in accordance with the CHARTER OF THE UNITED NATIONS and the Principles of International Law, the sovereign right to exploit their resources pursuant to their own environmental policies and the responsiblity to ensure that activities within their jurisdiction or control do not cause damage to the environment of other States or of *areas beyond the limits of national jurisdiction.*

States' liabilities are thus extended outside their natural and historical frontiers. Principle 22 develops this point:

> States should cooperate to develop further the International Law regarding liability and compensation for victims of pollution and other environmental damage caused by activities within the jurisdiction of the control of such States *to areas beyond this jurisdiction.*

States have therefore the basic duty to *consult and inform* other states about activities which could affect areas beyond their national jurisdiction. The American Clean Air Act of 1978 is derived from those principles and includes the necessary provisions to prevent and/or control 'the release of pollutants' emitted in the United States... which 'may reasonably be anticipated to endanger public health and welfare in a foreign country...'

World cultural and natural heritage

When developing this concept further, it soon became apparent that 'environmental protection and conservation' was not only a technical problem involving physical, chemical or biological parameters, but that it included also a vast amount of socio-cultural and ethical issues, as well as many intangible values such as 'quality of life' or 'cultural identity'. At that point, the legislators felt that this 'civilization patrimony' had also to be protected and, in 1972, UNESCO proposed to the international community to sign a 'Convention for the Protection fo the World Cultural and Natural Heritage'. This Act was made to protect natural sites of international significance as well as:

> Works of man or the combined works of nature and man and areas including archaeological sites which are *of outstanding universal value* from the historical aesthetic, ethnological or anthropological points of view.

and:

> Whilst fully respecting the sovereignty of the states on whose territory the cultural and natural heritage... was situated and without prejudice to property rights provided by

national legislations, the States Parties to the Convention recognized that such heritage did constitute a world heritage for whose protection is was the *duty* of the *international community as a whole to cooperate.*

It even identified, within the World Heritage List, 'properties of outstanding universal value', which could be condsidered 'in danger' and for the conservation of which major operations were necessary, for which assistance might be requested under the Convention. At last, international law considered the world as a global issue which had to be managed and developed under the joint sponsorship and responsiblity of all states, in due respect of their diverse historical patrimonies and future potentialities.

The arctic Haze problem in legal perspective

To assess how international law could apply to long-distance atmospheric pollution in the Arctic, we need to consider the scientific hard facts which are available and put them in perspective with the basic requirements of international law. Along those lines, it is a useful exercise to return to the court ruling in the Trail Smelter Case of 1935 (see above):

> No State has the right to... cause injuries by fumes in or to the territory of another one... when the case is of *serious consequence* and the *injury is established by clear and convincing elements.*

The facts

The occurrence in crystal-clear arctic air, during the winter months, of a dense, pervasive, pollution-derived, aged aerosol of anthropogenic origin has been clearly established (Rey and others 1983, Rahn and McCaffey 1983. Ottar and Pacyna 1986, Shaw 1985).

The interfacing of air-borne pollutants with terrestrial (and marine) ecosystems has been equally demonstrated and the remarkable work carried out by Hanson on radionuclides in Anaktuvuk Pass has shown how airborne chemicals are incorporated ito the food web over years and finally hit the human consumers.

The remote origin of the polluted air-masses and their anthropogenic nature is now beyond doubt; there is ample evidence that urban transport, smelting operations, and the generation of electricity and steam by oil- and coal-fired power plants in lower latitudes result in the massive atmospheric release of sulphur compounds, nitrogen oxides, trace metals, soot and graphitic carbon, which are driven to the Arctic by global atmospheric circulation. It is also highly probable that agrochemicals, sprayed in excess over the tropical belt, reach high latitudes by successive distillation-redeposition mechanisms and concentrate there in the fat of living animals (Ottar 1982).

Careful analysis of air-masses and synoptic meteorological studies also demonstrate that the pollution aerosol follows particular pathways around and across the Arctic, and these can often be traced back with precision to particular source areas in western and central USSR, Europe, and occasionally central and western USA and Canada. However, we know also that a substantial part of the polluted air-masses which enter the Arctic eventually leaves before being scavenged or deposited, and might even come back again, but we ignore today the relative importance of this mechanism.

In summary, we can definitely claim that the Arctic regions are impacted by *foreign fumes* resulting from man-made activities in *distant* mid and low-latitude areas. It is this fair to assume that the states where those fumes are generated bear the responsiblity of such activities carried 'within their jurisdiction and control *which extend to areas beyond their jurisdiction*'. We know, however, that today this global responsiblity is denied unless the *exact* origin of the pollution fumes is specified.

Unfortunately, in the present state of development of chemical analytical techniques, and without the capability of following the pathways right to the emission sites (which are often numerous sources spread out over a large territory, or located in forbidden areas where regular access is currently denied, as in northern Siberia) we are still unable to establish the specific origin of long-distance atmospheric pollutions *by clear and convincing elements*.

The Effects

The occurrence of such atmospheric pollution aerosols in arctic and subarctic regions definitely interferes with the local environment.

Climatological effects appear to be the most striking, since the presence of thick, absorbing layers or extended patches of haze changes the radiation balance between the surface and outer space, thus inducing an overall cooling of the area, enhances cloud formation by nucleation effects and modifies the albedo of ground and/or ice by deposition of dark particles. These combined effects can be considered as 'environmentally harmful' in the light of the Nordic Environment Convention which stipulates that 'environmentally harmful operations shall mean... air-pollution... *changes in temperature...* '

Incorporation of foreign chemicals into the arctic ecosystem, particularly into pollution-accumulating lichens, is an established fact as we said earlier, but until now scientists and public health officers have been unable, at least in the case of radio-nuclides, to demonstrate any *injury* to animals and man. As such, arctic atmospheric pollution does not fall into the general ruling of the Trail Smelter Case, which implies that, in order to be punishable, the effects of fumes should have *serious consequences*.

Quite clearly, this statement is based only on our present knowledge, and there is growing evidence that the long-term effects of remote pollution by pesticides, organo-chlorinated compounds and even soot (a complex mixture of graphitic carbon, anthracenic and heterocyclic compounds, and diverse chemicals) might be detrimental to plants and animals and impair human health. Multi-year longitudinal epidemiological studies are absolutely needed to investigate and document this case. However, today, we have *no clear and convincing elements* to claim that arctic atmospheric pollution has *serious consequences* on living ecosystems and man.

Suggestions for action

This rapid review of the present situation shows that we are not yet in a position to assign precise liabilities and, as a consequence, to take action to prevent or control arctic atmospheric pollution. However, there are different lines of approach that could be followed to help in settling this case.

Global responsibility of mankind

One way to attack the problem would be to stress the 'universal importance' of the arctic regions for the international community and, as such, under the UNESCO Convention, to try and claim that they constitute, in given areas, a *'World Heritage in Danger'*. This is certinly true for the arctic native cultures altogether, but it seems very unlikely that the international community would be prepared to consider the whole Arctic, and even the North American Arctic, Spitsbergen or North Greenland, heavily impacted by arctic haze, as a *'property of outstanding universal value... for whose protection it is the duty of the international community as a whole to cooperate'*. The area is by far too vast, the impacts still nebulous and the natural and cultural heritage properties too imprecise and spread out to allow any such proposition to meet the standards for international recognition. However, this does not mean that some actions in that direction shoud not be undertaken.

International solidarity

Another possiblity would be to acknowledge the fact that arctic air pollution is the result of a large pool of activities carried out by a number of wealthy, industrial, developed countries, impacting at long distance isolated communities deprived of real economic and political power; accordingly it falls within the agreed principle of international solidarity to try and help them solve this difficult problem.

We could, then, identify a few selected actions to document the case with adequate, in-depth international research and, on that basis, propose scavenging and/or pollution control measures to be applied at the main emission sites in order to reduce the long-distance effects. The cost of such research and protection measures would then have to be borne collectively by the 'polluter group' under both the *Principle of Solidarity between Nations* and assistance to the more exposed and less developed ones, and to the *polluter pays* principle. Obviously, Western Europe, the eastern European countries, western and central USSR and to a certain extent mainland China, Japan, Korea, the USA and Canada, would by implication be involved in this international environmental protection programme. An international fund would have to be identified and appropriations made to service the programme.

Joint development of chemistry, meteorology and international law.

A third approach, which might prove more practical, would be to foster a joint development of analytical chemistry, dynamic meteorology and international law so that their efforts merged in a solution. Advanced methods in analytical chemistry should be capable of providing a reliable 'fingerprint' of the main polluted air-masses and, with them, of the original pollution plumes. Recent research in that field (Rahn 1985) shows that it is possible to characterize quite accurately any air-mass by its contents in arsenic, antimony, selenium, non-crustal vanadium, zinc, non-crustal manganese and indium, and to identify this 'signature' all along its route, despite its dilution and admixing with other air-masses on the way, and internal changes by ageing processes.

Similarly, combining data from satellites, ground station networks, soundings and drifting constant-altitude balloons gives an increasingly accurate picture of the evolving synoptic situation, and allows us to trace the polluted air-masses back to their sites of emission with increased accuracy. This is however still in the development stage, and has not yet reached that point of sophistication which would allow a given polluter to be identified with certainty, though it can delineate with high probability a source area of a few tens of thousands of square miles. There is little doubt that dramatic improvements in this technology are bound to occur in the years to come.

Thus, the overall trend in the field is to achieve more and more accurate identification of the emission areas which, in the long run, might very well be boxed into quite restricted geographical settings.

In parallel, we should progressively adapt national and international laws to extend the responsiblity of the states where atmospheric pollution is generated, so that the collective liability of individual operators can be established, at least at the regional level. Thus, instead of having a collection of anonymous isolated polluters, we would be able to identify particular industrial areas responsible for long-distance pollution, which would together bear the consequences of their action.

Ideally, this two-way process should facilitate assignment of collective responsiblity to polluting areas of a size which would be compatible with the accuracy of the analytical and meteorological identification techniques. On that basis it would then be much easier to develop legal instruments capable of indicting polluting groups, and enforce de-pollution and scavenging at the emission sites with ad hoc international monitoring and control. This would equally allow the target communities to claim equitable compensation, and give them legal means for protecting their environment.

We are quite convinced that such an evolution would take place if we were able to involve in this process *all circumpolar countries* without exception. Indeed, as we have underlined many times, the Arctic Ocean is a 'Polar Mediterranean' (Rey 1982, 1984), a world of its own. Circumpolar arctic lands and seas have to be considered in global terms as a single entity. Any attempt to solve arctic environmental problems in a sectorized way is thus bound to complete failure.

It is our sincere hope that our 20th century scientists, lawyers and politicians will understand this situation, and, rising above self-interest and partisan ideologies, they will pool their efforts to safeguard the last frontier of the great north, and protect the Arctic brotherhood.

References

HANSON, W. 1986. Ecological processes involved in cycling of radionuclides within Arctic ecosystems. In STONEHOUSE, B. (editor). *Arctic Atmospheric Pollution*. Cambridge, Cambridge University Press: 221–8.

THEUTENBERG, B. J. 1976. *International environmental law*. Stockholm, Liber Forlag.

OECD. 1977. *Legal aspects of transfrontier pollution*. Paris, OECD.

OECD. 1981. *Transfrontier pollution and the role of states*. Paris, OECD.

OTTAR, B. 1982. Long-range transport of airborne pollutants with accumulation and re-emission problems of mercury and chlorinated hydrocarbons. In REY, L. (editor). *The Arctic Ocean*. London, MacMillan: 197–210.

OTTAR, B. AND PACYNA J. M. 1986. Origin and characteristics of aerosols in the Norwegian Arctic. In STONEHOUSE, B. (editor). *Arctic Atmospheric Pollution*. Cambridge, Cambridge University Press: 53–67.

RAHN, K. A. AND LOWENTHAL, D. H. 1986. Who's polluting the Arctic? Why is it so important to know? An American perspective. In STONEHOUSE, B. (editor). *Arctic Atmospheric Pollution*. Cambridge, Cambridge University Press: 85–95.

RAHN, K. A. AND MCCAFFEY, R. J. 1979. Long-range transport of pollutants to the Arctic: a problem without borders. In *Symposium on the long-range transport of pollurants*. Sofia, Bulgaria (WMO No 538): 25–35.

REY, L. 1982. The Arctic Ocean: a polar Mediterranean. In REY, L. (editor). *The Arctic Ocean*. London, MacMillan: 3–38.

REY, L. 1983. (editor). *Arctic energy resources*. Amsterdam, Elsevier.

REY, L. AND OTHERS. 1983. Arctic environment issues of energy production at mid-latitudes and the role of long-distance air-transport of pollutants. In *12th World Energy Conference, New Delhi*, 1983. London, World Energy Conference: 3,2,06.

REY, L. 1984. (editor). The polar regions: economic and strategic issues. *Geopolitique*, Paris, 5: 80–110.

REY, L. 1985. Environmental issues of industrial development in Arctic regions. In *POAC 85*, Narsarsuaq, Greenland.

REY, L. 1985. The Arctic world: an exclusive frontier. In GERWICK, B. C. (editor). *Arctic Ocean engineering for the 21st century*. Marine Technology Society, Washington: 15–30.

SHAW, G. E. 1982. Evidence for a central Eurasian source area of arctic haze in Alaska. *Nature*, 299: 815–18.

SHAW, G. E. 1985. Alaskan and Arctic-scale air pollution. *Northwest Environmental Journal*, 1(2): 85–95.

COOPERATION OF CHINA, JAPAN AND THE UNITED STATES IN THE STUDY OF LONG RANGE AEROSOL TRANSPORT TO THE PACIFIC AND ARCTIC

JOHN W. WINCHESTER, MING-XING WANG
AND YOSHIKAZU HASHIMOTO

ABSTRACT. Three laboratories in China, Japan, and the United States have initiated cooperative investigations of atmospheric aerosol formation and transport to the Pacific and Arctic. Both desert dust and air pollutants from urban or industrial areas have been sampled near their sources in China, at sites in Japan, above the marine boundary layer on Hawaii, and over the Arctic Ocean. Aerosols, sorted by particle size or temporally resolved using cascade impactors or time sequence filter samplers, were analyzed for elemental composition by proton-induced X-ray emission (PIXE). Dust storms in China greatly raised concentrations of very coarse particles, but particles finer than 1 μm (those of greatest human health concern) are raised much less. Dust carried over Japan during spring contains three components—soil particles, sea salts and air pollution sulfur—recognizable by element concentrations in filter samples collected sequentially. Hawaiian springtime aerosol contains particulate sulfur temporally coherent with soil dust, suggesting trans-Pacific Asian pollution and desert sources. Arctic aerosol, sampled in spring by aircraft-mounted cascade impactor, contains both fine pollution sulfur and soil elements mainly in ultra-coarse particles more than 16 μm in diameter, with Si/Al approximating that of shales or clay minerals but less than in average earth crust or Asian loess. Dust from a Chinese desert area shows Si/Al close to average earth crust or loess. These results suggest that, if coarse Arctic aerosol is derived from fine Asian soil dust, silica is depleted during transport (eg by fractional removal of quartz from clay) and particles coagulate into coarser aggregates. Alternatively the coarse Arctic aerosol may be of eruptive origin within the Arctic basin. Both natural and anthropogenic pollution components of the Arctic aerosol will be studied in our continuing cooperative program.

Contents

Outline of cooperative program

Informal cooperation has commenced between three laboratories in China, Japan, and the United States for investigating chemical aspects of aerosol formation and long range transport through the atmosphere. Desert, urban, and industrial locations in Asia are

sources of aerosol particles which, under certain atmospheric conditions, may be raised
high above ground and carried over great distances. In spring dust storms occur in
northern China. Their clouds, often appearing over Japan as 'kosa' (yellow sand), also
reach Hawaii and can be carried to the Arctic. Fine sulfate aerosol has also been detected
at Mauna Loa, Hawaii. Correspondence of its periodic concentration variation with that
of soil dust suggests an Asian air pollution origin. The extent that Asian pollution sulfur
sources also contribute to haze in the Arctic is a subject of ongoing investigation.

Our three laboratories have cooperated in atmospheric chemical research over a
number of years. Joint research between the US and Japanese (Keio University)
laboratories dates from the 1960s; cooperation between the US and the Institute of
Atmospheric Physics in China began in 1979. The work has concentrated on measurement
of elemental composition of aerosol samples, to identify sources, track long range
atmospheric transport, and obtain evidence of chemical transformations of pollutants in
the atmosphere.

A special opportunity for three-nation cooperation was presented in April 1980 when
a dust storm passed over northern China and reached Japan, an event for which our three
laboratories had planned in advance. In Beijing during the storm, elemental composition
as a function of particle size was determined by cascade impactor sampling and PIXE
analysis of the size fractions. In Japan, cascade impactors and time-sequence streaker
filter samplers were operated at five locations which were expected to receive mainland
dust, according to weather forecasts provided by Professor A. Ono, Nagoya University.
Our cooperative program has since been extended by sampling in the Arctic from aircraft
during 1983, and further international investigation of the Arctic atmosphere is planned.
Continuing objectives are identification of major sources of both terrestrial dust and
particulate sulfur from polluted air.

Coarse and fine particle concentrations in a Chinese dust storm

Dry lands of northwestern and western China are the largest source of airborne soil
particulate matter in the north temperate latitude zone. Their dust, raised especially in
spring, darkens the sky over Japan, supplies a substantial amount of deep sea sediment
in the North Pacific, and is transported as far as Hawaii and the Arctic. Depletion of soils
by dust storms contributes to desertification, a problem of current concern in China.
Each spring clouds of dust blow across Beijing, elevating especially the concentrations
of coarsest particles which are normally not very abundant in ambient air. An immediate
problem in China is to set air quality standards that distinguish coarse natural soil dust
(which does not easily enter the respiratory tract) from finer particulate matter which can
be inhaled deeply, and may present a relatively greater public health risk.

During the dust storm that passed over Beijing on 19 April 1980, we operated cascade
impactors and fractionated particles according to size, ranging from greater than 16 μm

Table 1. Dust concentrations, μg m^{-3}, during Beijing dust storm. Calculated from
measured major elemental abundances, 19 April 1980. Compared to Xinglong
background (XL bkg), near Beijing, March 1980.

Sample	Over 16	16–8	8–4	4–2	2–1	1–0.5	0.5–0.25
A	452	38.1	127	197	11.6	6.83	1.48
B	124	5.2	79	172	35.4	3.03	0.18
D	338	26.8	120	387	97.5	3.43	1.41
XL bkg	3.7	8.7	15.4	14.6	6.3	1.50	0.19

Aerosol particle aerodynamic diameter range, μm

to 0.25 μm aerodynamic diameter (μmad), and determined elemental concentrations by PIXE analysis (Wang and others 1982). The results, summarized in Table 1, show that elevation of concentrations was greatest for the coarsest particles, 100–fold for over 16 μmad, compared to only about 5–fold for the fine (less than 1 μmad) range. Thus, although concentrations in all particle size ranges were elevated by the storm, respirable particle concentrations were affected much less than those of coarser or of total suspended particulate matter.

Transport of aerosol from the Asian mainland to Japan

Kosa dust storms are under active investigation by Japanese atmospheric scientists. The April 1980 storm that we sampled in Beijing, an especially strong event, was sampled cooperatively when it appeared as a kosa in Japan, at Nagasaki, Okayama, Nagoya, Yokohama and Niigata (see acknowledgements). Both streaker samplers and cascade impactors were used, and elemental concentrations were determined by PIXE analysis (Hashimoto and others 1984). Advance weather information allowed the samplers to be started before the kosa arrived, so samples were obtained before, during and after the event at each site. Time sequence samples at Niigata showed the aerosol cloud to be a mixture with components from desert dust, sea salt, and air pollution sources, based on differences in arrival times for their indicator elements. Let us examine these differences in some detail so as to illustrate how relative elemental composition measurements help to infer long range aerosol transport.

The record of variation with time over a 6–day period at Niigata (Figure 1) is shown for three elements: Fe, a mineral constituent representative of several elements derived from desert soil; Cl, a major constituent of sea spray originating near Japan; and S, a common air pollutant from fuel combustion, contained also in both soils and sea spray. All three elements show large variations of concentration with time, with considerable difference in detail. Yet there is a general similarity; an initial period on 19–20 April is followed by a storm period for 3–4 days starting 20 April and then by a final period starting 23 or 24 April. This sequence corresponds to the movement into Japan of a dusty air mass, displacing other air, and its later disappearance when atmospheric conditions returned to normal. The dust, which darkened the sky with a yellow sandy appearance for three days, corresponded to the period of high Fe concentrations.

Detailed differences between Fe, Cl, and S indicate that the dusty air mass was not of uniform composition, but contained different aerosol components. The first to arrive in the early afternoon of 20 April was Cl, probably from the sea as the storm crossed from the continent. At about the same time S decreased, from a pre-storm concentration of sulfate in polluted ambient air to one expected for cleaner marine air. Early on 21 April Fe increased abruptly, and S also increased, but Cl decreased. The increases indicate transport of desert dust and some accompanying sulfate to the Niigata sampling site; the decrease in Cl may signify that the dusty air mass had descended to ground level from a higher altitude where sea salt was less abundant.

In the afternoon of 21 April Fe reached very high concentrations, and maxima also occurred in S and Cl. Although the elements were all present in the dusty air, temporal variations indicated that the air was not completely well mixed. Fe was mainly in soil particles transported from Asian deserts, and Cl was derived from the sea surface. The S was much too abundant in relation to Cl to be mainly derived from the nearby sea, and too abundant in relation to Fe to be mainly derived from natural desert dust; both components no doubt contributed, together with further S of air pollution origin, from urban areas on the mainland, swept up by dusty air passing over its sources and

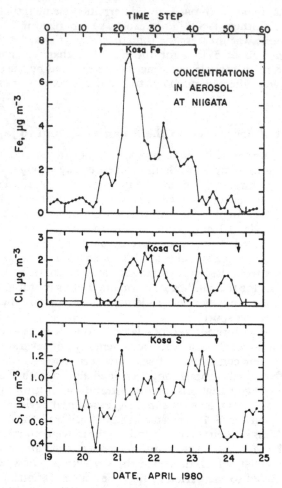

Fig 1. Concentrations of aerosol Fe, Cl and S at Niigata, 19–25 April 1980, during a dust storm from Asia. The times of increase and decrease of the three elements are not coincident, indicating that the air mass was not uniform in composition as it passed over the sampling site. Fe may be mainly due to soil dust of desert origin, Cl may be from sea spray picked up en route to Japan, and S may be mainly from Asian air pollution sources.

transported to Japan. Toward the end of the dust storm on the afternoon of 23 April, Fe was the first to decrease to low concentrations. S persisted at high values until early on 24 April, when it also decreased to the low level at which it had been late on 20 April. Cl did not decrease until the afternoon of 24 April, marking the end of the marine air episode which carried the dust constituents. When the Cl had decreased, S began to rise toward its initial concentration, a level perhaps typical of ambient air quality normally found in Niigata.

This example illustrates how possible long range pollution transport may be inferred from measurements made in time sequence. The physical processes suggested by the data may then be validated by further meteorological analysis.

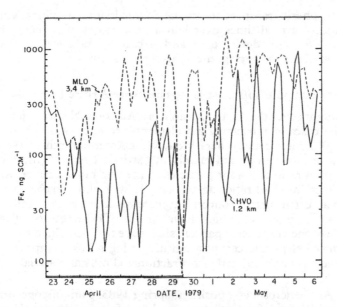

Fig 2. Concentration of aerosol Fe sampled concurrently at the Mauna Loa Observatory (MLO) and the Hawaiian Volcano Observatory (HVO), 23 April to 6 May 1979. Daily rainfall at HVO exceeded 1.3 mm only 23–29 April, indicated by bar along abscissa; no rain fell at MLO. At HVO Fe concentrations show a regular 24 hour periodicity, due to daily wind variations on the island. At MLO a prominent period about 10% shorter is found which cannot be driven by local winds, but must be due to larger-scale air motions, eg an Asian diurnal lifting modified by wind speed convergence during transport across the Pacific.

Transport of dust and pollution from Asia to Hawaii

Wind-blown dust from Asia has long been known to be an important component of Pacific sediments (Arrhenius 1963), but only recently have atmospheric scientists come to realize that it may be the most abundant aerosol component of the springtime haze over Hawaii. This aerosol may also contain sulfur from Asian air pollution; thus acid rain measured on Hawaii may also be due in part to long range transport of air pollution (Miller 1981).

During April 1979 dust storms in China resulted in kosa events in Japan (Iwasaka and others 1983) and increased haze intensity over Hawaii (Shaw 1980, Bodhaine 1982). Surface and satellite observations showed that dust clouds were carried over the Pacific at two different altitudes, travelling up to 10,000 km from source areas in north and northwestern China. In a study using streaker time-sequence aerosol samplers (Darzi and Winchester 1982a, 1982b) the dust was intercepted on Hawaii at Mauna Loa Observatory (3.4 km altitude) and the Hawaiian Volcano Observatory (1.2 km altitude) between 23 April and 6 May 1979. PIXE analysis for elemental concentrations showed periodicities that indicated a probably Asian source for both soil dust and air pollution constituents. A comparison of concentrations of Fe at the two Hawaiian sites (Figure 2) shows a diurnal pattern at the lower site, explained by local winds and soil particles, but a prominent period averaging 10% less than 24 hours between maxima at the upper site. In addition at the upper site, particulate sulfur, probably of air pollution origin, corresponds in time with solid dust elements. Diurnal lifting of dust and pollutants over Asian source areas, and long range atmospheric transport across the Pacific in a nonuniform wind field, can cause foreshortening of arrival times over Hawaii consistent with these observations. Thus Asian sources of both desert dust and urban or industrial fuel combustion pollutants are indicated for the springtime aerosol over Hawaii.

Coarse particle soil dust in springtime Arctic aerosols

In 11 sampling flights over the Arctic Ocean from Alaska to Norway, aerosol particles were collected using cascade impactors, and elemental concentrations were determined by PIXE analysis (Winchester and others 1984, Winchester and others 1985). In six of the flights very coarse dust was found, mainly in particles over 116 μm diameter, with a median concentration ratio of 5.4 for particles in the size range greater over 16 μm to 16–8 μm. Coarse fraction Si/Al ratios agreed closely with those expected for shales or clay minerals from an earth surface source, but were much lower than for average earth crust material or loess, a soil of wind-blown origin. These results (Table 2) were unexpected because of the very coarse particle size range that contained most of the dust, yet with relative major elemental composition indicating a dust composed mainly of a mineral fraction, such as clays, instead of unfractionated crustal material.

Table 2. Coarse Arctic aerosol composition, spring 1983. Sum of concentrations in over 2 μm diameter fractions compared to reference compositions.

Sample	Si/Al	Weight ratios Si/Ca	Si/Fe
Median Arctic	0.61	1.34	1.46
Shales	0.91	3.30	1.55
Earth crust	3.41	7.64	5.54
Loess	3.14	50.0	10.15

There are two possible sources. The first is dry Asian lands where soil particles could be formed and transported nothward, though they would have to be fine in size to avoid loss due to deposition during transport. Under atmospheric conditions in the Arctic, a particle size increase by coagulation would have to occur to explain the measured preponderance of aerosol particles over 16 μm diameter. A second possibility is a localized Arctic eruptive source of coarse aerosol with clay mineral composition. Volcanic activity was not known in the Arctic during the measurement period. However, a non-volcanic release of visible plumes from the vicinity of Bennett Island in the Soviet Arctic continental shelf occurs periodically, and was occurring during the measurement period. These plumes may contain coarse sedimentary debris that could be dispersed widely throughout the Arctic atmosphere.

Dust characterization study in the interior of China

Measurements of elemental composition signatures at source areas, and comparison of signatures at receptor sites in the Arctic and elsewhere, can be combined with meteorological analysis to establish likely transport trajectories. Our program included intensive sampling in the Bogda Shan range of central Xinjiang, in order to characterize the aerosol from the Dzungarian Basin, an important part of the Asian arid lands system. PIXE analysis of these samples is in progress as part of the continuing cooperative program. A preliminary result of the study (Wang and Winchester, to be published) can be used to test the hypothesis that some of the coarse dust found in the Arctic originates in Asia. Dust from this Chinese desert area was sampled sequentially during two weeks

Table 3. Aerosol composition in northwest China, August 1980. Measured in normal, dust event, and normal time periods.

Period	Dates	Si/Al	Weight ratios Si/Ca	Si/Fe
Normal	8–12	2.9	3.5	4.0
Dust event	13–17	2.9	4.2	15.0
Normal	18–22	2.7	3.3	4.4

in August 1980 when a dust event occurred between two normal periods. Results in Table 3 show Si/Al much greater than we have measured in the Arctic and close to average earth crust and to the composition of loess in China. This finding suggests that, if coarse Arctic aerosol is derived from Asian soil dust, only much finer mineral grains could be transported so far and silica depletion must have occurred during transport (eg by fractional removal of quartz grains from clay) prior to coagulation into coarser aggregates in the Arctic atmosphere. Whether this is more likely than a non-volcanic Arctic eruptive source must await further investigation. We intend to pursue this question, as well as others, in future phases of our cooperative study of long range aerosol transport to the Pacific and Arctic.

Acknowledgements

We thank Kimio Arao of Nagasaki University, Yusaku Nogami of Okayama Prefecture Center for Environment and Health, Akiro Ono of Nagoya University, Shigeru Tanaka of Keio University, and Takaaki Yanaka of Niigata Prefecture Kogai Center for carrying out sampling operations during the 1980 kosa investigation. The cooperation described here has depended on exchanges of personnel among the three laboratories through short term visits that have been assisted financially by the China Committee of the US National Academy of Sciences, the Institute of Atmospheric Physics of the Chinese Academy of Sciences, and the Faculty of Science and Technology of Keio University. Research funding within the three nations has been obtained through our respective laboratories and agencies.

References

ARRHENIUS, G. 1963. Pelagic sediments. In HILL, M. N. (editor), *The sea*, vol 3: 655–727. New York, Interscience.

BODHAINE, B. A. 1982. Background aerosol measurements at four remote sites. In preprint volume: FISHMAN, J. L. (editor) *Second symposium on the composition of the nonurban troposphere*. Boston, American Meteorological Society: 6.3: 154–58.

DARZI, M. AND WINCHESTER, J. W. 1982a. Aerosol characteristics at Mauna Loa Observatory, Hawaii. *Journal of Geophysical Research*, 87: 1251–58.

DARZI, M. AND WINCHESTER, J. W. 1982b. Resolution of basaltic and continental aerosol components during spring and summer within the boundary layer of Hawaii. *Journal of Geophysical Research*, 87: 7262–72.

HASHIMOTO, Y. AND OTHERS. 1984. Detection of an Asian dust storm event in Japan during April 1980. *Nuclear Instruments and Methods*, 231(B3): 526–28.

IWASAKA, Y. AND OTHERS. 1983. The transport and spacial scale of Asian dust storm: a case study of the dust storm event of April 1979. *Tellus*, 35B: 189–201.

MILLER, J. M. 1981. A five-year climatology of back trajectories from the Mauna Loa Observatory, Hawaii. *Atmospheric Environment*, 15: 1553–58.

SHAW, G. E. 1980. Transport of Asian desert aerosol to the Hawaiian Islands. *Journal of Applied Meteorology*, 19: 1254–59.

WANG, M. X. AND OTHERS. 1982. Chemical elemental composition of wind blown dust, Beijing, April 19, 1980. *Kexue Tongbao* [Science Bulletin], 27: 419– 22 [in Chinese], 27: 1193–98 [in English].

WINCHESTER, J. W. AND OTHERS. 1984. Coarse particle soil dust in Arctic aerosols, spring 1983. *Geophysical Research Letters*, 11: 995–98.

WINCHESTER, J. W. AND OTHERS. 1985. Particulate sulfur and chlorine in Arctic aerosols, spring 1983. *Atmospheric Environment*, 19: 2167–73.

INTERNATIONAL SCIENTIFIC COOPERATION IN THE ARCTIC: PROBLEMS, OPPORTUNITIES AND US RESPONSIBILITIES

JUAN G. ROEDERER

ABSTRACT. Many arctic research problems do not conform to national or geographic boundaries, but require simultaneous observations in several different places. Other problems require expensive facilities or logistic operations that demand the pooling of resources from different institutions and different countries. The need for cooperative research in the Arctic is discussed, and related problems and opportunities are reviewed, listing examples of research questions that require an international cooperative approach. Different possible modes of multinational cooperation are analyzed. The idea of a non-governmental 'SCAR of the North', recently advanced at international meetings, is discussed, and related scientific and political implications, including the possibility of organizing an 'International Year of the Arctic' are examined. Recent developments in the formulation of United States Arctic Policy and related research policy are discussed, with special emphasis on US commitments and responsibilities in international cooperative activities in the Arctic.

Contents

International cooperation in science

Economic and social development of all nations is becoming increasingly dependent on advances in technology, which in turn depend on the unpredictable results of science. Scientific research is becoming complex and expensive, requiring sophisticated instruments, extensive logistic support bases, huge computer centers and laboratory installations. It demands knowledge of the integral earth, the global environment, global ecological and population trends and global change, so that measurements must be taken virtually everywhere, requiring costly expeditions, multi-satellites missions, and collective data repositories—all endeavours that no single nation, let alone individual research groups, can afford on their own.

Measurements must be made simultaneously in many places, and in such a way that they can be fed into a common data base available to all participating scientists. This requires international agreements on what to measure, on instruments and measurement standards to be used, on data formats, and on how the data are to be analyzed. To disseminate and apply state-of-the-art instrumentation and data handling techniques, channels of communication and modes of scientific discourse must be established, demanding special information services and special international meetings, symposia and workshops.

A particular factor that today presses for international cooperation in science is the growing trend toward interdisciplinary research. Traditional disciplinary boundaries are being demolished. Physics and physical research methods are pervading all disciplines, including the social sciences; interactions between biological and physical systems are targets of intensive research; space is routinely used for research in astronomy, geophysics, and biology. All this involves megaprojects, and an application of knowhow that is not evenly distributed among nations, not even among the advanced ones.   There are other, non-scientific reasons for international cooperation in science. Programs of international scientific cooperation have served through informal, unofficial contacts as catalyzers, preludes or overtures to diplomatic intercourse. They have led directly or indirectly to international treaties such as the Antarctic Treaty of 1961, the Test Ban Treaty of 1963, the Space Treaty of 1967, and to current international discussion on 'nuclear winter' and the effects of nuclear war. Indeed, science transcends cultural, ideological and nationalistic narrowness. Programs of international scientific cooperation have given participating scientists the possiblity of seeking common denominators that encourage cooperation despite political or ideological differences; they have made it possible for scientists to voice and publish their concerns about international affairs, even when the concerns were at variance with official or government attitudes.

Problems in arctic scientific cooperation

Having reviewed scientific cooperation on a world-wide scale, it is important to examine arctic cooperative research in the world-wide context, even though many fewer nations are involved. First let us define the Arctic. Normally it is considered as a geographical region defined by geographic boundaries—for example the latitude of the Arctic Circle. However, in the context of human activity, technology and science the concept of 'arctic' is much broader, and should be defined in an operational way. In this arcticle the term Arctic is applied to those high northern latitude regions in which the effects of cold climate, a large-amplitude photocycle, short and intensive growing seasons, sea ice, scarcity of fast surface transportation routes, geomagnetic and auroral perturbations, etc, individually or in concert impair or prevent the straightforward transplantation of 'traditional' technologies and lifestyles from lower latitudes. It is clear that such an operational definition will necessarily extend to a fringing sub-arctic region. It must also be pointed out that in today's space age, the Arctic is a three-dimensional entity: its upper environment stretches out thousands of km into space, with the earth's magnetic field playing the role of the frame that sustains this three-dimensional physical structure. Indeed, many important phenomena occurring in the Arctic upper atmosphere, and many disturbance effects detected on the ground, result from processes occurring in this overlying space structure.

The importance of large-scale international cooperation in polar regions has long been recognized; indeed the First Polar Year of 1882 was the first major, carefully planned international scientific enterprise ever to be carried out. This was followed by the Second Polar Year 50 years later. Their successor was the International Geophysical Year (IGY) of 1957/58, with a strong arctic research component; in that year were launched both the space age and large-scale international cooperative research in Antarctica.

Antarctic research cooperation is often pointed out as an ideal to be emulated for the Arctic. However, direct comparisons need to be made with great care. Firstly, while there exists an international treaty on Antarctica, the Arctic is one notable area as yet largely untouched by existing or prospective international agreements (the Svalbard Treaty is geographically constrained). Secondly, there are the facts that no indigenous people exist in Antarctica, there is as yet no resource development there, and military activities are banned and territorial claims suspended as long as the Antarctic Treaty remains in force. Thirdly, while Antarctic research is mainly driven by scientific curiosity, much of Arctic research is directed toward solving specific problems related to the inevitable expansion of human activity and interests toward the north.

Many research needs in the Arctic are linked to political, economic and military interests. This makes international cooperation difficult; in certain geographic areas or fields of research cooperation becomes impossible, when research findings are classified as sensitive or proprietary by governments or industrial firms. Even within a single country arctic research issues may prove socially or politically sensitive. The research needed to resolve such domestic conflicts as the impact of industrial development on arctic environments, or the impact of western influences on native culture and subsistence lifestyles, cannot fail to stir strong political controversy in the arctic communities most concerned.

In several countries including the United States, research in the Arctic competes directly or indirectly with research in Antarctica for funding. Although logistic costs of a given research operation are usually much higher for Antarctica than for the Arctic, they are usually borne directly by the government agency responsible for Antarctic research; those for the Arctic must be budgeted for in individual research proposals. This puts Arctic research at competitive disadvantage wih projects in Antarctica or in less costly lower latitudes. In several countries fundamental data-gathering and interpretation of information on the Arctic are the responsibility of public agencies whose missions are separate and whose budgets do not reflect the priorities of Arctic issues. Industrial and urban development in the Arctic often has preceded the acquisition of necessary scientific knowledge, and scientific research has frequently been supported only as a corrective measure after serious mistakes have made the need for such knowledge apparent.

Another problem is that physical and biological phenomena in the Arctic tend to cross disciplinary boundaries, partly because of the strong interactions and delicate equilibria, both physical and ecological, that we find in polar regions. The results are that Arctic science is eminently interdisciplinary in nature, that Arctic phenomena and processes cannot be studied in isolation, that these studies are very expensive, that the need for coordination is dictated more by logistic consideration than by strictly scientific arguments, that these studies have to compete with more traditional science endeavours at lower latitudes, and that they often fall between the cracks of agencies' programs and peoples' frames of mind.

Opportunities for international cooperation

There are numerous scientific questions, the study of which will benefit from, or even necessitate, international cooperation. In applied science and technolgogy, international cooperation and/or coordination may offer important economic advantages to the participants, for it can help in avoiding costly duplication, promote the acquisition of relevant data banks, and prevent the need for a 'reinvention of the wheel' in the development of many industrial projects. Pertinent research goals can be grouped into the following general categories:

(1) Development of a scientific data base necessary for the search, assessment, extraction and transportation of natural resources in cost-effective manner with minimum adverse effects on the local environment and its residents;

(2) Assessment and prediction of natural and anthropogenic hazards, and development of mitigation techniques and related policies;

(3) Prevention and cure of diseases and social ills.

In the case of basic research, advantages of international cooperation are mostly scientific. Examples of fundamental research questions that would benefit from international cooperation are listed below. Most of these questions are from publications of the US National Academy of Sciences (1985) and the University of Alaska Foundation (1985). Some were recommended at recent public meetings held by the US Arctic Research Commission.

Atmospheric and space sciences
Long-term monitoring of atmospheric constituents.
Study of Arctic haze: sources, pathways, conversions, fingerprints and effects.
Study of ocean-ice-atmosphere interactions and their effect on the variability of the polar heat sink.
Study of upper atmosphere phenomena in the polar cap and their relations to magnetospheric tail and boundary processes.
Study of solar-variability-controlled effects on the neutral upper atmosphere.

Glaciology, oceanography and solid earth
Study of impurities, such as acid precipitation, in snow.
Long-term oceanographic and sea ice measurements in the Arctic by SAR satellites, moorings, buoys, floats and chemical tracers; sea ice prediction.
Development of improved methods to clean up oil spills in broken ice.
Paleoclimatic data from ocean cores in the Arctic Ocean, glacial deposits and ice cores.
Seismic measurements from ice islands to define the configuration and origin of major structures underlying the Arctic Ocean.

Life sciences
Integrated long-term marine ecosystem and food-chain studies on major Arctic shelves.
Study of migratory species: energy and nutrient transfer, adaptations for migration, routes and timing and interference by man.
Marine mammal research: population size, habitat use, life histories, physiology, effects of noise and toxicants.
Comparative study of social and cultural factors in community health.
Study of cold regions problems, including cold injury and chronological and photogenetic effects on human adjustment.

Social Research and Economics
Comparative study of the impact of imported lifestyles, diets, diseases and social ills on native populations in the Arctic.
Study of native languages and oral traditions.
Study of prehistorical migration of Native populations.
Comparative study of new political and economic units in the Arctic.
Comparative economic analyses of the subsistence element in the mixed economies of Native villages in the Arctic regions.

International scientific cooperation: levels and modes

International cooperation in research can occur at many different levels and in many different modes of organization. It may involve exchange of information on who is doing what, where and when; exchange of data taken with some specified type of instrument, at specified places and during specified time intervals; the organization of short-term experiments or expeditions; the establishment of permanent data repositories and data analysis centers; the foundation of new specialized journals and newsletters; the organization of workshops, symposia and conferences, and so on. All this is desirable for cooperative research in the Arctic—but who would organize the planning and the actual coordination and implementation of the cooperative work?

Most cooperative ventures in the Arctic have been conducted in the past on an ad hoc basis. In terms of multidisciplinary bodies, there is no Arctic equivalent at non-governmental organization (NGO) level to the Scientific Committee on Antarctic Research (SCAR). The idea of a 'SCAR of the North' has been advanced several times recently; it has been suggested as an inter-union committee of the International Council of Scientific Unions (ICSU) dealing with Arctic research. It might be premature to create such a body at this time, however, especially in view of the problems mentioned above. The strong links of many Arctic research topics to defense, proprietary industrial or native interests have no parallel in the south, so SCAR has worked with noteable scientific objectivity since its inception. It should be noted, however, that increasing pressures of non-scientific nature are currently developing within SCAR, brought about by commercial or political interests in Antarctic resources and tourism.

Perhaps a gradual, step-wise establishment of an ICSU body dealing with the Arctic would represent a more realistic approach. It might initially involve the appointment of a small ad hoc committee to coordinate Arctic aspects of Union activities between themselves and with those of such international programs as the World Climate Program, or the proposed International Geosphere-Biosphere Program. A fundamental function of such a committee would be to provide the necessary visibility to Arctic research issues in bodies that are usually dominated by scientists interested in problems of mid- and low latitudes. As the next stage, ICSU might promote the establishment of an Arctic data center and information clearing house, oversee the organization of selected inter-disciplinary research projects involving the Arctic as a whole, and sponsor related scientific meetings and symposia. As a further step, a major international program such as a 'Year of the Arctic' or a 'Third Polar Year', with strong interdisciplinary emphasis, could be envisaged. Just as the IGY led to the creation of SCAR, a major internationally coordinated effort such as an 'International Year of the Arctic' during the early part of the next decade (perhaps around the 1991 solar maximum to allow for optimum disciplinary participation) could well consolidate the establishment of an international committee on Arctic research under ICSU sponsorship.

In the immediate future bilateral and multilateral research programs, especially at an institute-to-institute level like AGASP, or at mixed private and governmental level like those promoted by the Arctic Ocean Sciences Board, will probably remain the most successful modes for cooperative research in the Arctic, in which major political problems can be circumvented. Signatories of the Svalbard Treaty will no doubt continue to conduct research in that particularly important area of the Arctic.

Special-interest groups or organizations, such as the Inuit Circumpolar Conference (ICC) and North Atlantic Treaty Organization (NATO), and private scientific associations such as the Comité Arctique International and the Arctic Institute of North America will continue to play an important role in cooperative research on subjects of their own specific interest and choice. It should be pointed out, however, that the international character of these organizations is limited in the sense that eligibility to membership is restricted either legally by charter, or de facto by the individual scientist's ability to pay the dues in convertible currency.

US responsbilities for cooperative research

Until a few years ago the United States had no coherent arctic research policy. General arctic policy guidelines had been issued by the executive branch of the federal government, but were only occasionally referred to by the Departments of State and Defense, and more or less ignored by other federal agencies. Several attempts to establish a research policy for the Arctic were made during the 1970s, but none ever reached the state of implementation. By and large, responsiblity for planning, implementing, and funding arctic research remained divided between several federal agencies, the State of Alaska, and private groups whose mandates or objectives were often unconnected.

Several important developments concerning arctic research policy took place in recent years, involving the National Academy of Sciences, some federal agencies, and Alaska Senator Frank Murkowski who in 1982 introduced a bill in Congress that was passed in final form as the Arctic Research and Policy Act of 1984, and signed into law by President Reagan in July of that year.

Prior to this the US President had issued a directive reaffirming earlier statements on general arctic policy, and declaring that this policy will be based on (i) protection of essential US security interests in the Arctic; (ii) support of a sound and rational development in the Arctic region, while minimizing adverse effects on the environment; (iii) promotion of scientific research on the Arctic or on subjects which can most advantageously be studied in the Arctic; and (iv) promotion of mutually beneficial international cooperation in the Arctic.

The Arctic Research and Policy Act of 1984 established an Arctic Research Commission and an Interagency Arctic Research Policy Committee. The Commission is directed to develop and recommend an integrated arctic research policy, and the National Science Foundation is designated as the lead agency responsible for implementing this policy with other federal agencies. The Act mandates the formulation of a comprehensive five-year Arctic Research Plan to be updated biennially and requires the preparation of a single, integrated multi-agency yearly budget request for arctic research.

The Arctic Research and Policy Act states inter alia that 'expanded knowledge of the Arctic increases the opportunities for international cooperation among Arctic-rim countries' and declares that 'industrial pollution of non-Arctic origin that collects in the polar air mass must be controlled through international cooperation and consultation'. The Act further directs the Interagency Arctic Research Policy Committee to coordinate and promote cooperative arctic scientific research programs with other nations, subject to US foreign policy guidelines and to the integrated research policy to be devloped by the Arctic Research Commission.

Thus it follows from both the US Arctic Policy established by the President, and the Arctic Research and Policy Act passed by Congress, that the United States has a commitment to and responsibility for the promotion of international scientific cooperation in the Arctic. The commitment has been declared, but the responsibilities are only now beginning to be formulated by the Arctic Research Commission and the Interagency Arctic Research Policy Committee. It will take several years to propose and implement coherent plans for US initiatives in international cooperation in the Arctic.

We sincerely hope that all Arctic rim nations, without exception, will join in international research ventures. A better knowledge of the Arctic is not only necessary for scientific and technological reasons: it can contribute effectively towards an increased insurance for world peace.

References

NATIONAL ACADEMY OF SCIENCES/NATIONAL RESEARCH COUNCIL, POLAR RESEARCH BOARD. 1985. *National issues and research priorities in the Arctic*. Washington DC, National Academy Press.

UNIVERSITY OF ALASKA FOUNDATION. 1985 *A challenge to Alaska*. Fairbanks, University of Alaska Foundation.

ARCTIC HAZE AND BUSH ALASKAN VILLAGES

HAROLD SPARCK AND DAVID FRIDAY

Contents

Introduction

Within the past two decades, industrial man has brought significant changes to indigenous hunting and gathering communities of Bush Alaska. The changes were political, social, economic. Little understood during this period of limited western observations of aboriginal socities were environmental and ecological changes that occurred at the same time.

Once all Arctic communities were self-sufficient. Now most of the indigenous communities in the north depend on government transfer payments to maintain the imported western portion of their economy. This portion is called 'the other village' (Sparck 1977). The basis of the local economy and society remains however 'subsistence' (Napoleon and Davidson 1973). Stable ecosystems are mandatory for these communities in transition to maintain their marginal economy (Sparck and Gaber 1981).

The most recent boom of industrial growth in the north has taken place during a period of long-term cyclic weather change. Industrial change and man-made discharges into the arctic environment have accelerated disruption of long-standing ecosystems, threatening the modern way of life of Arctic indigenous communities (Hunt and Sparck 1977). Today's Bush village cannot control its resource dependence to adjust to rapid change in wildlife population or migration (Sparck and Barker 1980).

Political changes

Only recently has the United States begun to address the 100–year-old unsettled issues of title, sovereignty and resource allocation in its American Arctic, resulting from its purchase of Russia's claim to Alaska. In 1958 Congress legislated statehood for Alaska, and in 1971 enacted the Alaska Native Claims Settlement Act (ANCSA) which provided for 44 million acres and $M962 to be settled on Alaska's native peoples as reparation for aboriginal lands taken by the State of Alaska prior to Settlement Act. Congress intended

ANCSA to satisfy all aboriginal claims to the landmass. Recent decisions of the judiciary have concluded that ANCSA settled all aboriginal claims to the continental shelf of Alaska as well.

In 1980, Congress determined how ANCSA lands would be managed, with an act called the Alaska National Interest Lands Conservation Act (ANILCA). This landmark of environmental legislation grew out of Section 17(d)(2) of ANCSA. Congress set aside 132 million acres of unreserved federal land as conservation units, including wildlife refuges, parks and monuments, forests, and wild and scenic rivers. The Congress described in ANILCA compatible-use standards of State and private owners with the dominant Federal land owner.

The federal executive had been very active in Arctic industrialization since the discovery of Prudhoe Bay oil in 1968. A resource development policy with heavy federal subsidy was outlined by the first National Security Council Decision Memorandum 144 in 1971 (National Security Council 1977).

In response to growing federal budget deficits, and a failure to locate additional arctic world-class mineral deposits, federal subsidies to arctic industrialization were first reduced, then reconsidered as an on-need basis only, and to be reimbursable (US Department of State 1982). Federal agencies with arctic missions explained in their 1982 ANILCA's Title X Arctic review to Congress that all goals were being achieved at the current rate of funding (US Depts of Interior, Defence and Energy 1982). The US Congress recently passed the 1984 Arctic Science Policy Act to encourage scientific cooperation in the Arctic.

In the 1983 NSC Decision Document 90, federal market intervention in arctic exploitation was explicitly curtailed. The executive's new arctic goal was to generate national profit from the arctic to reduce federal budget deficits. Federal arctic revenue-producing programs would be increased. Future federal Arctic support would be offered on a cost-reimburseable basis only. All direct federal costs in the arctic would be reduced (National Security Council 1984). In this federal policy, any third-party social costs would be the responsibility of developers. Arctic industry has refused this new social tax burden. The national debate on social responsibility between government and industry continues. At this time, no party is responsible for social costs to arctic residents or the arctic environment.

Societal changes

The aborginial peoples of Alaska's Bush have seen many changes. Social tinkering in indigenous societies took place on an unprecedented scale after the world-class oil find in Prudhoe Bay. Village society was overwhelmed. Bush people have had difficulties responding to immediate change mandated by the sovereign (Hamm 1982).

Subsistence still remains as the foundation of Bush life. Vast tracts of unsettled arctic lands and waters, called 'cuilkuq' in Cup'ik Eskimo, allowed the wildlife to replenish their populations even during low cycles, and feed indigenous populations (Sparck and Barker 1978). Once Alaskan land title was settled, lines began to appear on maps to carve out potential market riches. Extremes formed between those wishing Bush Alaska to remain undisturbed, and those seeking to maximize wealth through western values. During this period Bush people were either patronized or ignored, while their future as a viable people in balance with the natural arctic environment was at stake (Kairaiuak and others 1975).

Ecology and Bush economics

But the most unusual change in Bush Alaska during this period of temperate-zone man's tinkering has been in the weather. Arctic environmental services have been disrupted. The Bush society and economy developed over 3,000 years on winter and summer time and temperature are now distinctly different (Sparck 1979). Recent weather changes have significantly modified the aboriginal economy. The cash sector of the modern village economy is quite limited. Available cash is consumed by today's village luxuries like home heating fuel, transportation equipment, gasoline, electricity, and windows. A survey by Nunam Kitlutsisti (NK) revealed that 52% of the disposable income of village families was in 1984 being spent on imported fuels (Sparck 1981).

An economy and technology to wander in search of reduced and remote wildlife is not affordable (Paniyak and Sparck 1984). Bush people were unsettled when their fur garments splattered with mud kicked up from all terrain vehicles at Christmas time. Nor are people capable of moving far offshore with their limited technologies to look for random ice and its associated marine mammals. Winds that move fresh water plumes seaward altered salmon migration patterns. Unseasonal rains rot or harden air-dried fish, reducing edible local protein. Coastal villages located on narrow fresh water lenses within permafrost zones are searching for alternate fresh water sources now that their snow and rain are often tainted with dirt and sulphur-smelling compounds.

A recent survey by Nunam Kitlutsisti revealed that the social costs of converting the subsistence hunting and gathering economy of Bush Alaskan villages would be very costly. Denied traditional wildfoods, the costs to government to maintain rural settlements with a western imported food economy without basic infrastructure would be greater than $1.7 billion dollars in the initial years (Friday and Sparck 1983). The alternative of relocating Bush residents involuntarily to urban centers for maintenance levels of living would be equally costly if total social costs were calculated, and was not considered.

Conversion of villages to imported food economy

1974–76 subsistence calendar survey

NK surveyed the 48 occupied villages of the Yukon-Kuskokwim Delta in southwestern Alaska in 1974 and 1976. The study instrument was to first develop a subsistence harvest calendar for each of these villages. Research identified three distinct regions and seven patterns of subsistence food harvest. These data are available on request from Nunam Kitlutsisti. Each household in the villages of the seven regions was sent a calendar. Village officials explained the calendars to the households, and regional information messages via public radio supplemented this information.

Householders were requested to fill out the calendars. Similar calendars by government agencies had had limited success: NK believed that as an Alaskan native organization, supported by local officials, it would receive better reports. Village families mailed the calendars back to NK, and tonnage values were compiled for each species by village. As a condition of receiving voluntary information, some of it involving violation of hunting regulations, NK agreed that specific family totals would be held confidential, and results would be listed as village-wide totals only.

The calendars were divided into five classes of local food: fish, land mammals, sea mammals, vegetation and waterfowl. Totals were produced in each village by extrapolating reported family's totals by the number of households in the village. Net weights were ascribed to each species of wildlife based on average weights provided by the Alaska Department of Fish and Game for the given season of harvest; for example mature females were ascribed higher body weights during breeding seasons.

Once gross tonnage was totalled, each category of food harvest was expressed as a percentage to indicate its significance to to the village's yearly protein economy. Coastal villages had higher percentages in marine resources than inland villages, which had higher percentages of land mammal tonnage. Fish protein accounted for 70–80% of gross tonnage. The percentage of fish as the principle protein source did not vary between coastal and inland communities. In 1974, a total of 14.26 M lb of local protein were reported harvested by participating villages, and 16.66 M lb in 1976.

Subsistence harvest interruption study

Concern for expanded and expedited oil and gas development in the Bering Sea and the uplands of Alaska has precipitated regional demand for a comprehensive study of potential local hunting for meat, in view of possible changes arising from man-made disturbance to the ecology. As a basis for a survey, NK identified four types of habitat each with villages that maintained traditional ways of life, and the village councils were invited to participate. One sample village was selected from each habitat for in-depth study. The four villages were: Newtok, representing marine wet-tundra habitat (28 families); Lower Kalskag, representing interior, tree-line river habitat (55 families); Marshall, representing interior river-alpine tundra habitat (64 families); and Atmauth-luak, representing interior lake-moist tundra habitat (47 families).

NK staff visited the villages to see if there had been significant changes in the bi-weekly harvest calendar first developed in 1974. No changes were identified in any of the four villages, although modifications in harvest techniques had occurred. Gross total were obtained for the most significant species taken during the test year for each species. As expected, coastal villages or those with river-access to the sea had higher harvests of marine mammals, while inland village took more land mammals.

NK made incidental studies of family income, employment, prices of selected stores items and other factors affecting standards and costs of living in the test villages: details and results are available from NK on request. The major effort, however, was to estimate the amount of local protein consumed. In each village three groups, each of five families, were identified to represent families with high (90%), medium (50%) and low (25%) consumption of local protein. Each family was surveyed during a test year (1983) to determine sources and quantities of protein consumed, from the subsistence harvest calendar taken that year. The councils then identified the total number of families in each of the three consumption classes in each village, and mean tonnage reported by each class was weighted to arrive a total protein tonnage consumed in the village for the test year. Survey family tonnage per sample village was also calculated. Comparing survey results for 1974 and 1976 with test village harvest tonnages for 1983 (Table 1), differences are

Table 1. Protein harvests for 1974, 1976 and 1983 in four Alaskan villages. Figures show amounts in lbs taken by an average family in each village.

Sample village	1974	1976	1983
Newtok	3,356	11,020	8,797
Lower Kalskag	5,046	4,408	8,031
Marshall	10,122	10,570	9,027
Atmauthluak	4,178	6,580	18,080

immediately apparent in the accuracy of reported figures. Those for 1974 and 1976 were based on voluntary reporting, and totals were averaged, then extrapolated for the entire village. Rather than chance large differences within the village, NK calculated 1983 tonnages from average tonnage in each class of survey family.

Identification of harvesting areas
NK staff visited representative hunters in each village to identify areas of seasonal harvest. Although the tundra has no visible boundaries, villages observe traditional harvest areas

and respect resources of neighboring villages. Efficient transport is limited to the few with seasonal jobs of high value. Although all protein sources would appear to be available to a mobile hunter, unique species recently introduced and actively managed by government, moose for example, are subject to capture only by highly mobile village takers. Most species of local protein are managed by Yup'ik Eskimo traditional hunting practices. Down-river trappers would not seek beaver in the interior mountains, nor would interior hunters seek seals on the coast, unless invited. In times of need, social barriers are dropped, and the intra-communal sharing that identifies the Yup'ik Eskimo way of life becomes sub-regional. Villages with surpluses invite those without to take, or harvest additional protein for consumption by neighboring villages. Maps from this survey are on file and available for consultation through NK.

Costs of annual subsistence harvest

Private sector expenditures for annual subsistence harvest were determined for each sample village, indicating the cost of protein produced by local harvest. Costs were generally low in all the villages (Table 2, line 4), indicating that local protein is still important and

Table 2. Annual costs of local hunting for protein in four Alaskan villages, compared with estimated capital and variable costs of importing an equivalent amount of protein; for explanation see text.

	Newtok	Lower Kalsag	Marshall	Atmauthluak
1. Edible protein costs (per lb dry weight)	$0.72	$0.69	$0.69	$0.31
2. Mean fixed costs of hunting, per family	$3,251	$2,949	$3,141	$2,841
3. Mean variable costs of hunting, per family	$1,195	$ 915	$1,191	$1,024
4. Food obtained (lbs dry dressed), per family	6,158	5,622	6,319	12,656
5. Importing protein; fixed capital cost	$M5.72	$M4.85	$M8.95	$8.05
6. Importing protein; annual variable costs	$M0.65	$M1.10	$M1.43	$2.05

available for domestic consumption in the subsistence economy. In coastal and river villages costs were inflated by the expense of maintaining reliable boats and accessories to reach distant hunting grounds; they were lowest in Atmauthluak where villagers commuted short distances to the hunt. Costs were derived by taking into account fixed and variable costs of production, and subsistence output, as follows.

Fixed costs. Table 2, line 2 shows mean costs of essential hunting equipment used by the villagers, based on costs to the 15 families, and annual depreciation costs based on the mean lifespans of the equipment. Villagers were questioned on the proportion of time spent in using each type of equipment for subsistence harvesting, as a portion of the cost from income-generating activities, to determine yearly cost in subsistence production. Families then calculated annual repair costs in 1983, and a final fixed cost was determined for each type of gear, from which total fixed costs for each village were calculated.

Variable costs (ie running costs of hunting) were calculated from the mean total consumption of gasoline and oil per family, from which was deducted amounts used in non-hunting activities (Table 2, line 3).

Subsistence output for 1983 (Table 2, line 4) was derived from total tonnage per test village, reduced by 30% losses in dressing to give lbs of edible products.

The cost per lb of subsistence protein production output is equal to fixed plus variable costs divided by subsistence output.

Public costs of replacing protein

The cost of completely replacing this locally-hunted protein by imported equivalents (should that become necessary because of a natural or man-made catastrophy) was calculated for each village. The calculations took into account capital costs of constructing airport and road extensions, distribution centres, freezer facilities, storage capacity for 14 days, and increased electical power capacity; also running costs of freight and additional services, as well as the cost of replacement meat. Details of these calculations are available on request from NK; rounded totals are given in Table 2, lines 5 and 6.

Conclusions and recommendations

The survey indicates that to replace local protein with its equivalent in imported meat, the average cost per family in the four sample villages would amount to $169,069 per family. In Alaska alone, where 211 Bush villages exist, subsidized conversion would absorb $M8.20 for first-year fixed capital costs, plus variable costs of $M1,730.00 per village. These figure exclude operation and maintenance costs, and the human costs inherent in rapid acculturation from social shock. Clearly, it is the social, and economic interest of the US, and of other arctic states on whose lands live the indigenous peoples of the North, to determine whether or not long-term changes in arctic weather and ecology have been accelerated by arctic haze.

If man-made aerosols are found to be a cause of changing climate, can the nations that contribute to arctic haze find quick solutions, to prevent further modifications to weather and environment? The process of US-Canadian negotiations on trans-boundary 'acid rain' offers no remedies for solving arctic haze problems. As with 'acid rain', only international cooperation and speedy resolution will lessen the social costs of arctic haze. To wait is merely to pass on to future arctic generations the costs of future solutions.

The potential exists for temperatures in the entire northern hemisphere to be affected by a variety of arctic atmospheric events, including Arctic haze. Failure by arctic states to address the issue could lead to peoples in temperature zones of the earth becoming involved, at a greater political cost. 'Common heritage of mankind' claims need not be applied solely to seabed mining.

References

FRIDAY, D. AND SPARCK, H. *Harvest interruption study of Bush Alaskan villages.* Juneau, Ak, Nunam Kitlutsisti, Bureau of Indian Affairs, US Dept. of the Interior.
HAMM, K. 1982. *The issue is survival.* Bethel, Ak, Nunam Kitlutsisti.
HUNT, C. AND SPARCK, H. 1977. *Oil spills in rural Alaska.* Bethel, Ak, Nunam Kitlutsisti.
KAIRAIUAK, C. AND OTHERS. 1975. *I feel like I'm just wasting my breath.* Bethel, Ak, Nunam Kitlutsisti.
NATIONAL SECURITY COUNCIL 1977. *National Security Decision Memorandum 144.* Washington, NSC. (Declassified memo ASD, ISA, 1–5948/77, 25 May).
NATIONAL SECURITY COUNCIL 1984. *National security Decision Document 90.* Washington, NSC.
NAPOLEON, H. AND DAVIDSON, A. 1973. *Does one way of life have to die so another can live?* Bethel, Ak, Association of Village Council Presidents.
PANIYAK, J. AND SPARCK, H. *Just a small fishery: the Cape Romanzoff District commercial herring fishery.* Bethel, Ak, Nunam Kitlutsisti.
US DEPARTMENT OF STATE. 1982. *Arctic policy evaluation.* Interagency Arctic Policy Group. Washington, USDS. (2 March).
US DEPARTMENTS OF INTERIOR, DEFENSE AND ENERGY. 1982. *A study of US Arctic research policy and the possible roles of the Naval Arctic Research Laboratory.* Washington, US Government.
SPARCK, H. 1977. The other village. *Tundra Times,* 15, Nos 14, 15, 16, 17.
SPARCK, H. 1979. The other village revisited. *Tundra Times,* 16, Nos 47, 48, 49, 50.

SPARCK, H. 1981. *Directions in growth: a study of capital projects in three bush villages.* Anchorage, Ak, RuralCap.
SPARCK, H. AND BARKER, J. 1978. *A special relationship with the land.* Bethel, Ak, Nunam Kitlutsisti.
SPARCK, H. AND BARKER, J. 1980. *We need time!.* Bethel, Ak, Nunam Kitlutsisti.
SPARCK, H. AND GABER, S. 1981. *You don't forget hunger.* Bethel, Ak, Nunam Kitlutsisti.

ARCTIC AIR POLLUTION AND ALASKA'S CONCERNS

RICHARD A. NEVÉ

As Senior Science Advisor to the Governor of Alaska, one of my responsibilities is to acquire from the various state agencies a list of research needs, and to identify who within the universities and industries can resolve those needs. Another task is to interact with the recently established federal Arctic Commission, which is the vehicle for national and international scientific cooperation. The Arctic Research Policy Act, passed by President Reagan in 1984, has made the National Science Foundation the lead agency for coordinating arctic studies within Alaska—the only arctic state in the Union.

The newly appointed Commissioners and I had the opportunity to discuss with many citizens their concerns and hopes for the state, beyond just resource development and defense. Specifically, because of the accelerating pace and scale of developmental activities in the Arctic, concern has been expressed by Alaskans, especially native groups, about the impact these activities will have on the health of all residents in the Arctic. The current status of conducting and supporting arctic health research is inadequate. There are no medical or health research units in existence which can adequately assess problems related to occupation safety and health, or long-term environmental health consequences of life in the Arctic.

As indicated by the funding of this conference, the Alaska state government had concerns with air pollution which does not originate in the Arctic. As a result of data presented at this symposium, we see there is minimal danger to the Arctic from air pollution. However, we still have concern with pollution which is generated in local environments, eg from industrial activity at Prudhoe Bay and Fairbanks. That activity is becoming more significant, and points the need for long-term monitoring to measure and predict increases in atmospheric pollution from CO_2, NO_x gases, particulates, and any other substances which have an effect on the climate and life in earth. Long-term monitoring is not glamorous and *is* costly. A coordinated state, federal, and international effort in resolving this problem is mandatory. Alaska is interested in cooperating not only with the US Arctic Commission, but also with the other circumpolar nations.

How will we achieve this? Alaskan state research needs must be coordinated at this time in our state through the Governor: that is why I am here. Science in Alaska has more than a window into policy making. It has a Senior Science Advisor and a Science and Engineering Advisory Committee. Very few states have this kind of science support. Keep in mind that dollars for research *compete* with other budget requests, all of which is an opportunity for scientists in Alaska to participate in formulating state research policy. At this symposium we are dealing with atmospheric pollution. Other scientists in other disciplines will be making their needs known. The important point is that science research needs (eg atmospheric pollution) must be made known. Then, and then only, will they be competetive in acquiring any level of funding.

MODERN LIFE SYSTEMS OF INDIGENOUS PEOPLE OF BUSH ALASKA

MATTHEW BEAN

The Alaskan native people have gone through a great deal of change within the last two generations. Their remote location in the Arctic no longer isolates them from the modern world. Natural resources of the Bush—fish and other forms of wildlife, locatable and leasable minerals, and tourists attracted to the Bush way of life (called 'subsistence')—have all brought change to the way that people live, and the air, lands and waters they live with and depend on.

The majority of Alaskan native people still live a hunting-gathering way of life. They do not sit in office buildings. They do not wear coats and ties. They do not work eight hours a day. However, the modern world of the Alaskan native does include labour for cash. Our villages are no longer self-sufficient. Framed houses have replaced the sod houses built with drift wood. Our villages now require imported energy for space heating and electricity production. We watch television from New York, and see new movies from Hollywood right in our homes.

Zimbabwe in south central Africa is now very important to us. Why would Bush Alaskan natives care about that place? Zimbabwe produces cobalt. We need cobalt to produce hard alloys for our snow machines and outboard engines, to get us to remote locations to harvest the food we eat.

Even with all these modern luxuries, we still need wildlife resources. The harvest of wildlife is the key to our culture and to the maintenance of our health. In the Arctic, a person needs protein and fat. Our wildlife foods give us that. Our body systems have not adjusted totally to Western foods. Wild foods are low in the poisons that modern society injects into domestic animals to make them grow faster and larger.

Changes in the environment, air pollution for example, can cause many problems for our Bush communities. When pollution causes air temperature changes during the winter, our ways of travel—dogs and snow machines—do not work. Our winter foods are few in number and hard to get to. Changes from normal temperatures in spring produce unusual winds, which lock the ice to our shores and prevent our harvest of seals. When the winds are unseasonably warm they blow the ice away, so that no marine mammals can be found. We do not have the funds or technology to search the coastline for these mammals. Our villages are located in places where our elders have said the animals show up every year.

The more I am hearing of arctic haze and its actions, the more confused I become. Questions continually pop up in my mind. For example, if the sulphur particles are deteriorating in a certain length of time, where are the spent particles going? Are they going out into space, or are they going into the water and the ground? If they are falling to the ground, what effect is the plant life getting from the fallout; if they fall in the sea, how are the particles affecting the food chains leading up to the sea mammals? Since we have to depend on plant life to survive, we in Alaska are constantly wondering what is happening to our health. Our bodies are not adapted to domestic livestock, therefore we have to depend on the wildlife and the plant life in our immediate area. And the marine mammals that are in our areas are of great importance to us, as they provide the fats and the protein to keep warm in the harsh winters.

Every year this arctic haze is more noticeable in our area. The sky is never deep blue any more. It is very pale in colour. In this present day we never hear or see anyone get snowblind, and it seems on some days— even on semi-cloudless days—we have to keep our lights on. Our plants are not so healthy it seems. The leaves are not as green as they used to be. In recent years the leaves are turning pale green. Some are withering, such as Labrador tea and the water lilies; also in some areas our spruce trees are red in colour.

And the behaviour of our sea mammals is getting odd. Ordinarily we don't see sea mammals dead on the beaches when they die a normal death. Now we are seeing more mammals dead on the beaches, such as Bowhead whales, Killer whales, walruses. Clams are floating, washed ashore with empty stomachs. Our wildlife are constantly being found with some of the fur missing from their pelts, and their behaviour is odd. Some of the wildlife are not afraid of humans. In fact some go after humans, which was never seen like that in early times.

In early times our forefathers would go outdoors in early mornings and were able to read the weather forecast, and on certain days they would warn us not to travel. They read the stars, the moisture and the visibility—how clearly they could see an object as visible. Now we can't see in the distance because of the haze obstructing visibility, nor can we predict what the day will be like.

Our forefathers were able to tell what kind of winter to expect and prepare for by the behaviour of the plant life. They knew by the growth of the plants whether the marine mammals and wildlife would be scarce, whether there would be times of plenty or times of want. Now the weather is very unpredictable. Storms appear without warning or indications of coming, and our rain barrels have grey sediment at times. In some areas it is reported that the snow has some kind of ash with it. I try to tell our people to take samples and send them to me, but as this is not natural, on the spur of the moment they don't think to take samples.

How much of the sulphates, aerosols and carbons are in our snows in the winter is unknown, as there is no research done in our area at this time. We can only guess as to how these are present now. We know there is a huge presence because of our deteriorating visibility each year.

With your expertise and knowledge of this, if we could get some help in getting some research done we can all better understand why our wildlife, marine mammals and our waterfowl are dying, and why suicide, drugs and alcohol, and sexual violence are on such a dramatic rise. Perhaps with better understanding we can control these behaviours and correct them. Our small bush villages are built on a social and economic system that requires predictable weather. We do not have the funds to wait to subsist, nor do we have alternative foods that we can gather to feed our families.

The elders of our region believe that something must be done about air pollution. We have begun to ask elders in the other Bush Alaskan regions and in the Inuit Circumpolar Conference to join us to work to clean up our air.

We are very pleased that the State of Alaska Government and our State Senator Frank Ferguson have founded this international meeting. We look forward to all of the Arctic nations, and those other industrial nations whose air pollution is settling in the Arctic, to join with us in seeking solutions to this problem of Arctic Haze. Thank you.

CONCLUSIONS FROM THE ARCTIC ATMOSPHERIC POLLUTION CONFERENCE

C. C. WALLÉN

ABSTRACT. There is clear evidence that an atmospheric transport of pollutants to the Arctic takes place from the industrialized areas of surrounding continents, particularly in late winter when the anthropogenic component may rise to levels ten times higher than in summer and SO_4 concentrations are between 1.5 and 4.0 $\mu g/m^3$. In Alaska pH in snow has changed from 5.5 to 4.9 in the last 20 years. The basic meteorological cause for high levels of local and long-range pollutants in the Arctic in late winter is the lack of solar radiation, which contributes to high stability of air. Emission inventories, statistical analyses of trace elements and trajectory model calculations indicate that most winter aerosol at levels below 2 km originates in Europe and western and central Asia. The origin of pollution at higher levels is still unknown. Proposals for further monitoring and research are presented.

Studies of ice cores have indicated a marked increase in air pollution levels in this century, particularly since the mid-1950s.

The radiation balance over the Arctic could be significantly modified by aerosols like graphitic carbon particles which are common in arctic air pollution. Aerosols increase absorption of solar energy in the atmosphere and decrease absorption on the ground. In March they cause a mean atmospheric heating rate of the order of 0.20 K day^{-1} at latitudes 70° to 75°N. Proposals are presented for monitoring and research to determine more precisely the impact of changes in the radiation balance on regional climate in the Arctic.

The conference agreed that arctic air pollution is currently below levels that should trigger public health alarms or cause human illness. It is important however that further data be collected to determine whether pollution in the Arctic is increasing or decreasing. As northern ecosystems are very sensitive to pollutants, there is a significant risk of ecological degradation. Interdisciplinary studies and integrated monitoring of arctic ecosystems are urgently needed.   The conference agreed that existing international mechanisms such as UNEP, WHO, WMO and the Comité Arctique should continue to be used for monitoring and assessment of atmospheric pollution in the Arctic. Special activities are listed in this regard. Proposals are also presented for national activities in general and for Alaskan activities in particular

Contents

Session 1: arctic haze; composition, sources and pathways

The conference brought clear evidence that there is an atmospheric transport of pollutants to the Arctic from industrialized and densely populated areas of the surrounding continents. In late winter, peak concentrations in the arctic atmosphere may reach levels comparable to annual mean concentrations over the industrial urban regions at mid-latitudes. The most typical manifestation is a haze which reduces visibility from its clear-air limit of about 300 km to 30–70 km. The aerosol mass in the arctic atmosphere is clearly concentrated in the lower 5 km. Aerosol scattering coefficient data from Barrow, and other data relating to arctic air pollution, show a strong annual cycle with a maximum in winter and spring and a minimum in summer and fall.

Statistical analyses of chemical composition show that arctic haze aerosol contains 3–5 independent source-related components of distant origin. Two of the component which dominate in summer are of natural origin, one a crustal and the other a marine component. The most important of the anthropogenic components has a strong sulfur element and may in winter rise to levels 10 times higher than in summer. Mean SO_4 concentrations in January-April are typically 1.5–3.9 $\mu g/m^3$ in the Norwegian Arctic and 1.2–2.2 $\mu g/m^3$ in the North American Arctic. During the same period the sulfur dioxide precursor shows mean concentrations of 0.4–2.0 $\mu g/m^3$ in the Norwegian Arctic. Ground level concentrations of graphitic carbon particles in the haze are typically three to four times lower than those found in urban areas at mid-latitudes, ie about 0.3 $\mu g/m^3$. Concentrations of graphitic particles in upper levels can be as large as over polluted centres at mid-latitudes.

Arctic air pollution also contains a number of heavy metals which are characteristic of fuel combustion and different industrial products, as well as many polychlorinated hydrocarbons which, after deposition and enrichment, could have serious effects. Deposition in snow-covers leads to an increase in acidity, and the mean pH of Alaskan snow has changed from 5.5 to 4.9 in the last twenty years.

The basic meteorological cause for the existence of high levels of both local and long-range air pollution in winter and spring in the Arctic is the lack of solar radiation, which contributes to high stability of the air and reduces scavenging as well as mixing of air masses. Polluted air become stagnant over the Arctic, either by general subsidence of air masses brought by general circulation from lower latitudes into the polar region, or after more direct episodic transport of polluted air from specific source regions.

In winter air pollution from sources north of the polar front often moves into the Arctic at low level, and the resulting haze normally appears below a height of 2 km. In summer when the polar front is north of the main sources, polluted air from lower latitudes must be lifted and cooled adiabatically before entering the Arctic. The result is isolated layers of polluted air at various levels above about 3 km. Such isolated layers of polluted air may also be observed in winter at different levels up to the tropopause. Aircraft observations indicate that these layers may have an extension of a few hundred km, confirming their episodic nature.

Several atmospheric circulation factors favour a eurasian origin of the arctic haze in winter at levels below 2 km. Emission inventories, statistical analysis of trace element composition of arctic aerosol, and trajectory model calculations indicate that most winter aerosol originates in Europe and in western and central Asia. The time taken to transport particles across the Arctic from the Kola Peninsula [Kol'skiy Poluostrov] to Barrow has been calculated to be of the order of 10 days. Transport from the North American continent have also been observed.

In summer arctic air below 2 km is generally very clean. Occasionally low level transport of pollutants from western Europe may reach the Svalbard area across the Norwegian Sea. The origin of the polluted layers above 2 km is however still unclear; there are indications that sources may be spread all over the northern hemisphere. It seems also that higher levels may be a source region for lower levels. Although the above indications exist, observations of air pollution above 2 km are at present too few to establish the origin with certainty. For this to be done, current techniques will have to be applied more extensively to examining how air pollutants are transferred to the Arctic at greater heights. We must also understand more clearly the chemical and physical transformations that take place during long-range transport of air pollutants.

Due to lack of solar radiation, very high stability of the air and other special circumstances in winter, local air pollution in cities and industrialized areas can be very strong, and may add to imported pollution. These two kinds of pollution must be clearly distinguished and studied separately.

Acidity and conductivity measurements of ice cores have provided the first historical record of arctic air pollution, indicating a marked increase in air pollution levels in this century, in particular since the mid-1950s, and reflecting an increase in northern hemisphere emissions at least up to the mid-1970s. Further records from ice-cores are needed to establish the trend over the last 10 years, now that emissions have decreased in many source areas.

Session 1: further monitoring and research

In air pollution monitoring, identification of emissions and source areas, detection of long-term trends and related fields of enquiry, the following actions are recommended.

Monitoring

(1) Increase routine monitoring of atmospheric pollution at ground level by adding at least ten national and international stations in the Arctic and sub-Arctic, for monitoring precipitation chemistry, aerosols, soot and trace elements. Details should be agreed by national and international programmes. Use of automatic stations and satellite information for monitoring air pollution in the Arctic should also be considered.

(2) Add monitoring of CH_4, CO, O_3, freons and chlorinated hydro-carbons to routine monitoring programmes of existing stations whenever this is feasible and not already done.

(3) Increase aircraft observations through national or international programmes like AGASP. More information on the vertical distribution of pollutants is essential. Airborne monitoring campaigns are needed during normal 'background' periods, and even in periods of extremely clean arctic air. Every effort should be made to coordinate aircraft programmes with ground observations.

Emissions

(4) Obtain better information about emissions of pollutants in mid-latitude source areas of the northern hemisphere, particularly in Eastern Europe and Asia. Emission data should include haze-producing aerosols (sulfate, nitrate, graphitic carbon, trace elements) and gaseous compounds affecting radiation balance (CH_4, N_2O, CO_2, CO, O_3, freons and chlorinated hydrocarbons). From areas where some data are already available, for example Western Europe and North America, further information is needed on variations, particle size and chemical specifications of emissions.

Source areas

(5) Further develop techniques for statistically evaluating chemical data through the use of elemental signatures, principal component and discriminant analysis, which have already proved useful in tracing origins of arctic air pollution. There is a special need for more information on changes of source signatures, by chemical transformation, selective deposition and mixing, during the journey from source to Arctic.

(6) Make special efforts to model long-range transport of air pollution into the Arctic. In particular there is need for new modelling of transport occurring above 2,000 m, as sources of upper-air pollutants are not well known. Aircraft observations mentioned above should provide useful data for this purpose. Back-tracking of pollution transport by use of suitable models should be considered.

(7) Use organic trace gases as signatures for identifying source regions. Ratios of F_{11}/F_{12} and CCl_4/C_2HCl_3 may be applied.

(8) Study in detail synoptic climatology governing the transport of pollutants into the Arctic. It is essential to know the frequency and duration of the different types of circulation and air masses involved in transporting pollution from different mid-latitude regions.

(9) Gradually achieve agreement between results from the various techniques used in tracing origins of arctic air pollution, such as statistical evaluations of chemical data, modelling long-range transport, use of organic trace gases and studies of atmospheric circulation climatology.

Other research

(10) Carry out special studies of wet and dry deposition of pollutants in the Arctic. Budget materials may be useful tools when the total wet and dry deposition cannot be determined by measurements.

(11) Carry out further research on chemical transformation and physical properties of arctic pollutants. In this context contributions from natural sources must be considered, particularly in summer when both sea-salt and organic sulfur may play important roles. In particular, aerosols with particles in the size-range of 0.02 to 0.5 μm should be useful to study.

(12) In parallel with long-range air pollution, study local air pollution and meteorological conditions that favour the trapping near the ground of both local and long-range pollutants. Ratios between local and long-range air pollution should be quantified and monitored at suitable locations.

Trend studies

(13) Apply ice-core data from the Arctic and sub-Arctic to study pre-industrial references, long-term trends in pollution, overall contamination and deposition problems.

(14) Use local and regional observations in snow packs for assessing seasonally-integrated depositions of air pollutants in the Arctic.

Session 2: pollution, radiation balance and climate

There are clear indications that radiation balance over the Arctic could be significantly modified by combustion-generated aerosols. Over highly reflecting arctic ice regions, aerosols cause more solar energy to be absorbed by the earth-atmosphere system. The most efficient absorbing material is graphitic carbon particles, which may absorb 5–10% of incoming solar radiation. The absorption optical depth associated with vertical profiles of graphite particles is large enough to cause a substantial change in solar radiation balance over the reflecting surfaces of arctic sea-ice. Vertical profiles can be strongly-layered, or almost uniform to the top of the troposphere. Graphite particles show both large vertical inhomogeneities and large horizontal variations with characteristic scales of 50–100 km. Their residence time is estimated at about two weeks.

Arctic aerosols increase the solar energy absorbed by the atmosphere and decrease the amount absorbed at the ground. They cause mean atmospheric heating rates of the order of 0.20 K day^{-1} at latitudes between 70 and 75°N in March. Haze layers over sea-ice reduce effective planetary albedo by 9%; over water albedo increases by about 2.5%. Spring proliferation of algae within and on the surface of sea-ice may contribute to a change in its albedo.

To determine the impact of changes in radiation balance on regional climates would require more observations and research on spatial and temporal properties of haze layers. These could be provided by a further AGASP program or its equivalent. The importance of aerosols for the creation of clouds in the Arctic was mentioned as an area for future research. Anthropogenic aerosols are of little importance for cloud creations in summer; some influence may exist in winter, but cloud amount in winter is small.

Energy budget studies for the whole arctic basin have shown that infra-red cooling by the atmosphere-surface system amounts to 180 W/m² annually, while heat transport to the basin amounts to about 110 W/m². The difference between these terms is made up by solar heating. If seasonal budgets could be obtained, the impact of arctic haze in spring could be calculated. It is also important to verify experimentally the calculated heating rates.

Regarding trends of particles in arctic air, it has been found at Barrow that condensation nuclei show a half-annual cycle but no significant trend since 1974. Light-scattering index shows an annual cycle but no trend. Further studies of trends during the last 10–20 years are important.

The detection of signals for impending climatic changes due to arctic air pollution is an urgent matter which may be tackled by studies of long-term records, by modelling, or by studies of ice-core data. Long-term climatological records should be considered, together with such other geophysical data as sea-ice, glaciers, vegetation and perma-frost. New climatic models are needed, especially designed for the Arctic to include air pollution data, and existing models should be modified to include arctic conditions.

Session 2: proposals for further action

In the fields of radiation balance and climate research the following actions are recommended.

Impact on radiation balance

(1) Detailed study of graphite particles in arctic air pollution.   (2) Study of variations in albedo due to deposition of graphite particles and development of ice-surface biota in the Arctic.

(3) Introduce into aircraft programmes sun-photometer observations and infra-red radiation measurements.

(4) Expand particle-size measurements at monitoring ground stations in the Arctic.

(5) Introduce more measurements of total/diffuse radiation and albedo at ground station in the Arctic and the sub-Arctic.

(6) Detailed studies of cloud physics and chemistry in the Arctic, particularly in winter. Cloud formation as it depends upon availability of particles needs special consideration.

Impact on climate

(7) Obtain better information about horizontal and vertical extension of arctic haze layers and of their temporal variability in time scales of months.

(8) Study the detection of signals of climatic change through analysis of (i) air and sea-temperature records, (ii) cloudiness records, (iii) albedo and extension of sea-ice and glaciers, (iv) bio-indicators such as lichens, or micro-organisms in pack ice and underlying layers of ocean, (v) pollen deposits in lake sediments, (vi) permafrost extension, (vii) naval records of marine parameters, (viii) oral information from natives.

(9) Develop a model especially designed for arctic climate, coupling ocean, ice, and boundary and free tropospheric conditions. The key optical parameters needed to incorporate arctic haze into climate models should be determined (eg aerosol absorption and scattering) and the vertical, horizontal and temporal variability of optical properties monitored in sufficient detail to obtain a semi-empirical model of the arctic troposphere.

(10) Incorporate arctic haze in climate models to establish its possible effects on arctic climate and on general atmospheric circulation, applying present concentrations as well as scenarios for possible future concentrations.

(11) Develop energy-economic models to predict future concentrations of arctic air pollution within various scenarios.

(12) Apply ice-core data for understanding climatic changes and to detect change signals. New arctic or sub-arctic locations for collecting ice-cores may be found on Svalbard and in Norway. The representativeness of ice-core data with regard to air masses should be considered.

(13) Study sulfate in snow and ice using experience from Antarctic studies, and taking into account marine organic sulfate, which should be analysed and monitored.

Session 3. Health and ecology issues

As background information a lecture was presented which discussed physiological changes and acute or chronic human health consequences that may be associated with air pollution in general. It also dealt with monitoring methods and research strategies needed for assessing health effects from air pollution. Against this background it was agreed that arctic air pollution is currently below levels that should trigger public health alarms or directly cause human illness; we cannot conclude that there is no impact, but present evidence gives no basis for expecting public health impacts. It is important, however, that further data be collected so that we can determine whether pollution in arctic air is increasing, decreasing or remaining at stable levels.

Possible health effects of low-level exposure to ionizing and non-ionizing radiation were also considered as background information. Estimates from non-threshold dose-response models suggest that health risks associated with natural background ionizing radiation are too small to be detectable by current methods. Validation of risk estimates will require further studies of the dose-effect relationships in populations exposed to higher dose levels, as well as of biological mechanisms underlying the effects. Exposure to natural background levels of ultra-violet radiation is known to be involved in causing skin cancers.

A special study of fall-out from tropospheric nuclear weapons testing in the early 1960s, in relation to cancer risk in Alaska natives, had shown maximum annual dose rates of ^{137}Cs and ^{90}Sr comparable to normal background radiation levels. There is agreement between cancer risk assessment data (assuming plausible risk models to predict accumulated impact) and observed cancer rates among affected populations. Risk cannot be said to be zero, but the number of cancer cases, the number of exposed individuals, and the radionuclide dosage to exposed individuals are so small as to constitute no basis for public health concern. Cancer cases occurring among Alaskan natives are not on present understanding due to radionuclides. There are no reasons for thinking that present models of radiation-induced carcinogenesis do not apply to arctic residence.

Ecological studies of long-term build-up of radionuclides in arctic ecosystems between 1959 and 1980 have shown effective concentrations of several radionuclides such as ^{90}Sr and ^{137}CS. Radiation exposures from such concentrations, particularly through the lichen-caribou-man-food chain, were two to three times greater for arctic populations than for most other world populations. As several of the radionuclides of prime importance are isotopes of elements present in arctic haze, further investigations are needed into the role of arctic pollution in high radiation exposure in arctic ecosystems.

Insights into Arctic pollutants that are dangerous to human health are given by studies of heavy metal exposures in the population of Greenland. Due to high consumption of fish, seal, and whale meat, Inuit of northwest Greenland were found to have blood concentrations of mercury as high as 200 μg/l. Lead concentrations in blood are similar to those in industrial communities of West Europe. No correlations have been found between lead levels and alcohol comsumption, smoking, or eating habits. Samples of human hair have revealed that significant increases in mercury and lead have occurred since the 15th century; cadmium levels have shown no change. Further investigations are required to determine the etiology of heavy metals in human blood among the Greenland population.

The roles of urban and indoor air pollution as health hazards need further study and better documentation; in particular data from simultaneous indoor and outdoor urban measurements are needed. There is a high potential for adverse health effects from indoor pollution, particularly from stoves and poorly-ventilated buildings.

Morbidity and mortality in northern populations are likely to change due to underlying changes in characteristics and composition of the population. Changes in age, diet, smoking habits etc, may be falsely attributed to particular causes; more precise demographic data are needed to take these factors into account in evaluating individual risks and risk factors.

Ecological risks associated with arctic air pollution were discussed in relation to the biogeochemical characteristics that make northern ecosystems sensitive to pollutants. Risk of significant ecological degradation appears to be high. Ecosystems impacts may therefore be more sensitive indicators of arctic air pollution effects than human health parameters. Arctic haze may effect ecosystems in three ways: (i) direct effects of pollutants on ecosystems; (ii) indirect effects by pollution-induced climate change; (iii) long-term ecological and climatic changes induced by bioclimatic feed-back processes. The first has been demonstrated by effects of sulfate deposition on lichens, which extend to such grazing populations as caribou. An example of the third category of impact is the risk of a pollution-induced surface warming of the Arctic, which might cause an increase of organic matter decomposition and add to a future warming from an increase of CO_2 and other greenhouse gases in the atmosphere. To understand the complex consequences to ecosystems that may occur from increased arctic air pollution, it is necessary to emphasize interdisciplinary studies combining ecology, chemistry, climate and toxicology. More stations will be needed for integrated monitoring in air, water, soil and living organisms.

Session 3: health and ecology; proposals for further action

To investigate impacts of arctic pollution on human health and the ecological environment the following actions are recommended.

Human health risks in general

(1) Obtain precise data on changes in population with respect to age, diet, smoking habits, etc, so that such factors can be taken into account in evaluating individual risks and risk factors from air pollution.

(2) Obtain precise information on morbidity and mortality among different population groups, to establish trends and possible etiologies for observed changes.

(3) Reactivate periodic monitoring of ecological human health parameters among individuals exposed to radionuclides in the 1950s and 1960s.

Impact of urban and indoor air pollution

(4) Obtain data from simultaneous indoor/outdoor air measurements to establish to what extent measured air pollutants from outdoors sources find their way indoors.

(5) Obtain more precise exposure data (both outdoors and indoors) than are currently available, as pollutant levels are not uniformly distributed, and individual exposures may be intermittent, of brief duration and highly variable.

(6) Take governmental action to prohibit the use of unvented space heaters in densely populated areas.

(7) Investigate to what extent radon and radon daughters are a health problem in the Arctic. Monitoring of radon in buildings is essential.

(8) Carry out surveys to establish to what extent excessive humidity with mould growth and dispersion is a serious problem in arctic conditions.

Impact on ecosystems from air pollution

(9) Carry out integrated monitoring for assessment of ecological impacts by the use of such selected parameters as acidity, sulfur, nitrogen, heavy metals, and chlorinated hydrocarbons, which should be measured in air, water, soil and biota. By such data, baselines and trends may be established.

(10) Carry out interdisciplinary studies in various ecosystems combining ecology, chemistry, climate and toxicology.

(11) Investigate the role played by the Arctic Ocean regarding long-term ecological and environmental effects from deposition of arctic air pollutants.

(12) Place considerably more research emphasis on the sub-Arctic, both to determine levels of pollution there and to understand what climate change in the sub-Arctic is likely to be connected with arctic haze.

(13) Direct more effort towards understanding hydrological changes caused by climatic fluctuations, which are likely to be as important to ecosystems as are direct climate changes.

(14) Obtain better understanding of the likely synergistic effects on ecosystems of the combined stresses of climate change and pollutant deposition.

(15) Carry out studies to determine the extent and amount of deposition in the Arctic of atmospheric pollutants.

Session 4. International cooperation and state responsiblities

The usefulness of international co-operation in arctic air pollution studies was demonstrated by the presentation of a joint paper about current, informal co-operation between three laboratories in China, Japan and the USA, to study long-range aerosol transport over the Pacific into the Arctic. Both desert dust and air pollutants from urban and industrial areas are sampled near their sources in China and Japan, as well as over Hawaii and the Arctic Ocean. Dust particles from the interior of China often reach Japan, and it has been shown that in spring they sometimes reach Hawaii. Whether Asian sulfur pollution sources also contribute to arctic air pollution is at present under investigation, through sampling in the Arctic from aircraft during major field programmes of dust transport. Preliminary results show that arctic aerosols sampled during spring contains both fine sulfur and soil elements with a relation of Si/Al less than the average in Asian loess. As dust from a China desert area shows Si/Al close to average of loess, depletion of silica during transport into the Arctic must have occurred if arctic aerosols are derived from Asian soil. The cooperative programme continues.

A special lecture dealt with US federal arctic policy and specifically referred to international co-operation as an important means promoting scientific research of the arctic environment. In the Arctic Research and Policy Act passed by the US Congress in 1984, it is stated that industrial pollution of non-arctic origin that collects in polar air mass should be controlled through international co-operation and consultation. A US Interagency Arctic Research Policy Committee, newly established, should co-ordinate and promote co-operative arctic research programmes with other nations in this area. A review of international co-operation and co-ordination was also presented. It was concluded that arctic science is eminently interdisciplinary in nature, that arctic phenomena and processes cannot be studied in isolation, that such studies are very expensive and that the need for coordination, therefore, is often dictated by logistic considerations. It would be desirable that all nations bordering the arctic basin should join in such co-ordinated activities. The possibility of establishing an international non-governmental scientific committee for the Arctic under ICSU (similar to SCAR for the Antarctic) was mentioned, though it was pointed out in the discussion that Comité Arctique International already to a large extent filled this role.

The fact that many of the large-scale concerns in the arctic environment call for international cooperation does not by any means exclude the need for local and state responsiblities regarding many environmental problems. A variety of such problems was reported to the Conference as, for instance, how the arctic haze may be blamed for having a serious impact on the ecological social and economic life in Bush Alaskan villages. It was stated that fluctuations in winter and spring temperatures have changed in Alaska over the last decades and that these changes may be coupled with increasing arctic pollution. Recent weather conditions have had a harmful impact, particularly on the possiblities both for fishing and hunting marine mammals, and has modified the economic and social circumstances for the population in Alaskan villages. It was recommended that possible inter-relationship between arctic pollution and weather in Alaska in winter and spring be investigated in detail and that, if a causal relationship were found, it would further enhance the need for mitigating arctic air pollution.

In the context of the discussion of national and international activities to mitigate arctic air pollution, various options for application of international legal instruments were discussed. It was noted that possibilities are indeed quite limited as only few instruments exist to prevent long-distance trans-frontier pollution, the reason being that individual country responsiblities cannot easily be identified. At the international level it was agreed that the only possibility would be to apply the principle of international solidarity between countries to the problem of arctic air pollution, basing oneself on the Declaration of the UN Conference on the Human Environment of 1972.

The conference expressed appreciation at the interest in the problem of arctic air pollution shown by the international organizations UNEP, World Meteorological Organization, World Health Organization and the Comité Arctique. It was agreed that those organizations could play an important and decisive role in the implementation of proposals for future action to improve understanding of arctic air pollution. In particular, the organizations were called upon to participate in the following ways in future activities: (i) stimulate and co-ordinate between them activities for monitoring of arctic air pollution; (ii) stimulate and support activities aimed at assessing arctic air pollution and its effects on weather, climate, human health and ecosystems. In discussing the participation of international organizations in the monitoring of and research into the effects of arctic air pollution, attention was drawn to the on-going joint activities under GEMS/UNEP with special reference to BAPMoN carried out by WMO, and to studies of effects carried out by WHO and the Comité Arctique.

Session 4: proposals for further action

The following international and national actions, and special measures for the State of Alaska, are recommended.

International activities

(1) Make use of already existing international mechanisms for monitoring and assessing atmospheric pollution.

(2) Encourage co-ordination of national programmes to study arctic air pollution as a precursor to future international arctic-wide multidisciplinary campaigns.

(3) Foster international exchange of data relating to arctic air pollution, and encourage international symposia to consider scientific and practical aspects of the problems.

(4) Actively promote the development, both nationally and internationally, of legal instruments for protecting the arctic environment, based on existing environmental legal principles.

National activities

(5) Develop national programmes for studies of arctic air pollution in all countries bordering on the arctic basin.

(6) Encourage the establishment of national networks for monitoring arctic air pollution.

(7) Encourage the establishment of an appropriate national data base for use in research of arctic air pollution.

(8) Encourage active consultation with and involvement of native residents in scientific programmes to ensure the collection of relevant information about environmental development.

(9) Encourage the development of science education programmes at national universities, to ensure development of expertise necessary to carry out long-term research of the arctic enviroment.

Alaska activities

In view of the strong interest shown by the State of Alaska in calling the conference on arctic atmospheric pollution, the conference in particular called upon the State to develop and conduct a research and monitoring programme that can be taken as a model by other Arctic Rim nations in a concerted international effort to understand the large-scale behaviour and potential impact on arctic residents of pollutants transported by the atmosphere.

The programme should address a selection of the research problems proposed above, definitely including the following:

(1) Establishing a network of air pollution monitoring stations at appropriate locations.

(2) Documenting environmental change symptoms reported by natives, and conerting it where possible into quantitative information.

(3) Studying the dynamics and chemistry of urban and indoor pollution.

(4) Studying health problems specific to pollution inhalation and ingestion, and their mitigation.

(5) Strengthening science education programmes to develop the necessary expertise for long-term research efforts.

INDEX

The numbers in italic represent figures or tables in the text.